Electronic Processes on Semiconductor Surfaces during Chemisorption

Electronic Processes on Semiconductor Surfaces during Chemisorption

T. Wolkenstein

Translated from Russian by
E. M. Yankovskii

Translation edited in part by
Roy Morrison

CONSULTANTS BUREAU • NEW YORK AND LONDON

Library of Congress Cataloging-in-Publication Data

Vol'kenshteĭn, F. F. (Fedor Fedorovich)
    [Ėlektronnye protsessy na poverkhnosti poluprovodnikov pri
khemosorbt͡sii. English]
    Electronic processes on semiconductor surfaces during
chemisorption / T. Wolkenstein ; translated from Russian by E.M.
Yankovskii ; translation edited by Roy Morrison.
        p.   cm.
    Translation of: Ėlektronnye protsessy na poverkhnosti
poluprovodnikov pri khemosorbt͡sii.
    Includes bibliographical references.
    ISBN 0-306-11029-6
    1. Semiconductors--Surfaces. 2. Chemisorption. I. Title.
QC611.6.S9V6513 1990
621.381'52--dc20   90-46366
                                                               CIP

ISBN 0-306-11029-6

The Russian edition of this work was published by Nauka in
Moscow in 1987. Professor Wolkenstein died while the English translation was
being prepared for publication.

© 1991 Consultants Bureau, New York
A Division of Plenum Publishing Corporation
233 Spring Street, New York, N.Y. 10013

PREFACE

"Hands are useless if there are no eyes to see what is obvious."
—M. V. Lomonosov

Dear Reader, I invite you to open this book and step on the semiconductor surface, where the processes that form the subject of the book come into play. The surface of the semiconductor is attracting more and more interest among researchers, in fact researchers in two different fields.

These are notably the physicists and engineers engaged in research in semiconductor physics and the making of semiconductor devices. The entire industry of semiconductor instruments hinges on the problem of the surface. The quality of semiconductor devices, whose use is growing steadily, depends essentially on the properties of the surface. The instability of these properties and their uncontrollable alterations with temperature and under the influence of environmental conditions result in a lack of stability in the performance of semiconductor devices, hence the high percentage of waste in their industrial production. The methods used in factory laboratories to prevent such waste are largely empirical. The properties of the surface, the nature of the physicochemical processes that take place on it, and the role of environmental factors still remain obscure. A major task of the semiconductor industry is to learn to control the properties of the surface.

The semiconductor surface is also attracting the attention of researchers in quite another field. These are physical chemists and chemists specializing in adsorption and catalysis. For them the semiconductor surface is where the processes of adsorption and catalysis take place. As a matter of fact, most semiconductors are catalysts in chemical reactions. Indeed, in catalytic research the semiconductor surface is dealt with far more often than may appear at first sight. The fact is that many metals are usually clad in a semiconductor sheath, so that the processes that appear to take place on the surface of the metal are really occurring on the surface of a semiconductor. Industrial chemistry is faced with the problem of producing catalysts sufficiently active for specific reactions, but, as a rule, the methods used are still nothing more than empirical. It is most essential

v

for the chemical industry to learn to control the chemisorptive and catalytic properties of the surface.

So we see that the surface is of interest both for semiconductor physics and semiconductor devices and for chemisorption and catalysis in the chemical industry that makes use of catalytic processes.

The surface of the semiconductor is the boundary of two phases. The main personages of this book meet on this boundary. Some come to the surface from the gaseous phase. These are gas molecules. Others come from within the depths of the semiconductor. These are free electrons and holes. Being the boundary of two phases, the surface of the semiconductor interacts with both phases. Hence, the problem of the surface is not the two-dimensional problem that it might seem at first sight. It is three-dimensional by its very nature. Researchers who ignore this fact and see the surface as a two-dimensional world will never solve its riddles.

To understand the surface, we must consider it together with the two phases on which it borders. This is the approach of this book. The book is mainly theoretical. An attempt has been made to reveal the mechanism of the processes taking place on the semiconductor surface. At the same time the book contains a review of experimental data. This review does not lay claim to being exhaustive, however. It has no independent status in the book and serves only to illustrate the theory. For reasons beyond the author's control, the review does not include the latest work in the field; but the author believes that this is of no great importance for a book of this type.

The semiconductor surface is the border not only between two phases but also between two sciences, physics and chemistry. Here "physics and chemistry are so closely interwoven that one cannot do without the other" (M. V. Lomonosov). The problems that face modern engineering and industry often border on the two different sciences, and it is these problems that hold out most promise theoretically and practically. The problem of the semiconductor surface, which confronts semiconductor physics on the one hand and catalytic chemistry on the other, is an excellent example. The surface is a two-faced Janus, with one face turned to physics and the other to chemistry. For this reason the book should appeal to both physicists and chemists. In view of this dual readership, the book starts with two introductory chapters, the first dealing with some problems of semiconductor physics and aimed at chemists unacquainted with this field, and the second containing information about the theory of adsorption and intended for semiconductor physicists not versed in the subject. Section 5.1, devoted to the basic concepts of the theory of catalysis, is also intended for semiconductor physicists.

Though the book describes the investigations of the author and his collaborators, it may also be used as a textbook for those studying the physics and chemistry of the semiconductor surface for the first time. Considerable attention is given to processes of chemisorption on the semiconductor surface. Special chapters are devoted to heterogeneous catalysis, photoadsorption and photocatalytic effects, and the effect of the surface on the luminescent properties of semiconductors. All these problems, however, are discussed from the standpoint of the electronic theory of chemisorption and catalysis, the theory on which the ideas and concepts of this book center.

In 1960 the author's book *The Electronic Theory of Catalysis on Semiconductors* came out in the USSR. It was then translated into French (1961), Polish (1962), English (1963), German (1964), and Japanese (1965). The last section of that book was partly devoted to the unsolved problems of the electronic theory of catalysis. Since then most of the problems have been solved and many of the results obtained are reflected in the present book. But this does not mean that the number of unsolved problems has diminished. New problems have arisen, since each advance in our knowledge poses more problems than it solves. As Louis de Broglie once remarked, "each new land brings to light the possibility that there may be continents as yet unknown." Notwithstanding the brilliant conquests of our mind and perhaps owing to these conquests, we remain surrounded by mysteries. These mysteries are born while others are solved, which is the reason why science is inexhaustible.

Recent years have seen a number of books devoted to the physical chemistry of surfaces. Two of these, *Reaktionen in und an festen Stoffen* by K. Hauffe and *The Chemical Physics of Surfaces* by S. R. Morrison, are veritable encyclopedias (the latter has more than a thousand references).

The semiconductor surface is as yet not entirely understood. In the catalytic industry, as well as in the industry of semiconductor devices, semiempirical methods continue to be used in the treatment of the surface. Three hundred years have passed since Robert Boyle declared that he despised researchers whose eyes and mind were choked with soot from their own ovens. His words still hold as far as the semiconductor surface is concerned.

Science, that "daughter of wonder and curiosity" (Louis de Broglie) poses crossword puzzles. Each empty square awaits its letter, the one letter that offers the clue to other words. Many squares have yet to be filled before it will be possible to decipher the word "surface." The present book tries to help in this. If we solve this crossword puzzle, we will acquire absolute power over the surface, the power to control it.

A number of books and articles written by the present author alone or in collaboration with colleagues form the core of the book. Among these are the *Physicochemistry of Semiconductor Surfaces* (Moscow, 1973), *Radical-Recombination Luminescence of Semiconductors* (in collaboration with A. N. Gorban and V. A. Sokolov, Moscow, 1976), and *Effect of Irradiation on the Surface Properties of Semiconductors* (in collaboration with V. G. Baru, Moscow, 1978).

The author expresses his gratitude to E. M. Yankovskii, who prepared the translation into English. He also thanks G. M. Zhidomirov, G. Ya. Pikus, and O. Peshev, who wrote several sections (the contribution is noted by footnotes), and also V. G. Baru and A. I. Loskutov, who participated in the writing of several chapters. And special thanks to E. V. Kulikova for invaluable assistance in preparing the book for publication.

T. Wolkenstein

CONTENTS

Chapter 1

ELECTRONS AND HOLES IN A SEMICONDUCTOR

1.1. ORDER AND DISORDER IN CRYSTALS

1.1.1. Types of Defects

All the macroscopic properties of crystals (semiconductors, for one) can be divided into two classes. To one class belong all properties that are determined by the periodicity of the crystal, and for which the defects present in any real lattice play the role of a small correction term. Such properties are known as *structure-independent*. The other class contains properties determined by local violations of periodicity of the crystal lattice. In this case the defects are of paramount importance. Such properties are known as *structure-sensitive*.

An example of a structure-sensitive property is the electrical conductivity of a semiconductor at not too high a temperature (impurity, or extrinsic, conductivity). At high temperatures the conductivity loses its sensitivity to impurities and structural defects and becomes a structure-independent property (intrinsic conductivity). Another example is the absorption spectrum of a crystal. At relatively low frequencies the absorption bands are structure-sensitive (impurity absorption), while absorption bands in the high-frequency region prove to be structure-independent (intrinsic absorption). Chemisorptive and catalytic properties of semiconductors constitute another typical example of structure-sensitive properties; they depend on the prehistory of the sample and can be changed by introducing impurities into the lattice.

Theoretical interpretation of structure-independent and structure-sensitive properties requires various approaches. In the first case we can proceed from the theory of the ideal crystal lattice. In the second we are forced to deal with the notion of a real crystal. Within the ideal crystal theory all structure-sensitive properties retreat completely to the background. In the theory of chemisorption

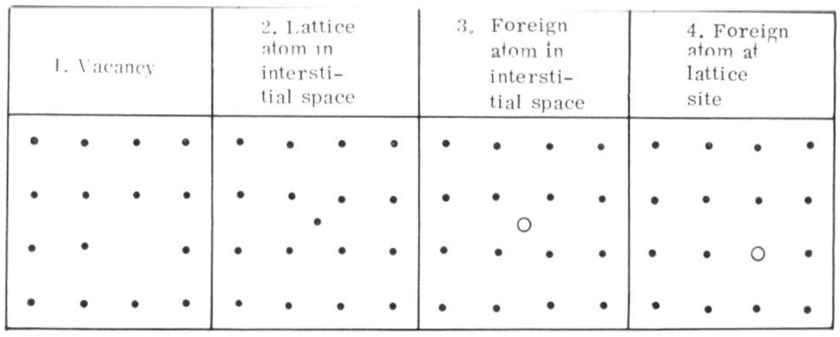

1. Vacancy	2. Lattice atom in interstitial space	3. Foreign atom in interstitial space	4. Foreign atom at lattice site
. . . .	. . . . .	. . . . .	. . . . .
. . . .	. . . . .	. . . . .	. . . . .
. . .	. . . . .	. . o . .	. . o . .
. . . .	. . . . .	. . . . .	. . . . .

Fig. 1.1. Types of microdefects.

and catalysis, as in semiconductor theory in general, we are clearly dealing with a real crystal instead of an ideal.

A real crystal differs from the ideal in that it has defects, i.e., local imperfections in the periodic structure of the lattice. Among the defects present in any real lattice we must distinguish the *macroscopic* and the *microscopic*. A macroscopic defect is an imperfection in the periodic structure that involves a region considerably larger than the lattice constant. These include cracks, pores, and various macroscopic inclusions. We will not deal with such defects at present. A microscopic defect is an imperfection whose size is of the same order of magnitude as an individual primitive cell. Here are the main types of such defects:

   (1) an unoccupied lattice position (a vacancy) in a crystal formed as the result of extracting an atom or ion from the ideal lattice;
   (2) a lattice atom or ion forced into an interstitial position;
   (3) a foreign atom placed in an interstitial position;
   (4) a foreign atom placed at a lattice site, i.e., replacing a lattice atom.

These types of microdefects are shown schematically in Fig. 1.1.

Defects of the first two types may be called *structural defects*. In the case of a one-component lattice (i.e., consisting of atoms of one type) these defects do not change its chemical composition, while in a multicomponent lattice they may change the stoichiometry. Defects of the other two types may be called *chemical defects*. In the narrow sense of the word these are impurities, which change the chemical composition of the crystal. Chemical defects are introduced into the crystal from outside—they occur due to a particular processing of the sample. Structural defects may not only be introduced from outside but appear as a result of heating the lattice. Note that each microdefect causes a certain distortion in the lattice around it. Strictly speaking, a defect is a region inside which the lattice is deformed.

1.1.2. Properties of Defects

Microdefects have a number of general properties independent of their concrete nature.

(1) They possess a certain mobility that increases with temperature (i.e., requires activation energy). Indeed, as the defect moves in any direction, the system energy varies periodically—energy minima and maxima alternate. Therefore, the motion of the defect is associated with its overcoming potential barriers, whose height depends on the nature of the defect, the structure of the lattice, and the direction of the defect's motion. We see that defect motion requires a certain activation energy, generally depending on the crystallographic directions. Defects inside a crystal may be considered fixed only if the temperature is not too high.

(2) Another general property of microdefects is their mutual interaction, which manifests itself the more strongly the closer they are. Indeed, the energy of the system depends, generally speaking, on the mutual position of the defects, which is evidence that defects interact. Defects may attract or repel each other. For instance, in the ionic lattice MR, built of ions M^+ and R^-, where M stands for metal and R for metalloid, the metallic vacancies repel each other but attract metalloid vacancies or interstitial metallic ions.

(3) When defects meet, they are capable of forming compounds and various groups, which must be considered as new defects with generally different properties. For instance, in an MR lattice metalloid and metallic vacancies together constitute a new formation of a different nature and with different properties than its components considered separately. "Reactions" between defects, just as ordinary reactions, may be exothermic or endothermic and may proceed with activation or without, depending on the nature of the reacting defects.

(4) Defects of each given type, taking part in reactions with other defects, may be created or disappear in the process. In equilibrium conditions we may assign to each defect a certain lifetime. Besides, they may be absorbed or created by the lattice. An example of such a process is the dissociation of atoms (or ions) from lattice sites to interstices, as a result of which defects appear in two types (interstitial atoms or ions and vacancies). Another is the recombination of interstitial atoms (or ions) with vacancies, as a result of which the defects vanish, as if "absorbing" each other.

(5) An important property of defects is the fact that they usually serve as attractive centers for free electrons or holes, i.e., they localize these entities near themselves. For instance, a metalloid vacancy in an MR lattice serves as a trap for a free electron, while an interstitial metallic atom is a trap for a free hole. In this way defects reacting with each other also react with electrons and holes in the lattice; i.e., the defects are ionized or neutralized.

Thus, the presence of defects in a crystal is violated in a real crystal. Some primitive cells in such a crystal prove to be "faulty." The ratio of faulty cells to the total number of cells is what may be called the *degree of disorder* in the crystal.

The degree of disorder is determined first of all by the prehistory of the sample; i.e., it depends on how the sample was manufactured, and to what external stresses it was subjected prior to the experiment. It also depends, as a rule, on temperature. Indeed, heating leads to the dissociation of lattice atoms (or ions) from sites to interstices and results in additional disorder in the lattice. Consequently, lattice disorder has a twofold origin: disorder due to the sample's prehistory and thermal disorder. The first type forms the irreversible part of disorder—it is the part retained at absolute zero, given to the lattice from the start, so to say. On this is imposed thermal disorder, which has a temperature nature.

In some cases thermal disorder prevails over the first type. We can then neglect the defects due to prehistory in comparison with defects of a thermal nature. In other cases the situation may be the opposite. This happens in considerably "spoiled" crystals at not too high temperatures and is common in semiconductors.

Thus, each crystal lattice with defects impregnated into it is a single system with properties governed by two competitive factors—order and disorder. The first determines all structure-independent properties, while the second all structure-sensitive properties of crystals.

1.2. ELECTRICAL CONDUCTION OF NONMETALLIC CRYSTALS

1.2.1. Factors Influencing Conduction

We now consider the peculiarities of semiconductors that distinguish them from dielectrics, on the one hand, and from metals, on the other.

Semiconductors constitute a broad group of solids occupying an intermediate position between metals and dielectrics. The difference lies primarily in the electrical conductivity κ. We have

for metals $\kappa = 10^6$–10^4 mho/cm,
for semiconductors $\kappa = 10^2$–10^{-10} mho/cm,
for dielectrics $\kappa = 10^{-14}$–10^{-16} mho/cm.

Hence semiconductors include solids with a wide range of conductivities, differing by millions.

The difference between metals, semiconductors, and dielectrics is not only quantitative but qualitative. This lies in the fact that the conductivity of the three groups of solids reacts differently to the same factors. These factors are

(1) impurities;
(2) temperature;
(3) electric field;
(4) light.

Under the influence of these factors the conductivity changes within an extremely wide range. For the same semiconductor in different conditions, the conductivity may have values differing by several orders of magnitude. Let us study these factors separately.

(1) In relation to semiconductors the term "impurity" should not be understood literally. An impurity does not necessarily mean there are foreign atoms introduced into the lattice from outside. What is meant is any local fault in the periodic structure of the lattice, and this may be a microdefect of any type.

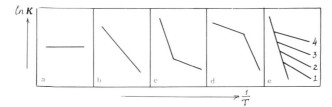

Fig. 1.2. Types of temperature dependence of conductivity: a) κ constant; b) slowly varying κ; c, d) varying κ and different A and B in different temperature intervals; e) different κ vs. $1/T$ dependences for different impurity concentrations.

When an imperfection is introduced, not only the electrical conductivity changes, increasing sometimes by a factor of a thousand or tens of thousands, but often the temperature dependence of the conductivity changes. This distinguishes semiconductors from metals, where an impurity at not too low a temperature, irrespective of its nature, reduces the conductivity but does not change its temperature dependence. Often the nature of the impurity determines the type of conductance in a semiconductor. For instance, semiconductors that are binary compounds (metal oxides or sulfides) where the metal is stoichiometrically in excess are n-type, while semiconductors where the metalloid is stoichiometrically in excess are p-type.

We note that at high temperatures (above a certain critical value) all semiconductors become insensitive to impurities. This critical temperature depends on the nature of the semiconductor and the number of impurity atoms.

(2) The temperature dependence of electrical conductivity differs from semiconductor to semiconductor. In many cases it grows with temperature according to an exponential law:

$$\kappa = A \exp(-B/kT), \tag{1.1}$$

just as it does in dielectrics (here T is the absolute temperature). Parameters A and B in (1.1) may be different in different temperature intervals, while for a given temperature interval they may prove to depend on impurity concentration in the semiconductor.

In other cases the conductivity may slowly drop with temperature, as in the case of metals.

We note that at high temperatures the temperature dependence of conductivity of all semiconductors has the exponential form (1.1). Parameters A and B in principle can be independent of impurity concentration and may be considered constants characteristic of the material.

The various types of temperature dependence of conductivity are shown in Fig. 1.2. In Fig. 1.2e the curves are numbered in order of increasing impurity concentration in the semiconductor.

(3) Strong electric fields are needed to find a dependence of the electrical conductivity of a semiconductor on the field strength. Below a critical value the conductivity does not depend on the field strength F, i.e., Ohm's law is valid:

$$\kappa = \text{const.}$$

Above this critical field strength we find ourselves in the *instability region*, where within certain intervals of F the conductivity becomes negative ($\kappa < 0$) and the current–voltage characteristic has an N or S shape. In still stronger fields, conductivity rapidly grows with F, and often we are dealing with the so-called Poole's law (Fig. 1.3):

$$\kappa = a \exp (bF) . \tag{1.2}$$

For sufficiently strong fields this law is valid for many semiconductors (oxides and sulfides) and is also valid for dielectrics.

(4) As a rule, the electrical conductivity of semiconductors is very sensitive to light, just as in the case of dielectrics. Light absorption usually leads to an increase in conductivity and changes the dependence of conductivity on various factors, so that conductivity in the presence of light differs from that without light, all other conditions being equal.

In addition to this so-called normal photoconductive effect, light sometimes does not affect conductivity (photoelectric nonactive light absorption) or even leads to a small decrease in conductivity (the anomalous photoeffect).

1.2.2. Types of Electrical Conduction

The electrical conduction of nonmetallic crystals may by nature be either *ionic* or *electronic*. The necessary and sufficient condition for ionic conduction is the Faraday effect, while an observable Hall effect is a sufficient but not necessary criterion for electronic conduction. Ionic conduction is provided by interstitial ions (the Frenkel mechanism) or vacancies in the ionic crystal (the Schottky mechanism), while electronic conduction is normally provided by free electrons (purely n-type conduction) or free holes, i.e., electronic vacancies (p-type conduction).

Strictly speaking, all types of conduction mechanisms are present in electrical conduction to a certain extent. However, these proportions depend on the conditions. By acting externally on a lattice we can change the role of a component. Usually one of these is predominant while others can be ignored. Note that the ion mobility is very small compared to electron mobility, whence the electron component becomes predominant over the ion component even when the number of free electrons is still very small compared to that of free ions.

In semiconductors electronic conduction prevails over ionic. On the other hand, in many dielectrics (e.g., alkali-halide crystals) the ionic conduction is the prevailing mechanism in ordinary conditions. This feature can be explained by

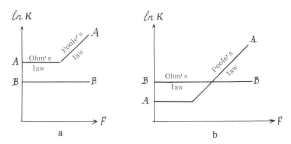

Fig. 1.3. Field strength dependence of conductivity:
a) in semiconductors; b) in dielectrics.

the fact that the energy needed to transport an electron from a bound state to a free state is much smaller in semiconductors than in dielectrics.

In the region of intense electric fields, dielectrics change their behavior in that the electronic component of the conduction becomes predominant. The reason for this is that the ionic component obeys Ohm's law (i.e., is independent of the field strength) up to very high fields, while as noted earlier the electronic component in high fields violates Ohm's law but obeys Poole's law (1.2). This is illustrated in Fig. 1.3, where AA' schematically depicts the electronic component as a function of field strength and BB' the ionic component. Figure 1.3a refers to semiconductors and Fig. 1.3b to dielectrics. With dielectrics, the transition from Ohm's law to Poole's law indicates that the nature of the electrical conduction has changed and that we are in the region of a much stronger field than in the case of semiconductors.

Let us return to semiconductors. The charge carriers are electrons and holes. Hence, the conduction consists of electron and hole components. At low temperatures one is greater than the other, and we have to distinguish between n- and p-type semiconductors. The type of semiconductor depends not so much on the nature of the semiconductor material as on the nature of the impurity in it, or, in the case of binary semiconductors, on the way in which the stoichiometry of the sample is distorted (the predominance of a metal or metalloid). By processing the sample in different ways, we can change the semiconductor type: transform an n-type semiconductor to a p-type, or vice versa. As the temperature increases, the contribution of the nondominant term in the conduction increases, too. At high temperatures both components are more or less equal in concentration and we are dealing with mixed conduction. The semiconductor is then of the intrinsic type (the i-type semiconductor).

1.3. THE MECHANISM OF n- AND p-TYPE CONDUCTION

1.3.1. Free Electrons, Holes, and Excitons in a Lattice

Let us examine the mechanism of n- and p-type conduction. If the crystal lattice possesses a free electron, one of the lattice atoms (or ions) must have an excess negative charge. This anomalous state may be transported from this atom (or ion) to a neighboring like atom (or ion) and in this way travel in the lattice.

For instance, in the lattice of the M_mR_r type, built of metallic ions with charge $+p$ and metalloid ions with charge $-q$ (with $mp = rq$), where M stands for metal and R for metalloid, the presence of a free electron indicates that there is an ion with the anomalous charge $+(p - 1)$ among the metallic ions. Similarly, the presence of a free hole means that one of the lattice atoms (or ions) has an excess positive charge that may be transported from atom to atom (or from ion to ion). In the case of an M_mR_r lattice, this may be a metalloid ion with the anomalous charge $-(q - 1)$ or a metallic ion with the anomalous charge $+(p + 1)$. For instance, in the lattice of ZnO, which we interpret here as an ionic lattice built on the Zn^{++} and O^{--} ions, the state Zn^+ corresponds to a free electron and the state O^- to a free hole. In the Cu_2O lattice built on Cu^+ and O^{--} ions, state Cu corresponds to a free electron and state Cu^{++} to a free hole, with the electron and hole migrating along the regular ions Cu^+.

When a one-component lattice is involved, e.g., the germanium lattice built on neutral Ge atoms, the presence of a free electron or free hole means that there is a Ge^- ion or, respectively, a Ge^+ ion among the neutral germanium atoms.

Electrons and holes remain free as long as they are far away from each other, so that their mutual interaction may be neglected. Then each electron or hole behaves as if no other electrons or holes exist. But if an electron and a hole are close to each other (with a separation of a few lattice constants), they are bound by the Coulomb interaction and as a whole constitute an electrically neutral formation. This is the *Mott exciton*, which may move inside the lattice while contributing nothing to charge transport. An electron and ion localized at neighboring lattice ions correspond to a Mott exciton in its ground (normal) state. The other various excited states of the exciton correspond to other separations.

An exciton is not long-lived. After it forms in the crystal, it sooner or later perishes. This happens in two ways: either by annihilation (recombination) of the electron and hole constituting the exciton or by dissociation of the exciton into a free electron and a free hole. In the first case the destroying of an exciton is accompanied by energy release, while in the second the process requires expenditure of energy.

In the limiting case where the electron and hole are localized at the same lattice atom or ion, the Mott exciton turns into the *Frenkel exciton*. The atom or ion does not then carry excess charge but is in an excited state, and this state may be transported to other like atoms or ions and in this way migrate through the lattice.

Free electrons and holes occupy the center of the stage in this book. The main subject matter is the role that these entities play in surface phenomena.

A lattice in its ground state contains no free electrons, free holes, or excitons. Their presence indicates lattice excitation. An electron and hole may be created at the expense of the inner resources of the lattice when an electron hops from one atom (or ion) to another atom (or ion). Two atoms (or ions) are created in the process, both with excess charges. We denote the energy required by this process by E. Obviously,

$$E = E^- + E^+ + \Delta E,$$

with E^- the energy required to introduce an electron into the lattice, E^+ the energy required to remove an electron from the lattice (or, in other words, to introduce a hole), and ΔE the interaction energy between electron and hole, where in the case of a free electron and hole $\Delta E = 0$.

Let us take as an example the MR lattice built on M^+ and R^- ions, where state M corresponds to an electron and state R to a hole. Then

$$E^- = \frac{e^2}{a} \mu - J, \qquad E^+ = \frac{e^2}{a} \mu + A, \qquad (1.3)$$

where e is the electron charge, a the smallest possible separation between two oppositely charged ions, μ the Madelung constant, J the ionization energy of atom M, and A the electron affinity energy of atom R. The term $e^2\mu/a$ in (1.3) is the energy needed to introduce an electron into a metallic site or a hole into a metalloid lattice site; in other words, it is the interaction energy of the given electron (hole) with all the lattice ions except the one at the given site. If electron and hole are localized at neighboring ions (the exciton in its ground state), then

$$\Delta E = -e^2/a. \qquad (1.4)$$

But if the two are far from each other (free electron and free hole),

$$\Delta E = 0. \qquad (1.5)$$

Thus, the energy needed to create a ground-state exciton, according to (1.3) and (1.4), is

$$E = \frac{e^2}{a} (2\mu - 1) - J + A, \qquad (1.6)$$

while the energy needed to create a free electron and a free hole (a hole–electron pair), according to (1.3) and (1.5), is

$$E = \frac{e^2}{a} 2\mu - J + A. \qquad (1.7)$$

1.3.2. Energy Levels of Electrons and Holes

Figure 1.4 shows the energy patterns for the free electron (level E_C) and the free hole (level E_V) in a lattice built on M^+ and R^- ions, with the electron energy plotted along the vertical axis upward and the hole energy along the same axis downward. In the ground state the level E_V is occupied by an electron while E_C is vacant. An electron transition from level E_V to E_C means creation of a pair consisting of a free electron and a free hole. If we consider this problem quantum mechanically, we will find (see Section 1.4) that the two levels spread out into more or less wide energy bands, which are usually called the *conduction* and

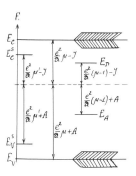

Fig. 1.4. Energy level patterns in a lattice built on M^+ and R^- ions.

valence bands (they are depicted in the right half of Fig. 1.4; the labels C and V, therefore, stand for "conduction" and "valence").

We note that on the surface of a crystal the Madelung constant μ' differs from that in the bulk, so on the surface instead of (1.3) we have

$$E_s^- = \frac{e^2}{a}\mu' - J, \qquad E_s^+ = \frac{e^2}{a}\mu' + A, \qquad (1.3a)$$

where $\mu' < \mu$ (for crystals of the NaCl type we have $\mu = 1.74$ and $\mu' = 1.68$). Here E_s^- is the energy needed to move an electron from infinity to the ion M^+ on the surface, while E_s^+ is the energy needed to move an electron from the ion R^- on the surface to infinity.

The corresponding levels are depicted in Fig. 1.4 and are denoted by E_C^s and E_V^s. Their common name is the *surface Tamm level (state)*, but we have not allowed for the quantum mechanical effect of the overlap of wave functions. This will be done in Section 1.5. The electron on the E_C^s level and the hole on the E_V^s level can move freely from one atom to another along the crystal surface, but they can only go inside the crystal if they acquire a surplus energy $e^2(\mu - \mu')/a$.

In a similar way we may speak of a surface exciton, which has less energy than a body exciton. Instead of (1.6) we then have

$$E_s = \frac{e^2}{a}(2\mu' - 1) - J + A. \qquad (1.6a)$$

This exciton can exist on the surface and freely move along it but cannot go inside the crystal.

We note in conclusion that in real crystals not only the regular atoms or ions of the crystal lattice but impurities and structural imperfections present in the crystal as well may be sources of free electrons and free holes. The energy required to form a free electron or free hole in this case may be lower. In this respect all defects can be divided into two groups: *donors*, which are defects that supply free electrons, and *acceptors*, which are defects that supply free holes. In the electronic "household" of a semiconductor, lattice defects play the main role.

As an example we take the metalloid and metallic vacancies in an MR lattice built on M^+ and R^- ions, where state M corresponds to a free electron and state R to a free hole. Such vacancies are equivalent to the presence in the lattice of a positive and negative charge, respectively.

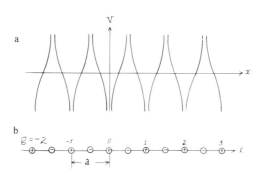

Fig. 1.5. One-dimensional model of a crystal lattice: a) periodic lattice potential; b) (infinite) chain of alternating M^+ and R^- ions.

Suppose that E_D is the energy needed to move an electron from infinity to the ion M^+ that is the nearest neighbor of a metalloid vacancy, and E_A is the energy needed to move an electron to infinity from the ion R^- that is the nearest neighbor of a metallic vacancy. We then obviously have from the Madelung model:

$$E_D = \frac{e^2}{a}(\mu - 1) - J, \qquad E_A = \frac{e^2}{a}(\mu - 1) + A.$$

The corresponding levels are denoted by E_D and E_A in Fig. 1.4.

The electron on level E_D is localized at the metalloid vacancy and neutralizes its charge. Such a defect, called an F-center, is a typical example of a donor. The hole on level E_A is localized at the metallic vacancy and together with the vacancy may be considered as an electrically neutral defect. Such a defect, called a V-center, is a typical example of an acceptor.

1.4. THE ENERGY SPECTRUM OF AN ELECTRON IN AN INFINITE CRYSTAL LATTICE

1.4.1. Statement of the Problem

How does a free electron behave in a crystal lattice? We will assume, for the sake of definiteness, that we have an ionic lattice MR built on ions M^+ and R^- that are point charges. Let us also assume that the lattice is ideal, i.e., without imperfections, and infinite. To make the problem simple, we will take the one-dimensional model, which is an infinite chain of alternating M^+ and R^- ions (see Fig. 1.5b). In our discourse we will gradually lift all these assumptions and arrive at the general picture.

We neglect the interaction of a chosen free electron and all other free electrons and holes and assume that the concentration of free electrons and holes is low in the lattice. We have thus reduced the problem to the one-electron one.

Let $\psi(x, y, z)$ be the wave function that describes the behavior of our electron and satisfies the Schrödinger equation

$$\hat{H}\psi = E\psi,\tag{1.8}$$

where the Hamiltonian $\hat{H}$ is

$$\hat{H} = -\frac{h^2}{2m}\Delta + V(x, y, z).\tag{1.9}$$

Here

$$V(x, y, z) = V(x - na, y, z)$$

is the periodic lattice potential depicted in Fig. 1.5a, a is the lattice constant (the distance between neighboring like ions), and n an integer.

The next step is to number the metallic ions M^+ in the lattice. Suppose that g is the ion number. Obviously,

$$g = 0, \pm 1, \pm 2, \pm 3, \ldots$$

Let us assume that $\varphi_g(x, y, z)$ is the atomic wave function that describes the electron behavior in an isolated gth metallic atom, with all other M^+ and R^- ions absent. This function, which we assume to be known, satisfies the Schrödinger equation

$$\hat{H}_g\varphi_g = E_0\varphi_g,\tag{1.10}$$

where

$$\hat{H}_g = -\frac{h^2}{2m}\Delta + U_g(x, y, z),\tag{1.11}$$

with

$$U_g(x, y, z) = U_0(x - ga, y, z),$$
$$\varphi_g(x, y, z) = \varphi_0(x - ga, y, z).\tag{1.12}$$

Here U_g is the potential energy of the electron in the field of the gth isolated metallic ion, the function φ_g is assumed nondegenerate (an s function), and the level E_0 is assumed to be sufficiently far from its neighbors (see below).

What we must do is find eigenfunctions and eigenvalues of Eq. (1.8), assuming that the eigenfunctions and eigenvalues of Eq. (1.10) are known. We look for the function ψ in the form of a linear combination of atomic functions φ_g (see [1]):

$$\psi(x, y, z) = \sum_g a_g \varphi_g(x, y, z),\tag{1.13}$$

where summation is over all lattice ions. The problem is therefore reduced to finding the expansion coefficients a_g in (1.13). These coefficients must be chosen in such a way that the function (1.13) satisfies Eq. (1.8) in the best possible way. This requirement minimizes the integral

$$J_E = \int \psi^*(\hat{H} - E)\psi \, d\tau \qquad (1.14)$$

Integration is over the entire volume where ψ is normalized. The minimization condition is expressed by a system of linear homogeneous equations:

$$\frac{\partial J_E}{\partial a_g^*} = 0.$$

We write this system explicitly, assuming that

$$\hat{H} = \hat{H}_g + V - U_g,$$

which follows from (1.9) and (1.11), and introducing the following notations:

$$S_{gg'} = \int \varphi_g \varphi_{g'} \, d\tau, \quad P_{gg'} = \int \varphi_g (V - V_{g'}) \varphi_{g'} \, d\tau, \qquad (1.15)$$

We have

$$\sum_{g'} \{(E_0 - E)S_{gg'} + P_{gg'}\} a_{g'} = 0. \qquad (1.16)$$

which is an infinite system of linear homogeneous equations with an infinite number of unknowns, a_g.

1.4.2. Eigenfunctions and Eigenvalues

Noting that the function φ_g, being an s function, is spherically symmetric and assuming that the overlap integral of the wave functions of two neighboring atoms is small, we may put

$$S_{gg'} = \begin{cases} 1, & \text{if} \quad g' = g, \\ 0, & \text{if} \quad g' \neq g, \end{cases} \quad P_{gg'} = \begin{cases} \alpha, & \text{if} \quad g' = g, \\ \beta, & \text{if} \quad g' = g \pm 1, \\ 0 & \text{otherwise.} \end{cases} \qquad (1.17)$$

In view of (1.17) the system of equations (1.16) takes the form

$$(E_0 - E + \alpha)a_g + \beta(a_{g-1} + a_{g+1}) = 0 \qquad (1.18)$$

for all g's. We look for the solution of this system of equations in the form

$$a_g = a_0 \exp(i\lambda g) \quad \text{or} \quad a_g = A \exp(i\lambda g) + B \exp(-i\lambda g), \qquad (1.19)$$

with λ any number (generally complex), and A and B arbitrary coefficients.

Substituting (1.19) into (1.18) and (1.13), we obtain

$$E = E_0 + \alpha + 2\beta \cos\lambda, \tag{1.20}$$

$$\psi(x, y, z) = a_0 \sum_g \exp(i\lambda g)\varphi_g(x, y, z). \tag{1.21}$$

Parameter λ can be taken as a quantum number, since it completely defines eigenfunction (1.21) and the corresponding eigenvalue (1.20). We note that in the case of an infinite lattice the number λ must be real, since otherwise the function (1.21) becomes infinite at infinity. We also note that (1.21) is periodic in λ with a period of 2π. This implies that we can restrict the values of λ to an interval 2π wide, e.g.,

$$-\pi \leqslant \lambda \leqslant +\pi. \tag{1.22}$$

Often we impose on ψ the periodicity condition (the Born–von Karman condition), conceiving the infinite crystal as an infinite repetition of the same crystal:

$$\psi(x + L, y, z) = \psi(x, y, z), \tag{1.23}$$

where $L = aN$, with N a large positive integer. If we substitute (1.21) into (1.23), we find that parameter λ takes discrete values:

$$\lambda = \frac{2\pi a}{L} j, \tag{1.24}$$

where, according to (1.22), $j = 0, \pm 1, \pm 2, ...$, with $|j| \leq N/2$.

Equation (1.20) shows that when we go over from an isolated atom M to a lattice, the level E_0 is shifted by α and broadens out into a band, as shown in Fig. 1.6. We have

top of band $(\lambda = 0)$ E_{max} $= E_0 + \alpha + 2\beta$,

middle of band $(\lambda = \pm\pi/2)$ E_{mid} $= E_0 + \alpha$,

bottom of band $(\lambda = \pm\pi)$ E_{min} $= E_0 + d - 2\beta$,   (1.25)

band width $E_{max} - E_{min}$ $= 4\beta$,

since in our case β is positive. The band is the wider the greater β is, which is the overlap integral of the functions φ_g and $\varphi_{g\pm1}$ of two neighboring atoms. This integral is the greater the closer the lattice atoms are to each other (the smaller a is), and for a given position of these atoms, the higher E_0 lies. [We note, however, that this overlap is a source of inconsistency because we assumed, following (1.17), that the φ_g are orthogonal, which is definitely not the case for overlapping functions.] If parameter λ is continuous, so is the band, but if this parameter is discrete [(the Born–von Karman condition (1.23)], the band splits into a system of

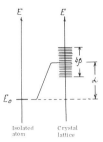

Fig. 1.6. Electron level broadening in a crystal lattice.

levels lying close together and condensing toward the top and bottom of the band. The j in (1.14) can serve as the number of a level in the band, while the total number of states contained in the band is N, i.e., the number of ions M^+ lying in the interval L.

It is possible to show that an electron in the periodic lattice field behaves like a classical particle under an external force but with an effective mass m^* instead of its true mass m (e.g., see [2, 3]):

$$m^* = \left(\frac{h}{2a}\right)^2 \left(\frac{d^2E}{d\lambda^2}\right)^{-1} \tag{1.26}$$

This combined with (1.20) yields

$$m^* = -\frac{h^2}{8a^2\beta\cos\lambda}.$$

Thus, the effective electron mass depends on what level the electron is on in the band. As we can see from (1.26), its absolute value is smaller, the wider the band. The effective mass is positive in the lower half of the band ($\pi \geq |\lambda| > \pi/2$), negative in the upper half ($\pi/2 > |\lambda| \geq 0$), and becomes infinite in the middle ($|\lambda| = \pi/2$). We see that in the upper half of the band the electron behaves very peculiarly—it accelerates in the direction opposite the acting force. Near the lower edge of the band (when λ is close to $\pm n$) and near the upper edge (when λ is close to zero) we have

$$m^* = \mp \frac{h^2}{8a^2\beta}, \tag{1.27}$$

where the minus corresponds to the upper edge and the plus to the lower edge.

1.4.3. A Three-Dimensional Lattice

Let us now go over from the one-dimensional model of a crystal to the three-dimensional cubic lattice. Instead of one integer g characterizing the position of a metallic ion M^+ in the lattice, we will need three integers g_1, g_2, and g_3, viz.,

$$g_1, g_2, g_3 = 0, \pm 1, \pm 2, \pm 3, \dots,$$

which can be taken as the components of a vector $\mathbf{g}$. Then (1.21) becomes

$$\psi(x, y, z) = a_0 \sum_{\mathbf{g}} \exp(i\lambda\mathbf{g})\varphi_{\mathbf{g}}(x, y, z), \tag{1.28}$$

where λ is a vector with components λ_1, λ_2, and λ_3, which serve as three quantum numbers. We may also put

$$-\pi \leqslant \lambda_i \leqslant +\pi,$$

where $i = 1, 2, 3$. Summation in (1.28) is over all values of g_1, g_2, and g_3. Then instead of (1.20) we have

$$E = E_0 + \alpha + 2\beta(\cos\lambda_1 + \cos\lambda_2 + \cos\lambda_3). \tag{1.29}$$

Here

$$\alpha = \int \varphi_{\mathbf{g}}(V - U_{\mathbf{g}})\varphi_{\mathbf{g}}\,d\tau, \quad \beta = \int \varphi_{\mathbf{g}}(V - U_{\mathbf{g}})\varphi_{\mathbf{g}'}\,d\tau, \tag{1.30}$$

where $\mathbf{g}$ and $\mathbf{g}'$ are the "number" of neighboring metallic ions. If we apply the Born–von Karman conditions, the parameters λ_1, λ_2, and λ_3 become discrete:

$$\lambda_i = \frac{2\pi a}{L} j_i,$$

where $j_i = 0, \pm 1, \pm 2, \dots$, with $|j| \leq N/2$ and $i = 1, 2, 3$.

The parameters E_0 and α in (1.29) have a simple physical meaning. Obviously,

$$E_0 = -J,$$

where J is the ionization potential of atom M, while α is the potential energy of the electron belonging to the $\mathbf{g}$th metallic atom in the field of all other ions, i.e., except the $\mathbf{g}$th ion. The electron is taken to be smeared out with a density $\varphi_{\mathbf{g}}{}^*\varphi_{\mathbf{g}}$. If we assume that the electron belonging to the $\mathbf{g}$th metallic atom is point-like and localized at the $\mathbf{g}$th metallic site, the quantum-mechanical parameter α becomes a classical quantity:

$$\alpha \to \frac{e^2}{a}\mu,$$

as $\beta \to 0$, and Eq. (1.29) takes the form (1.3) (see Section 1.3):

$$E = \frac{e^2}{a}\mu - J.$$

If the atomic wave functions $\varphi_{\mathbf{g}}$ are not s functions, which we have assumed them to be up till now, but p functions, i.e., threefold degenerate, then instead of one energy band (1.29) we have three bands superimposed on each other (in other words, a complex band with three times as many states as in the case of s-states [4]). In the case of d-functions, which are fivefold degenerate, we have five superimposed bands [5]. In all these cases and also with a noncubic lattice (up till now we have dealt only with cubic lattices), the dependence of E on the quantum numbers λ_1, λ_2, and λ_3 proves to be more complex than in (1.29). But in all cases E is a periodic function of λ_1, λ_2, and λ_3 with a period of 2π.

Thus, for a free electron in a lattice we have as the energy spectrum a system of bands corresponding to the levels of an isolated metallic atom. The lowest band is known as the *conduction band*. These bands usually overlap somewhat, so that we are dealing practically with a continuous spectrum.

1.4.4. The Energy Spectrum of a Hole

Everything we have said about an electron can be repeated for a hole. We also have an energy spectrum in the form of bands. If the holes are fixed *a priori* at negative metalloid ions of the lattice, these bands correspond to the levels of an isolated metalloid atom. The electron and hole spectra can be depicted in one picture if they are plotted in different directions, as in Fig. 1.4, since introducing a hole is equivalent to extracting an electron. The highest of all hole bands is known as the *valence band*. Let us assume that the presence of a free hole in a lattice means that one of the metalloid ions is neutralized. This neutral metalloid atom may be in either its ground state or one of its excited states, depending on whether the hole is in the valence band or one of the lower lying hole bands.

Hole energy bands may be considered as bands "filled" by electrons, i.e., bands that can be used as a source of electrons. In this sense electron energy bands are "empty," i.e., electrons can be introduced into the bands. As long as hole energy bands contain no holes and the electron energy bands no electrons, our lattice is an ideal dielectric (insulator). To create electrical conduction we must provide electrons for the conduction band (n-type conduction) or provide holes for the valence band (p-type conduction) or do both (mixed conduction).

It goes without saying that in the energy spectrum the highest of the hole bands lies below the lowest of the electron bands; i.e., the valence band is below the conduction band. Indeed, if this were not so, the electrons from the valence band would move to the conduction band, which would mean that our initial idea of a lattice built on M^+ and R^- ions was untrue. This would simply mean that energetically such a lattice structure is not advantageous.

We must specially note that the idea of energy bands has been introduced without any reference to the *collective electron method* (Bloch's method) and, therefore, is free from all deficiencies and restrictions inherent in this method [6]. The band picture of the energy spectrum of an electron or hole can be extended in certain conditions from ionic lattices, which have concerned us up till now, to homopolar crystals. The only difference is that instead of the electrons of the lattice atoms we have to consider either the excess electron on the specific atom or the hole produced by extracting an electron from the atom. The limits of the band picture will be discussed in Section 1.6.

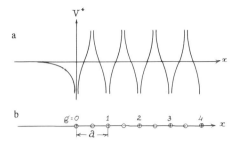

Fig. 1.7. Behavior of a free electron in a semiinfinite crystal: a) periodic lattice potential; b) (semiinfinite) chain of alternating M^+ and R^- ions.

In conclusion we note that the various microdefects present in a real lattice influence the energy spectrum. They have corresponding to them so-called *local* electron and hole levels (acceptors and donors, respectively) lying in the forbidden gap between the conduction and valence bands. While the levels lying inside the energy bands have corresponding to them wave functions whose squared modulus is periodic with the lattice period, the local levels have wave functions with pronounced maxima at the defects and that more or less rapidly fall off as the distance from the defects increases. An electron or hole in an energy band is spread out over the entire crystal, while an electron or hole on a local level is localized at the defect. The position of a local level in the spectrum is determined by the nature of the defect, while the extent to which an electron or hole is localized on such a level is determined by its position in the spectrum—the closer the electron (acceptor) level lies to the conduction band or the hole (donor) level to the valence band, the lower the degree of localization of the electron or hole on the respective level, i.e., the more the corresponding wave function is spread out. In *n*-type semiconductors the donor levels, particularly the ones of practical interest, often lie near the conduction band, while in *p*-type semiconductors the acceptor levels lie near the valence band.

1.5. THE ENERGY SPECTRUM OF AN ELECTRON IN A FINITE CRYSTAL LATTICE

1.5.1. Statement of the Problem

We would like to know how a free electron behaves in a semiinfinite crystal, i.e., a lattice limited on one side. Just as we did in the previous section, we consider an MR lattice built on M^+ and R^- ions. We keep to the one-dimensional model and consider a chain of ions unlimited from the right but limited from the left (Fig. 1.7b). It makes no difference whether the first ion in the chain is metallic (as shown in Fig. 1.7b) or metalloid; what is important is that in the first potential well we allow for the distortion introduced by the termination of the chain. Let us assume that g, the number of the metallic ion, takes the following values:

$$g = 0, 1, 2, 3, ...$$

The Schrödinger equation for the electron of such a lattice is

$$\hat{H}\psi = E\psi, \tag{1.31}$$

where

$$\hat{H} = -\frac{h^2}{2m}\Delta + V^*(x, y, z).$$

Here V^* is the sum of potentials of all ions M^+ and R^- in the lattice and, obviously, is not a periodic function of x. It may be considered periodic only approximately, at large values of x. But we can put

$$V^*(x, y, z) = V(x, y, z) - V'(x, y, z),$$

where V is a periodic potential corresponding to an ideal unlimited crystal lattice, and V' is an addition that distorts the periodicity and is due to the presence of a boundary. Obviously $V' \to 0$ as $x \to \infty$. We assume that V' is nonzero only near the first ion ($g = 0$), since already the second ion ($g = 1$) does not feel the boundary. In other words, we assume (and this is the approximation) that

$$V = \begin{cases} V & \text{for} & x > a, \\ V - V' & \text{for} & x \leqslant a. \end{cases} \tag{1.32}$$

and this is depicted in Fig. 1.7a.

We look for the solution of Eq. (1.31) in the form

$$\psi(x, y, z) = \sum_{g=0}^{\infty} a_g \varphi_g(x, y, z), \tag{1.33}$$

where φ_g is the atomic wave function corresponding to the gth metallic atom (we assume this is an s function). Function (1.33) satisfies Eq. (1.10) and possesses the property (1.12). If we build the variational integral (1.14) and find its minimum, we obtain, just as in Section 1.4, the system of equations (1.16):

$$\sum_{g'} \{(E_0 - E)S_{gg'} + P_{gg'}\} a_{g'} = 0, \tag{1.34}$$

which can be used to find the unknown coefficients a_g in the expansion (1.33). In (1.34), just as in (1.16), summation is over all values of g' (in our case $g' = 1, 2, 3, ...$), while the parameters $S_{gg'}$ and $P_{gg'}$ are given, in contrast to (1.15), by the formulas

$$S_{gg'} = \int \varphi_g \varphi_{g'} \, d\tau, \quad P_{gg'} = \int \varphi_g (V^* - U_{g'}) \varphi_{g'} \, d\tau. \tag{1.35}$$

1.5.2. Eigenfunctions and Eigenvalues

Let us study the integrals (1.35). Since the φ_g are spherically symmetric and their overlap integrals are small, using the approximation (1.32) we may put

$$S_{gg'} = \begin{cases} 1, & \text{if} & g' = g, \\ 0, & \text{if} & g' \neq g, \end{cases}$$

$$P_{gg'} = \begin{cases} \alpha', & \text{if} & g' = g = 0, \\ \alpha, & \text{if} & g' = g > 0, \\ \beta, & \text{if} & g' = g \pm 1, \\ 0 & \text{otherwise.} \end{cases} \tag{1.36}$$

Then (1.34) takes the form

$$(E_0 - E + \alpha')a_0 + \beta a_1 = 0 \qquad \text{for} \quad g = 0,$$
$$(E_0 - E + \alpha)a_g + \beta(a_{g-1} + a_{g+1}) = 0 \quad \text{for} \quad g > 0. \tag{1.37}$$

We seek the solution to this system in the form

$$a_g = A \exp(i\lambda g) + B \exp(-i\lambda g), \tag{1.38}$$

where A and B are arbitrary coefficients, and λ is any number (in general, complex). Two cases are of interest here: (a) both A and B in (1.38) are nonzero, and (b) one of them is zero. In the first case we arrive at the notion of an energy band, while in the second, which is of special interest, we arrive at the notion of a discrete surface level. We will study both cases.

(a) Suppose that $A \neq 0$ and $B \neq 0$. Then parameter λ must be real; if this were not so, the function ψ at infinity would become infinite, as we can see from (1.33) and (1.38). The second equation in (1.37) yields (1.20). We have an energy band that is exactly the same as for an infinite lattice (see Section 1.4). Substituting (1.38) and (1.20) into the first equation in (1.37), we have

$$A \exp(i\lambda) [\alpha' - \alpha - \beta \exp(-i\lambda)] + B \exp(-i\lambda) [\alpha' - \alpha - \beta \exp(i\lambda)] = 0.$$

We see that A and B in (1.38) prove to be coupled, while for an infinite crystal these parameters are independent (see Section 1.4).

(b) Now we assume that either A or B is zero. For the sake of definiteness we assume that $A \neq 0$ and $B = 0$. The parameter λ need not be real-valued for the function ψ to be finite everywhere. In this respect a finite lattice differs from an infinite. Indeed, as we will shortly see, the finiteness of ψ is ensured by the fact that

$$\lambda = n\pi + i\lambda', \tag{1.39}$$

where n is an integer, and λ' is positive. The second equation in (1.37) transforms once more into (1.20), while the first equation in (1.37) yields

$$E_0 - E + \alpha' + \beta \exp(i\lambda) = 0, \tag{1.40}$$

whence, substituting (1.20) into (1.40), we have

$$\exp(i\lambda) = \frac{\beta}{\alpha' - \alpha}. \tag{1.41}$$

Since α, α', and β are real numbers, Eq. (1.41) is valid only if λ is expressed via (1.39). Substitution of (1.39) into (1.41) yields

$$(-1)^n \exp(-\lambda') = \frac{\beta}{\alpha' - \alpha},$$

with n odd, since in our case $\beta > 0$ and $\alpha > \alpha' > 0$.

According to (1.41) and (1.38), the wave function (1.33) takes the form

$$\psi = A \sum_{g=0}^{\infty} \left(\frac{\beta}{\alpha - \alpha'} \right)^g \varphi_g. \tag{1.42}$$

The finiteness of the wave function is ensured only if

$$\frac{\beta}{\alpha - \alpha'} < 1, \quad \text{i.e.,} \quad \lambda' > 0, \tag{1.43}$$

a fact noted earlier. The wave function (1.42), which decays in the bulk of the crystal, corresponds to a local level, whose position in the energy spectrum may be found by substituting (1.39) into (1.20):

$$E = E_0 + \alpha - 2\beta \cosh \lambda'. \tag{1.44}$$

If we assume that

$$\exp(-\lambda') = \frac{\beta}{\alpha - \alpha'} \ll 1, \tag{1.45}$$

then substitution of (1.45) into (1.44) yields

$$E = E_0 + \alpha'. \tag{1.46}$$

We have a level lying below the conduction band.

We note that the numerator of the right-hand side in (1.41) is the band half-width, while the denominator is the distance between the band center and the surface level. Hence, according to (1.42), the closer the surface level is to the band, the slower its wave function falls off in the bulk of the crystal. If we assume the electron to be point-like, parameter α' turns into its classical analog:

$$\alpha' \to \frac{e^2}{a} \mu,$$

and, if we recall that $E_0 = -J$, formula (1.46) transforms into (1.3) (see Section 1.3):

$$E = \frac{e^2}{a} \mu - J.$$

We can repeat our reasoning for the case $A = 0$ and $B' \neq 0$. The finiteness of ψ is provided by the fact that

$$\lambda = n\pi - i\lambda',$$

where, as before, n is an odd integer, and λ' is positive. We arrive at the same eigenfunction (1.42) and (1.46).

However, if

$$\frac{\beta}{\alpha - \alpha'} \geqslant 1,$$

then there is no surface level (or, in other words, the level lies inside the band). In this case both coefficients A and B in (1.38) must be taken to be nonzero.

1.5.3. Tamm and Shockley Surface Levels

The idea of a surface level was first introduced by Tamm [7], who restricted his reasoning to the one-dimensional crystal model (just as we do here) and used the Kronig–Penney potential. Ryzhanov [8], Goodwin [9], and others generalized it to the three-dimensional case. The review of works devoted to surface state theory is given in [10, 11].

In the case of a three-dimensional crystal lattice, the discrete Tamm level (level $E_C{}^s$ in Fig. 1.4) spreads out into a more or less wide surface energy band (the surface conduction band). Introducing the periodicity condition (1.23) into the wave function in directions parallel to the surface, we split the band into a set of closely lying discrete levels. The number of states in the surface band is equal to the number of atoms on the surface (about 10^{15} cm^{-2}). The squared modulus of the wave function corresponding to a level in the surface band is periodic in directions parallel to the surface (the period is that of the lattice), while in the perpendicular direction it falls off with distance from the surface. This means that an electron belonging to the surface band can move freely along the surface but cannot go inside the crystal or leave it. Electrons in the surface band lead to surface electron conductivity of the crystal and may impart a negative charge to the surface with respect to the bulk.

What we have just said of electron surface levels can be repeated for holes. When we go over from a one-dimensional crystal to a three-dimensional, the hole surface level (level $E_V{}^s$ in Fig. 1.4) spreads out into a hole surface band (the surface valence band). Holes in this band lead to surface hole conductivity and may impart an excess positive charge to the surface with respect to the bulk. The fact that there is a potential difference between surface and bulk leads to important consequences, as we will see later.

Tamm surface states may not appear in a limited crystal lattice. This depends on certain conditions being fulfilled [e.g., condition (1.43)]. The surface levels may prove to be absorbed by the wide "bulk" bands [which is clearly demonstrated by (1.43), for instance]. The wave functions corresponding to these levels lose their damped nature and, hence, do not constitute surface states.

In conclusion we note that along with Tamm states, which exist due to the finiteness of the crystal, there may in certain conditions appear so-called Shockley surface states, which have quite a different nature. Shockley [10] dealt with a one-dimensional crystal lattice limited on both sides, and studied the behavior of the wave function and the energy spectrum of an electron as the lattice constant a drops from infinity (the case of isolated atoms) to small finite values. In the process the energy bands, which grow out of the levels of the isolated atoms, broaden, then overlap, but after a becomes smaller than a certain value, gaps appear in the continuous spectrum (forbidden sections). In these gaps one finds discrete levels (one from each band), each of which has corresponding to it a wave function that falls off as we depart from the crystal boundaries (Shockley surface states). These states, in contrast to Tamm states, appear only at small values of a, when the energy bands overlap.

Both Tamm and Shockley levels are characteristic of an ideal surface, where the potential is strictly periodic along the surface. A real surface differs from the ideal by the presence of surface defects, which cause local violations of this periodicity. This leads to the emergence of surface local levels with wave functions that have maxima near the defect (i.e., in surface plane or near it) but fall off as the distance from the defect increases. It is surface states of this type that will interest us in the following pages.

1.6. STATISTICS OF ELECTRONS AND HOLES
IN SEMICONDUCTORS

1.6.1. The Fermi–Dirac Distribution Function

In earlier sections we studied the behavior of an electron or hole in a crystal lattice and ignored the interaction between the specified electron or hole with other electrons and holes. Let us now study the assembly of electrons and holes populating a semiconductor, while still ignoring their interaction, i.e., assuming that the electrons and holes constitute an ideal gas.

Note, however, that all the results we will obtain are also applicable to the case where this interaction is taken into account by the *self-consistent field method*. In this method the interaction of the given electron (hole) with all the other electrons and holes is substituted by an effective field produced by the evenly distributed charge of all electrons and holes in the system. Our electron (hole) is then assumed to be moving in this effective (self-consistent) force field.

It is important that this self-consistent field possesses the same periodicity as the crystal lattice and that even with the interaction taken into account by the self-consistent field method the problem remains a one-electron one. From the standpoint of statistics this means that we are still in the framework of the ideal gas model.

With the energy spectrum of the electron or hole given, we have to find the distribution of the entire assembly over the levels of this spectrum. In other words: what is the probability of a given level E being occupied by an electron or remaining unoccupied (i.e., occupied by a hole)? We denote these probabilities by $f_n(E)$ and $f_p(E)$. Obviously,

$$f_n(E) = \frac{n(E)}{Z(E)}, \quad f_p(E) = \frac{p(E)}{Z(E)},$$

where $n(E)$ and $p(E)$ are the number of electrons and holes with an energy E, and $Z(E)$ is the total number of states corresponding to E or, in other words, the multiplicity of degeneracy of level E (per unit volume), with

$$n(E) + p(E) = Z(E)$$

and, hence,

$$f_n(E) + f_p(E) = 1.$$

According to the Fermi–Dirac statistics, which electrons and holes obey, in equilibrium

$$f_n(E) = \frac{1}{1 + \exp\dfrac{E - E_F}{kT}}, \quad f_p(E) = \frac{1}{1 + \exp\dfrac{E_F - E}{kT}}, \tag{1.47}$$

where energy E is reckoned with respect to an arbitrary (but fixed) reference point. The parameter E_F in (1.47) has the dimensionality of energy and is known as the *Fermi energy* (or *Fermi level*). This quantity, as we will see later, is a regulator of all chemisorptive and catalytic properties of a semiconductor.

The Fermi–Dirac distribution functions $f_n(E)$ and $f_p(E)$ are depicted in Fig. 1.8a and b, where the electron energy is indicated on the vertical axis upward and the hole energy on the same axis downward. The broken line in Fig. 1.8a belongs to the case of absolute zero ($T = 0$). From (1.47) it follows that at $T = 0$,

$$f_n(E) = 1, \quad f_p(E) = 0 \quad \text{for} \quad E < E_F,$$
$$f_n(E) = 0, \quad f_p(E) = 1 \quad \text{for} \quad E > E_F,$$

i.e., all levels below the Fermi level are occupied by electrons (i.e., contain no holes), while all levels *above* the Fermi level are empty (i.e., contain no electrons).

The smooth curve in Fig. 1.8b refers to the case $T \neq 0$ and is symmetric with respect to point $E = E_F$. From (1.47) we can see that

$$f_n(E_F + \Delta E) = f_p(E_F - \Delta E), \quad f_n(E_F) = f_p(E_F) = 1/2.$$

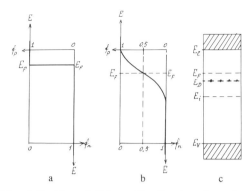

Fig. 1.8. Fermi–Dirac distribution functions
$f_n(E)$ and $f_p(E)$ and the energy spectrum of a
semiconductor with one type of defect: a) at
$T = 0$; b) at $T \neq 0$; c) the energy spectrum.

Hence, the Fermi level is one that is populated to an equal degree by electrons and
holes. This may serve as a definition of the Fermi level. From (1.47) it follows
that

$$f_n(E) = \exp\left(-\frac{E - E_F}{kT}\right) \quad \text{for} \quad \exp\left(-\frac{E - E_F}{kT}\right) \ll 1,$$

$$f_p(E) = \exp\left(-\frac{E_F - E}{kT}\right) \quad \text{for} \quad \exp\left(-\frac{E_F - E}{kT}\right) \ll 1,$$

(1.48)

i.e, the Fermi–Dirac distribution turns into the Maxwell–Boltzmann distribution.
For electrons this happens when the energies are considerably *higher* than the
Fermi level, while for holes it happens when the energies are much *lower* than the
Fermi level. In both cases the electron or hole gas is said to be *nondegenerate*.

Figure 1.8c depicts the energy spectrum of a semiconductor with one type of
defect. The extent to which the energy levels in Fig. 1.8c are filled with electrons
and holes is given by the curves in Fig. 1.8a and b, which correspond to a fixed
position of the Fermi level. This position changes with temperature (see below).

1.6.2. *n*- and *p*-Type Semiconductors

We wish to find the electron population of the conduction band and the hole
population of the valence band. In other words, we are interested in the charge
carrier concentrations in a semiconductor, which we denote by n and p, respectively. According to (1.47),

$$n = \int f_n(E) Z_n(E) dE, \quad p = \int f_p(E) Z_p(E) dE,$$

(1.49)

where $Z_n(E)\,dE$ and $Z_p(E)\,dE$ are the number of states inside the conduction band and valence band, respectively, in the energy range from E to $E + dE$ (per unit volume). In the formula for n the integration is carried over all the levels in the conduction band, and in the formula for p over all the levels in the valence band.

If we assume the electron and hole gases to be nondegenerate [see (1.48)], Eqs. (1.49) yield

$$n = N_n \exp\left(-\frac{E_C - E_F}{kT}\right), \quad p = N_p \exp\left(-\frac{E_F - E_V}{kT}\right). \tag{1.50}$$

Here E_C is the bottom of the conduction band and E_V the top of the valence band; the coefficients C_n and C_p depend on the nature of the bands: in the simplest case of an s-band (i.e., the band resulting from the s-level of an isolated atom) we have

$$N_n = 2\left(\frac{m_n kT}{2\pi h^2}\right)^{3/2}, \quad N_p = 2\left(\frac{m_p kT}{2\pi h^2}\right)^{3/2},$$

where m_n and m_p are the effective masses of the electron at the bottom of the conduction band and of the hole at the top of the valence band, respectively. In order of magnitude,

$$N_n \approx N_p \approx 10^{19} \text{ cm}^{-3} \ .$$

Equations (1.50) remain valid as long as

$$\exp\left(-\frac{E_C - E_F}{kT}\right) \ll 1 \text{ and } \exp\left(-\frac{E_F - E_V}{kT}\right) \ll 1, \tag{1.51}$$

i.e., as long as the Fermi level lies deep under the conduction band and high above the valence band. Then the conduction band can be considered as a discrete level coinciding with its bottom and N_n-fold degenerate, while the valence band can be taken as a discrete level coinciding with its top and N_p-fold degenerate.

A semiconductor with $n = p$ is called an *i-type semiconductor* (intrinsic semiconductor). We denote the position of the Fermi level for such a semiconductor by E_i and the electron (or hole) concentration by n_i. According to (1.50),

$$n_i = N_n \exp\left(-\frac{E_C - E_i}{kT}\right) = N_p \exp\left(-\frac{E_i - E_V}{kT}\right) \tag{1.52}$$

and, hence,

$$E_i = \frac{1}{2}(E_C + E_V) - kT \ln \frac{N_n}{N_p} \ . \tag{1.53}$$

At absolute zero ($T = 0$), or at any other temperature if we assume that $N_p = N_n$ (as a rule, $N_p \neq N_n$), the Fermi level of such a semiconductor lies exactly in the middle of the forbidden section between the bands.

On the basis of (1.52) we can write (1.50) as

$$n = n_i \exp \frac{E_F - E_i}{kT}, \qquad p = n_i \exp \frac{E_i - E_F}{kT}.$$

Obviously, if $E_F > E_i$, we are dealing with the n-type semiconductor ($n > p$). But if $E_F < E_i$, then we have a p-type semiconductor ($p > n$).

As the Fermi level moves upward, the electron concentration in the conduction band increases, as (1.50) implies, while the hole concentration in the valence band decreases; in the process (as long as both gases are nondegenerate),

$$np = n_i^2 = N_n N_p \exp\left(-\frac{E_C - E_V}{kT}\right). \tag{1.54}$$

1.6.3. Statistics of Local States

Now let us study the population of local levels by electrons and holes. In the energy spectrum these levels represent lattice imperfections (defects). We must distinguish between two types of local levels (corresponding to two types of defects): *donor levels* and *acceptor levels*. We deal with a donor or with an acceptor depending on whether the defect, being in an electrically neutral state, is depicted by a local level occupied by an electron or a hole. A donor may be either in the neutral state or (freeing itself of an electron) in the positively charged state, while an acceptor may be either in the neutral state or (accepting an electron) in the negatively charged state.

Suppose that a crystal has X donors and Y acceptors of a definite type per unit volume. We assume that X^0 and X^+ are the concentrations of neutral and charged donors, while Y^0 and Y^- are the concentrations of neutral and charged acceptors, so that

$$X = X^0 + X^+, \qquad Y = Y^0 + Y^-.$$

We then have

for donors $\qquad n = X^0 = X f_n, \qquad p = X^+ = X f_p,$

$$\tag{1.55}$$

for acceptors $\qquad n = Y^- = Y f_n, \qquad p = Y^0 = Y f_p,$

where now

$$f_n = \frac{1}{1 + \dfrac{g_p}{g_n} \exp \dfrac{E - E_F}{kT}}, \qquad f_p = \frac{1}{1 + \dfrac{g_n}{g_p} \exp \dfrac{E_F - E}{kT}}. \tag{1.56}$$

Here E denotes the position of the local level in the spectrum, and g_n and g_p are the statistical weights of the states occupied by an electron and a hole, respectively, i.e., the number of ways in which these states may be realized (if degeneracy is due only to the electron spin, then $g_n = 2$ and $g_p = 1$).

If we compare (1.56) with (1.47) we see that local and band states have somewhat different statistics. The reason is as follows. Let us assume that an energy level is g-fold degenerate. If this level lies inside a band, it can take on g electrons, which may be considered independent: for each electron there is the same probability of being on this level. But if an electron appears on a local level, all other electrons are excluded from it. In this case the interaction between the electrons cannot be neglected. In the final analysis the difference lies in the corresponding wave functions.

From the practical standpoint the difference between (1.56) and (1.47) is of little importance, and in what follows we will use (1.47) instead of (1.56). We can reduce (1.47) to (1.56) if we substitute E^* for E in (1.47):

$$E^* = E + kT \ln(g_p/g_n).$$

Hence, if we apply the Fermi–Dirac statistics to local levels, these levels shift by $\Delta E = kT \ln(g_p/g_n)$.

Here we will not discuss the often encountered case where one defect may take on several electrons or holes, i.e., an acceptor or donor capable of multiple ionization. Nor will we discuss the case (which can be reduced to the above one) where a defect can take on both an electron and a hole, i.e., act as an acceptor and donor simultaneously. Strictly speaking, such defects (just as excitons) cannot be depicted within the band picture, which is built on the one-electron approximation. To do this we would have to assume that local level population leads to another local level in the spectrum; i.e., each defect would have not one but several alternative levels.

1.6.4. The Position of the Fermi Level

The position of the Fermi level fixes the charge carrier concentration (the charge carriers are free electrons and holes) and the extent to which the defects in the semiconductor are ionized. It depends on two factors:

(a) the nature and concentration of the impurity in the semiconductor;
(b) the temperature.

These two factors make the Fermi level move in the energy spectrum. Introduction of a donor impurity moves the Fermi level up, while introduction of an acceptor impurity moves it down. Usually the Fermi level is found from the electrical neutrality of the crystal or, if the crystal is charged (a case we will deal with later), from charge conservation.

Let us find the position of the Fermi level for a neutral semiconductor with one type of donor as an impurity. We denote by n and p, just as we did before, the electron concentration in the conduction band and the hole concentration in the valence band, while X^0 and X^+ are the concentrations of the neutral and charged donors, respectively.

The electrical neutrality condition is simply that

$$n = p + X^+. \tag{1.57}$$

This means, among other things, that $n \geq p$ and, hence, $E_F \geq E_i$. According to (1.55) and (1.56),

$$X^+ = \frac{X}{1 + \exp \dfrac{E_F - E_D}{kT}} \tag{1.58}$$

[we have assumed that $E_D > (E_C + E_V)/2$; see Fig. 1.8c]. Substituting (1.58) and (1.50) into (1.57), we obtain a third-degree equation for $\exp[(E_C - E_F)/kT]$ or $\exp[(E_F - E_V)/kT]$. There are three cases for which we can easily solve this equation:

(1) When the temperature is sufficiently low, so that

$$\frac{N_p}{X} \exp\left(-\frac{E_F - E_V}{kT}\right) \ll \exp\left(-\frac{E_F - E_D}{kT}\right) \ll 1,$$

Eq. (1.57) yields

$$E_F = \frac{1}{2}(E_C + E_D) - kT \ln \frac{N_n}{X} \tag{1.59a}$$

and hence, substitution of (1.59a) into (1.50) yields

$$n = \sqrt{N_n X} \exp\left(-\frac{E_C - E_D}{2kT}\right). \tag{1.60a}$$

We see that at $T = 0$ the Fermi level lies in the middle of the forbidden gap between the conduction band and the donor levels, and as the temperature grows, the Fermi level moves down, and the greater X is, the slower it moves.

(2) When the temperature is intermediate, so that

$$\frac{N_p}{X} \exp\left(-\frac{E_F - E_V}{kT}\right) \ll 1 \ll \exp\left(-\frac{E_F - E_D}{kT}\right),$$

Eq. (1.57) yields

$$E_F = E_C - kT \ln \frac{N_n}{X} \tag{1.59b}$$

and, hence, substitution of (1.59b) into (1.50) yields

$$n = X. \tag{1.60b}$$

We see that in this case the free electron concentration is temperature independent, and the Fermi level proves to be under the donor levels and continues to move down as the temperature grows.

(3) When the temperature is high, so that

$$\frac{N_p}{X} \exp\left(-\frac{E_F - E_V}{kT}\right) \ll \exp\left(-\frac{E_F - E_D}{kT}\right),$$

Eq. (1.57) yields

$$E_F = E_i = \frac{1}{2}(E_C + E_V) + \frac{1}{2} kT \ln \frac{N_p}{N_n} \tag{1.59c}$$

and

$$n = n_i = \sqrt{N_n N_p} \exp\left(-\frac{E_C - E_V}{2kT}\right), \tag{1.60c}$$

i.e., the semiconductor becomes an i-type semiconductor [see (1.53) and (1.54)].

Following Eqs. (1.59a–c), we can depict the temperature dependence of E_F under the assumption that $N_p = N_n$ (see Fig. 1.9a). In Fig. 1.9b we depict the dependence of ln n on $1/T$ according to Eqs. (1.60a–c). The various curves in Fig. 1.9a and b correspond to different values of X (the numbers indicate an increase in X).

In the same manner we can determine the position of the Fermi level and its motion with temperature for any semiconductor and any system of local levels.

1.7. LIMITS OF THE BAND THEORY OF SEMICONDUCTORS

1.7.1. Characteristic Features of the Band Theory

Up till now in describing the behavior of electrons and holes in a crystal, we stayed within the scope of the *band theory of solids*. Basically, the subject matter of this book is confined to this theory. There are only a few cases when we have to go beyond the limits of this theory.

The band theory is very convenient and pictorial. Experimenters working in solid-state physics use it as their language and interpret the results of experiments in its terms. However, it is an approximate theory and like any approximation has its limits. Experimenters often go beyond these limits and use the language of the theory where the theory is no longer valid.

Let us examine the limits of the band theory. First we note some of its characteristic features.

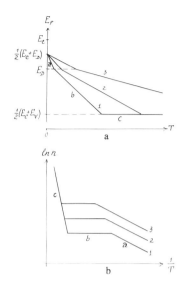

Fig. 1.9. Temperature dependence of position of Fermi level and electron concentration for a semiconductor: a) $C_p = C_n$; b) $\ln n$ vs. $1/T$ dependence.

The electrons populating a semiconductor are, strictly speaking, a system of interacting particles, and the problem is basically many-electron. The band theory makes this many-electron problem a one-electron problem. This is achieved by either neglecting the interaction or accounting for it via the self-consistent field method (Section 1.6) characteristic of the band theory.

The band theory is the theory of the behavior of a single electron. It gives us the right to speak of the behavior of each electron separately and to ascribe to each electron an individual wave function and individual energy. Within the band theory each electron lives its own life and pays no attention to the other electrons as if they did not exist. The only reminder of the other electrons is the Pauli principle, which forbids this electron from taking up quantum states already occupied by the others. This is one characteristic feature of the band theory.

Another feature is the assumption that the force field in which a single electron moves is periodic. The exact form of this field is unimportant to the band theory—it is only essential that the potential is periodic with a period equal to that of the lattice. The characteristic band picture of the energy spectrum, i.e., allowed energy bands generally alternating with forbidden sections, is the result of this periodicity. Hence, the band picture lies in the very prerequisites of the theory, in the assumption that the potential is periodic. The band theory ignores energy exchange between electron and lattice and the effect of lattice vibrations on the electron energy spectrum.

An important drawback to the band theory, as various authors have noted (e.g., see [6, 12]), is the fact that it takes no account of the tendency of electrons to stay far away from each other. Indeed, if we think of electrons as ignoring each other, there is the possibility of several gathering at one atom (or ion). It may so happen that several concentrate at one atom (or ion) simultaneously. Such states, characterized by overconcentration of electrons at one point, have an equal status with other states in the band theory.

This situation is well known from molecular theory, which uses a method equivalent to Bloch's method [1] in the band theory. This is the method of molecular orbitals (MO method), or the Mulliken–Hund method [13, 14]. If we use this method to solve the problem of the hydrogen molecule, we obtain, with nonpolar states, polar states, i.e., states in which both electrons gather at one of the two nuclei in the hydrogen molecule while the other nucleus is stripped of its electron. As a result the hydrogen molecule becomes heteropolar. In the MO method such heteropolar states have the same statistical weight as the homopolar states and correspond to the same values of energy. If we apply Bloch's method to the crystal lattice, we arrive at the same results.

1.7.2. The Validity of the Band Theory

The limits of the band theory were formulated by Pekar [15] (see also [12, 16]). Here are four conditions for the band theory to be valid:

(1) Each atom (or ion) in a crystal lattice has its normal set of electrons. Strictly speaking, the band theory does not work when applied to these "proper" electrons belonging to the atoms or ions of the crystal lattice. In other words, remaining within the framework of the band theory, we cannot give a correct description of such electrons.

But aside from proper electrons, an atom or ion in the lattice may hold an additional electron. In the case of an atomic lattice this additional electron placed on a neutral atom makes it a negative ion, while with ionic lattices such additional electrons produce ions with anomalous charge. These additional electrons may be introduced into the lattice from outside or borrowed from the lattice in the process of shifting electrons from one atom (or ion) to another. When we speak of the band theory being able to describe the behavior of lattice electrons, we mean these additional electrons.

(2) The band theory provides a correct description of the behavior of the additional (excess) electrons as long as the concentration of such electrons is low. In other words, for the band theory to be valid the number of atoms (ions) with an excess charge must be small compared to the overall number of atoms (ions) in the lattice.

(3) The above condition is necessary but not sufficient. The band theory can be applied to the excess electrons only if such an electron does not change the state of the inner electrons of the atom (ion) that holds such an electron. This is the case when, for instance, the inner electrons of an atom (ion) form a closed shell.

(4) Suppose that all the previous conditions are fulfilled. Still, the band theory ceases to be valid if the interaction between two or more of the excess electrons becomes significant. In other words, the band theory does not work for problems in which two atoms (or ions) carry excess charges, i.e., when two such atoms (ions) happen to be close and interact.

Thus, we see that while the band theory (Bloch's method) works when describing the behavior of a free electron, i.e., an electron in the conduction band, when we try to describe electrons that fill (partially or completely) the valence band, we immediately run into difficulties with Bloch's method. Indeed, the electrons that fill the valence band cannot claim to have individual wave functions. Their behavior is essentially a many-electron problem that cannot be reduced to a series of independent one-electron problems.

1.7.3. The Valence Band

Let us assume that the valence band contains a hole. The electrons in the valence band constitute a family whose behavior is described by a collective wave function that cannot be presented in the form of products of individual wave functions and depends on the position of all the electrons. By electrons we mean all "proper" electrons of the lattice atoms (ions) except one electron whose absence is equivalent to the hole. The position coordinates of this electron are not present in the collective wave function. We can thereby say that this function describes the behavior of the hole [6].

Thus, when describing an electron in the conduction band, we stay within the one-electron approximation, while when describing a hole in the valence band, we are dealing not with a one-hole problem, so to say.

The valence band, therefore, has a different meaning than the conduction band. The conduction band constitutes a system of electron levels and the valence band is effectively a system of hole levels. The relationship between the two is the same as between the optical terms of a one-electron atom and the x-ray terms of a many-electron atom (x-ray terms correspond to removal of an electron from an electron shell). For an atom we have a whole series of x-ray terms, i.e., the energy of the system depends on which electron is absent in the system. Figures in textbooks on atomic theory usually present electron levels with the energy increasing upward and hole (x-ray) levels with the energy increasing downward. The same is done in the band picture of a crystal, the only difference being that both electron and hole levels are combined in one figure.

Such an interpretation of the valence band differs from the Bloch interpretation. In Bloch's method the "proper" electrons have an equal status with the excess: valence band electrons are considered as independent as the conduction band electrons, and each is described by an individual wave function. This is as incorrect as trying to build a many-electron atom with n electrons by taking the system of hydrogen levels and placing the n electrons one after another on these levels, allowing only for the Pauli principle. For this reason, from the viewpoint of solid-state theory the method widely used in theoretical studies of heterogeneous catalysis in which the so-called d-electrons (i.e., electrons from the d-band) are considered as Bloch electrons must be seen as unjustified.

Thus, when we speak of the valence band of a semiconductor, we do not mean the behavior of separate and independent electrons. We mean the behavior of independent holes. In other words, we do not reduce the many-electron problem to a one-electron but to a one-hole problem. We see that the band theory cannot describe the ground state of a system, when the conduction band holds no free electrons and the valence band no free holes. But it can be used to describe the excited states characterized by the presence of electrons in the conduction band and holes in the valence band.

We know that quantum mechanically the excited states of a system can be considered as an ideal gas of so-called quasiparticles, *elementary excitations*. The system energy then splits into the ground state energy and the energies of individual elementary excitations. Free electrons and holes, with which the electron theory of solids is concerned, is nothing more than an example of such elementary excitations.

In conclusion we must note that there is a broad group of problems in solid-state theory that can be solved by the energy band approach. This method, if understood correctly, has every right to be further developed to explain ever more experimental data. Its use in the theory of luminescence or the theory of heterogeneous catalysis (to which the present book attests) is proof of the above statement.

But solid-state theory also has problems that go beyond the scope of the band theory. The recombination of a free electron with a free hole, a process that in the band theory is described as the drop of an electron from the conduction band to the valence band, is an example of such a problem. Here the band theory gives the initial and final states but is unable to disclose the mechanism of the process in which the hole and electron interact. The same refers, of course, to the inverse process of ionization (i.e., creation of a free electron and a free hole), where the band theory again gives only the initial and final states. Intermediate states, which are excitonic states, cannot be explained by the band theory. As one more example of such a problem we point to the interaction of a hydrogen atom with the lattice of a semiconductor. (In Section 2.6 we consider this problem in detail.) If here we stay within the one-electron approximation, we risk losing the most important part of the whole problem, the chemisorption bond that holds the atom on the surface.

Chapter 2

THE VARIOUS TYPES OF ADSORPTION

2.1. THE MAIN LAWS OF ADSORPTION

2.1.1. The Main Prerequisites for Langmuir's Theory

As soon as a semiconductor is brought into contact with a gaseous medium, its surface begins to be covered by the molecules of the gas, i.e., adsorption has set in. The process ceases when an equilibrium between the surface and the gaseous phase is established, i.e., when the number of molecules passing from the gaseous phase to the surface per unit time is equal (on the average) to the number of molecules leaving the surface for the gas over the same interval. The presence of the molecules adsorbed by the semiconductor surface changes the properties of the latter. Thus, adsorption is the agent by which the ambient acts on the surface and, indirectly, on some of the bulk properties of the semiconductor.

Sometimes adsorption proceeds very rapidly, so that the adsorption equilibrium sets in instantly. In other cases the process is fairly slow, so that considerable time must elapse before there is equilibrium between the surface and the gaseous phase. In such cases we speak of the kinetics of adsorption and introduce the function

$$N = N(t),$$

where N is the surface concentration of the adsorbed particles, and t is time.

When adsorption equilibrium is reached, the number of gas molecules N retained on a surface of unit area depends on the external conditions, i.e., pressure P and temperature T, or

$$N = N(P, T). \tag{2.1}$$

This is an equation of state. If we keep the temperature constant, we have an equation of an isotherm (the dependence of N on P for constant T), while with P constant we have an isobar (the dependence of N on T for constant P). The value of N for fixed values of P and T is the adsorption capacity, or adsorptivity, of the surface, and it depends on the quantities that characterize the nature and state of the surface and that enter Eq. (2.1) as parameters.

Let us now turn to the theory of adsorption proposed by Langmuir [1]. The theory, which we call classical, is based on the following assumptions:

(1) Adsorption takes place at separate adsorption centers, whose physical nature we will not specify for the time being. We assume that each center can hold only one gas molecule and that the surface has adsorption centers of only one type characterized by the adsorption heat q (the same for all centers), i.e., the same binding energy with respect to molecules of a definite type. Such a surface is said to be energy-homogeneous.

(2) The adsorbed molecules do not interact with each other; i.e., the strength with which a given molecule is coupled with a given center depends only on the type of molecule and center (we have assumed that all the adsorption centers are of the same type) and does not depend on whether there are molecules at the neighboring centers.

In addition to the above assumptions required for Langmuir's theory we introduce the following two assumptions, which are always taken for granted even if not formulated specifically:

(3) The number of adsorption centers on the surface is a constant characterizing the surface and depending on its prehistory. The overall number of centers does not vary with temperature and is independent of the degree to which the surface is filled by adsorbed molecules (i.e., the fraction of centers occupied with adsorbed molecules).

(4) Each given adsorption center may be coupled with a given molecule only in a definite way, so that the binding energy is a unique characteristic of the given center with respect to the given molecule. During the lifetime of a molecule in the adsorbed state (at a given adsorption center) the nature and strength of its bond with the center does not change.

In Section 6.4 we will analyze these assumptions in greater detail.

2.1.2. The Kinetics of Adsorption

We will now study the kinetics of adsorption in the framework of Langmuir's theory. Suppose that N^* is the number of adsorption centers per unit surface area. We then have

$$dN/dt = \alpha P(N^* - N) - \beta N, \tag{2.2}$$

where

$$\alpha = \kappa s/\sqrt{2\pi MkT}, \quad \beta = \nu \exp(-q/kT). \tag{2.3}$$

The first term on the right-hand side of Eq. (2.2) is the number of molecules from the gaseous phase arriving per $cm^2 \cdot sec$ onto the surface, while the second term is the number of molecules leaving the surface for the gaseous phase (from the same surface area and in the same time interval). Here M is the mass of an adsorbed molecule, s its effective surface area, κ the sticking probability, the probability that a gas molecule that has arrived at an adsorption center on the surface from the gas will become attached, β the probability of desorption for an adsorbed molecule per unit time (the quantity $\tau = 1/\beta$ is then the mean lifetime of a gas molecule in the adsorbed state), ν is a constant, and q is the binding energy of the adsorption heat).

At the beginning of the adsorption process, while the surface coverage is small ($N \ll N^*$) and the desorption can be neglected when compared with adsorption, the adsorption rate given by Eq. (2.2) is

$$dN/dt = \alpha P N^*. \tag{2.4}$$

Integrating Eq. (2.2) with the initial condition $N = 0$ at $t = 0$, we obtain

$$N(t) = \frac{N^*}{1 + b/P} [1 - \exp(-at)], \tag{2.5}$$

where

$$a = \alpha P + \beta, \quad b = \beta/a, \tag{2.6}$$

where for small t's (for $t \ll 1/a$) we again arrive at (2.4).

Langmuir's kinetics (2.2) or (2.4) is often observed in experiments. In some cases, however, the adsorption kinetics for a fairly wide range of surface coverages is given by the Roginskii–Zel'dovich equation [2]

$$dN/dt = C \exp(-\gamma N) \tag{2.7}$$

with $\gamma > 0$, whence

$$N(t) = \frac{1}{\gamma} \ln\left(1 + \frac{t}{t_0}\right). \tag{2.8}$$

Roginskii and Zel'dovich used Eq. (2.7) [or (2.8)] to describe the kinetics of adsorption of CO on MnO_2, while Taylor and Thon [3] have demonstrated its applicability to a vast body of experimental data. Stone [4] has given a survey of such data. On the other hand, Thuillier [5] in studying the kinetics of adsorption of O_2 on ZnO established that it obeys the following equation:

$$dN/dt = C \exp(-\gamma N^2), \tag{2.9}$$

to which we will return later. According to Schuttler and Thuillier [6], the kinetics of adsorption of O_2 on TiO_2 obeys Eq. (2.9), too. In many cases the adsorption kinetics for not very low or very high surface coverages obeys Bangham's law [7]

$$N(t) = Ct^n, \tag{2.10}$$

with $n < 1$.

2.1.3. Adsorption Equilibrium

Let us now examine the state of adsorption equilibrium. In Langmiur's theory an isotherm can be obtained by substituting $t = \infty$ for t in Eq. (2.5):

$$N = \frac{N^*}{1 + b/P}, \tag{2.11}$$

or

$$N = N^* \frac{P/b}{1 + P/b}, \tag{2.12}$$

where, according to Eqs. (2.6) and (2.3),

$$b = b_0 \exp(-q/kT). \tag{2.13}$$

In the region of low pressures ($P \ll b$ or $N \ll N^*$), Langmuir's isotherm reduces to Henry's law (or isotherm)

$$N = N^* \frac{P}{b}. \tag{2.14}$$

If, as is often the case, in adsorption a molecule splits into two particles (e.g., a H_2 molecule splits into two H atoms) each of which attaches itself to a definite adsorption center, we can show that Langmuir's theory instead of Eq. (2.11) yields

$$N = \frac{N^*}{1 + \sqrt{b/P}}, \tag{2.15}$$

where N is the number of occupied adsorption centers per unit surface area, N^* the total of adsorption centers within the same area, and q in (2.13) is the adsorption energy of the initial particle (allowing for its fragmentation). If the pressure is very low (Henry's regions), Eq. (2.15) assumes the form

$$N = N^* \sqrt{\frac{P}{b}}. \tag{2.16}$$

If the gaseous phase is a mixture of gases, each gas is adsorbed independently if it has its own adsorption centers, and for each gas we have a Langmuir's isotherm (2.12)

$$N_i = N^* \frac{P_i/b_i}{1 + P_i/b_i} ,$$ (2.17)

where i labels the specific gas in the mixture, and P_i is its partial pressure. But if all the gases in the mixture are adsorbed at the same centers, then, instead of Eq. (2.17), Langmuir's theory yields

$$N = N^* \frac{P_i/b_i}{1 + \sum_i P_i/b_i} ,$$ (2.18)

where N^* is the total number of adsorption centers. We see that here the adsorptivity of the surface with respect to a given gas is reduced due to the presence of the other gases in the mixture. This is the result of blocking of the adsorption centers by the molecules of the other gases. Note that actually this is not the only possibility. Below we will consider cases that cannot be studied in the framework of Langmuir's theory, when the adsorption of one of the gases in the mixture increases the adsorptivity of the surface with respect to the other gases instead of decreasing it.

Langmuir's adsorption isotherm (2.11) is far from being the only one observed in experiments. The so-called differential adsorption heat q in Eq. (2.11) [see also (2.13)] is often a function of N and decreases as N grows, which leads to a distortion in the Langmuir isotherm. Very often one encounters the so-called Freundlich isotherm [8]

$$N = CP^n ,$$ (2.19)

with C and n constant and $n < 1$, or the logarithmic isotherm [9]

$$N = C \ln \frac{P}{P_0}$$ (2.20)

(C and P_0 constant), both often being valid in a broad pressure interval.

These violations of the simple Langmuir laws point to the fact that the postulates underlying Langmuir's theory (see above) are not always realized in experiments. There are two ways out of this difficulty: to introduce interaction between the adsorbed molecules or to introduce the notion of an inhomogeneous surface with different types of adsorption centers that have different adsorption heats q. We can also modify the postulates of Langmuir's theory. We will return to this later.

Let us now discuss the adsorptivity of a surface with respect to a given gas. From the standpoint of Langmuir's theory, as demonstrated by Eqs. (2.11) and (2.13), adsorptivity is determined by the values of parameters N^* and q; the adsorptivity of a surface, i.e., the number of molecules N retained by it at given temperature T and pressure P, is the higher, the greater the number of adsorption centers N^* on the surface, and the stronger the bond q of the adsorbed molecule to the adsorption center. Actually the situation is more complex, since the adsorptivity often depends on factors that are not present in Langmuir's equation (2.11). For instance, it changes if we introduce impurities into the crystal or irradiate the crystal or place the crystal in an electric field.

In conclusion we will consider the kinetics of desorption. Let us assume that after the adsorption equilibrium between the surface and the gaseous phase has been established, the pressure suddenly drops. The adsorbed molecules begin to leave the surface or, in other words, desorption sets in. In Langmuir's theory this process is described by Eq. (2.2), where, however, we must put $P = 0$. Integration of Eq. (2.2) with the initial condition

$$N = N_0 \quad \text{at} \quad t = 0,$$

where N_0 is the equilibrium coverage, yields

$$N = N_0 \exp(-\beta t). \tag{2.21}$$

According to (2.3), the higher the temperature, the more rapid the desorption. Note, however, that with the temperature the same as for adsorption, it is not always possible to remove all the adsorbate from the surface by lowering the pressure since a fraction of the molecules remain permanently attached to the surface. If we wish to remove this fraction, we must raise the temperature considerably. This is why we often speak of partial irreversibility of adsorption. The concept of reversible and irreversible chemisorption is beyond the scope of Langmuir's theory.

2.2. PHYSICAL AND CHEMICAL ADSORPTION

2.2.1. The Difference between Physical and Chemical Adsorption

We must distinguish two types of adsorption, namely, physical absorption (physisorption) and chemical adsorption (chemisorption). On the other hand, we must distinguish activated adsorption from ordinary (nonactivated) adsorption. Note that the terms chemisorption and activated adsorption are often used interchangeably for no valid reason. Generally speaking, chemisorption and activated adsorption are not the same, and we will return to this later.

The difference between physisorption and chemisorption lies in the difference in the forces that retain the adsorbed molecules on the surface of the solid. Indeed, the forces that arise between a solid and a foreign molecule and produce adsorption can be of different nature. For instance, they may be of electrostatic origin, such as van der Waals' forces or forces of electrostatic polarization or image forces. In this case we speak of physical adsorption. But if the forces re-

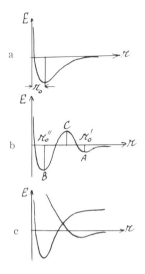

Fig. 2.1. Adsorption curves: a) with a single minimum; b) with two minima corresponding to physical adsorption ($r = r_0'$) and chemical adsorption ($r = r_0''$); c) the real adsorption curves corresponding to physisorption and chemisorption.

sponsible for adsorption are of a chemical nature (forces of the exchange type), we are dealing with chemisorption. Here adsorption constitutes a chemical combination of the molecule with the solid. As in every chemical compound, the forces making up the bond are in the given case covalent forces, but ionic interaction may be involved to a certain extent.

Figure 2.1a depicts an adsorption curve that represents the energy E of the system as a function of the distance r between the adsorbent surface and the particle being adsorbed. Here r_0 is the equilibrium distance. In contrast to physisorption, chemisorption is characterized by considerably smaller values of r_0 and considerably larger values of q. In other words, in chemisorption a molecule is bound to the surface more firmly than in physisorption. In the case of physical adsorption q is about 0.01–0.1 eV, while in chemisorption q is about 1 eV. However, we must note that the energy released during adsorption cannot serve as a guide in determining the true binding energy. Here is an example. Suppose that a molecule AB dissociates in the process of adsorption, which is often the case:

$$AB + L \rightarrow AL + BL,$$

where L stands for "lattice." The amount of energy released in this act we denote by Q. Obviously,

$$Q = (q_A + q_B) - D_{AB},$$

where q_A and q_B are the energies with which the atoms A and B are bound to the lattice, and D_{AB} the energy of dissociation of molecule AB. Here both q_A and q_B may be very large separately, while the total energy release Q may prove to be low.

The experimenter is not always able to draw a dividing line between physisorption and chemisorption. Along with the clear-cut limiting cases there are intermediate cases where the dividing line is fuzzy. It is difficult to find experimental criteria that would enable us to precisely distinguish physisorption from the so-called weak form of chemisorption (see Section 2.3).

In the theoretical context the approaches to physisorption and chemisorption must be radically different. In the case of physical adsorption the adsorbed molecule and the lattice can be treated as two independent systems. The influence of the adsorbent on the adsorbate can then be assumed to be a small perturbation and the problem can be solved by applying perturbation theory methods. In the case of chemical adsorption the adsorbed molecule and the lattice constitute a single quantum-mechanical system and must be considered as a united whole. Here the adsorption, as has already been pointed out, is the chemical combination of the molecule and adsorbent.

A simple example illustrating the difference between these two approaches is the problem of the interaction of an H atom with an H^+ ion placed at a distance r from one another. At large values of r the hydrogen atom may be considered as situated in the field of the H^+ ion, and this field can be treated as a small perturbation, which leads to polarization of the H atom and Stark splitting of its electronic levels. But as r decreases, this picture ceases to be correct, since we have to deal with the molecular ion H_2^+, which is a single system.

2.2.2. Calculating the Adsorption Minimum

In a number of theoretical papers devoted to physical (van der Waals') adsorption (e.g., see [10, 11]), the values of the equilibrium distance r_0 calculated from the position of the adsorption minimum on the E vs. r curve (see Fig. 2.1a) proved to be equal or even less than the sum of the atomic radii of the adsorbed particle and the adsorbent atom. We must bear in mind, however, that this sum is exactly the critical distance at which exchange interaction comes into play and van der Waals interaction (used in these papers) loses its meaning. Indeed, van der Waals' interaction, which in these papers produces the attraction of the adsorbed atom to the surface, is calculated in second-order perturbation theory. The concept of van der Waals' interaction, which in these papers produces the attraction of the adsorbed atom to the surface, is calculated in second-order perturbation theory. The concept of van der Waals' forces retains its meaning only as long as we remain within the framework of perturbation theory, i.e., at values of r so large that the wave functions of the adsorbed particle and the lattice of the adsorbent practically do not overlap. In the papers just cited the branch of the E vs. r curve lying to the right of the minimum in Fig. 2.1a is calculated from the van der Waals' attraction formulas, while the branch to the left of the minimum is calculated as a "repulsive potential," whose very nature is due to the overlapping of the wave functions. Thus, these two branches are calculated under assumptions that are contradictory and at the same time r_0 is determined from the interaction of these two branches.

The E vs. r adsorption curve is often plotted as in Fig. 2.1b, with two minima separated by an energy (activation) barrier. The shallow minimum at $r = r_0'$ is interpreted as physical adsorption, while the deeper minimum at $r = r_0''$ (with $r_0'' < r_0'$) is interpreted as chemisorption. Then the transition of the system from point A to point B over the energy barrier C means the transition of the particle from the state of physisorption to the state of chemisorption.

Note that the adsorption curve depicted in Fig. 2.1b cannot occur in reality, since there cannot be two minima on the same adsorption curve (corresponding to a given electronic state of the system). Indeed, the branch AC in Fig. 2.1b indicates the coming into play of the exchange interaction between the adsorbed particle and the adsorbent lattice, an interaction that leads to repulsion (the repulsive potential). At the same time the branch CB represents attraction due to the same exchange interaction. But an exchange interaction that gives rise to repulsion at large values of r cannot lead to attraction at small values of r. For this reason an adsorption curve with a minimum corresponding to physisorption cannot lead to chemisorption. The reverse is also true: there cannot be physisorption in an electronic state leading to chemisorption. In other words, physical and chemical adsorption necessarily correspond to two different adsorption curves, which express two different states of the system. An energy barrier that separates the two adsorption minima can only arise as a result of intersection of two such adsorption curves as shown in Fig. 2.1c and as occurs, for example, in the case examined by Lennard-Jones (see below).

In what follows we consider only chemisorption. It is the simplest type of a heterogeneous reaction in which a gas particle combines with a solid to form a single system.

2.2.3. Activated Adsorption

This is a type of adsorption that occurs only after a preliminary excitation (activation) is applied to the system. It requires a certain preliminary expenditure of energy (the activation energy), which, however, is later restored with excess as a result of the act of adsorption.

Usually chemisorption adsorption is activated adsorption, and the two terms are therefore often used interchangeably. Such confusion of concepts cannot be considered correct, as we have already remarked.

The presence of activation energy is not necessarily a criterion of chemisorption, since there are cases where chemisorption proceeds without activation energy. Indeed, not every chemical reaction requires activation. For instance, a substitution reaction of the type

$$AB + C \rightarrow AC + B,$$

in which the formation of new bonds requires breaking old bonds, is always characterized by a certain activation energy, while a reaction of the type

$$A + B \rightarrow C$$

may often proceed without any activation.

On the other hand, the absence of activation is not a necessary characteristic of physical adsorption (see below). We can say that the presence or absence of activation in adsorption says nothing about the nature of the adsorption forces.

Activated adsorption differs from normal (nonactivated) adsorption in the character of the kinetics. In the absence of activation energy adsorption proceeds extremely rapidly, so that equilibrium between the adsorbate and the gaseous phase is established almost instantaneously, and the lower the temperature the more rapidly the equilibrium is established. In the case of activated adsorption, however, equilibrium is established more slowly, adsorption proceeds with a measurable rate, and the higher the temperature the more rapid the adsorption.

At the beginning of the adsorption process, while the surface coverage is small, the rate of adsorption can be found from (2.4) and (2.3):

$$\frac{dN}{dt} = \kappa \frac{sN^*P}{\sqrt{2\pi MkT}}, \qquad (2.22)$$

where N is the number of adsorbed molecules of a given type, N^* the total number of adsorption centers (per unit surface area), s the effective surface area of an adsorbed molecule, M the mass of an adsorbed molecule, P the gas pressure, T the absolute temperature, and k the Boltzmann constant. The factor κ in (2.22) is the sticking probability.

The difference between activated and common (nonactivated) adsorption is reflected in the form of κ. In nonactivated adsorption κ is considered constant (temperature-independent) and is usually assumed to be equal to unity. The dependence of the adsorption rate on temperature is then given by (2.22), and we see that dN/dt slowly decreases as T grows. In activated adsorption it is assumed that

$$\kappa \sim \exp\left(-\frac{E_a}{kT}\right), \qquad (2.23)$$

where E_a is the activation energy. The factor κ given by (2.23) ensures a rapid growth of the adsorption rate with temperature.

The exponential factor (2.23) is characteristic of activated adsorption. It is the presence of this exponential factor in the kinetic formula (2.22) that is the criterion of activated adsorption and not the notions concerning the nature of the adsorption forces. The theory of activated adsorption must explain the origin of this factor. The most common way to do this is to introduce an activation barrier on the adsorption curve (see Fig. 2.1), with the height of the barrier being the activation energy E_a. The occurrence of such a barrier may be due to various causes.

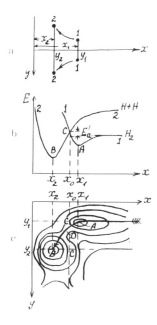

Fig. 2.2. Formation of activation barrier: a) dissociation of the H_2 molecule accompanying adsorption; b) energy diagram of the process; c) the corresponding Eyring diagram.

2.2.4. The Nature of the Activation Barrier

Lennard-Jones [12] showed that an activation barrier occurs when the adsorption of a molecule is accompanied by its decomposition into separate atoms or radicals (the mechanism of such dissociation in adsorption will be examined in Section 2.4.4). Following Lennard-Jones, let us imagine the adsorption of hydrogen accompanied by dissociation of the H_2 molecule into two atoms: $H + H$. Suppose that the H_2 molecule, when it approaches the surface, is parallel to the surface, as shown in Fig. 2.2a (the plane $x = 0$ is the adsorbing surface). The energy of the system, E, is a function of two parameters, i.e., $E = E(x, y)$, where x is the distance of the H_2 molecule from the surface, and y the distance between the two hydrogen atoms. Figure 2.2b, which is taken from [12], depicts the energy E as a function of x for two different values of y: for $y = y_1$ (curve 1) and for $y = y_2$ (curve 2), where y_1 is the distance between the two hydrogen atoms in the free H_2 molecule, and y_2 the lattice constant of the crystal (we assume that y_2 is greater than y_1). The shallow minimum on curve 1 (at $x = x_1$) corresponds to van der Waals' adsorption of the H_2 molecule, while the minimum on curve 2 (at $x = x_2$) corresponds to chemisorption of the H atoms. Transition from curve 1 to curve 2 indicates dissociation of the molecule. In Fig. 2.2a this process is represented by the transfer of the atoms from position 1–1 to position 2–2. The intersection of curves 1 and 2 in Fig. 2.2b leads to the formation of an energy barrier, which is usually interpreted as an activation barrier and its height E_a' as the activation energy.

We note in passing that the above interpretation of Fig. 2.2b is not completely accurate [29, 30]. Generally speaking, the barrier height E_a' in Fig. 2.2b may differ from the activation energy E_a corresponding to the given process. Curves 1 and 2 in Fig. 2.2b correspond to different values of the parameter y and, therefore, the barrier height cannot be interpreted as the activation energy. This becomes particularly clear on an Eyring diagram shown in Fig. 2.2c. Here a system of isoenergetic curves $E = E(x, y)$ is represented in the xy plane. The curves 1 and 2 in Fig. 2.2b are the intersections of this diagram with the planes $y = y_1$ and $y = y_2$. Point A corresponds to van der Waals' adsorption of the H_2 molecule and point B to chemisorption of two H atoms. Point C in Fig. 2.2c corresponds to the top of the barrier in Fig. 2.2b. However, the adsorbed molecule reaches point B along a path indicated by the solid bent arrow in Fig. 2.2c, passing on its way through a barrier whose summit is determined by point D instead of point C. Obviously, point D may lie either higher or lower than point C (Fig. 2.2c corresponds to the former case, i.e., with $E_a > E_a{}'$). Thus, the quantity $E_a{}'$ in Fig. 2.2b is fictitious and, generally speaking, differs from the activation energy E_a.

Let us now look at formula (2.22). In common theories of activated adsorption it is assumed that

$$N^* = \text{const}, \quad \kappa \sim \exp(-E_a/kT). \tag{2.24}$$

In other words, we assume that the number of adsorption centers (per unit surface area) is a constant (fixed once and for all) determined by the nature of the surface and temperature-independent. Such an assumption, however, may prove to be incorrect for certain types of adsorption centers. Later we will examine the case where the concentration of the adsorption centers does not remain constant but increases exponentially with temperature (see Section 6.1 and [13, 14]). If, instead of (2.24), we assume that

$$N^* \sim \exp(-E_a/kT), \quad \kappa = \text{const}, \tag{2.25}$$

we obtain all the characteristics of activated adsorption. In this case the activation energy E_a is determined by the nature of the adsorption centers and characterizes the energy of formation of these centers. If molecules of different types are adsorbed at centers of different types [differing in the value of E_a in (2.25)], then the activation energy proves to depend not only on the nature of the adsorbent but, in the final analysis, on the nature of the adsorbate. In general, we may put

$$N^* \sim \exp(-E_{a_1}/kT), \quad \kappa \sim \exp(-E_{a_2}/kT). \tag{2.26}$$

The activation energy is then made up of two terms:

$$E_a = E_{a_1} + E_{a_2}.$$

As we have already remarked, the existence of an activation barrier on the adsorption curve is not always proof that the adsorption forces have a chemical origin. In certain conditions such a barrier may also arise in physical adsorption, e.g., if there are repulsive forces between the adsorbed molecules. Let us examine this case in conclusion [15].

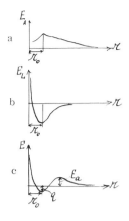

Fig. 2.3. Various branches of energy curves in chemisorption: a) representing the interaction of a molecule with remaining molecules of adsorption layer; b) representing the interaction of a molecule with adsorbent layer; c) representing the combined effect.

When there are repulsive forces between the molecules, each new molecule adsorbed is under the influence of two opposing forces, i.e., attraction toward the surface due to the atoms of the adsorbent and repulsion from the surface due to molecules already adsorbed. The interplay between these opposing forces changes as we move closer to the surface and in certain conditions (with certain assumptions concerning the nature of the repulsive forces) may lead to formation of an energy barrier near the surface.

Indeed, let us assume that one of the physically adsorbed molecules is removed from the adsorption layer and displaced in the direction normal to the surface by a distance r from the surface, while the remaining adsorbed molecules stay in their positions. The energy of the molecule can then be represented in the form

$$E(r) = E_L(r) + E_A(r),$$

where $E_L(r)$ is the energy of interaction of the molecule with the adsorbent lattice, and $E_A(r)$ the energy of interaction of the molecule with the remaining molecules of the adsorption layer.

Figure 2.3b gives a rough sketch of the E_L vs. r dependence, while Fig. 2.3a represents the E_A vs. r dependence. The positions of the maximum on the curve in Fig. 2.3a and the minimum on the curve in Fig. 2.3b coincide (and correspond to $r = r_0$). If we combine these two curves, we can arrive at the curve depicted in Fig. 2.3c. The greater the coverage of the surface, the higher the maximum of the curve of Fig. 2.3a, the greater the height of the barrier E_a, and the smaller the depth q of the well in Fig. 2.3c. Thus, the activation energy E_a and the adsorption heat q change in the opposite sense as the surface coverage increases. We see that the activation barrier can indeed arise in certain conditions in physical adsorption as a result of the interaction between the adsorbed molecules.

2.3. "STRONG" AND "WEAK" BONDS IN CHEMISORPTION

2.3.1. "Weak" and "Strong" Donor and "Strong" Acceptor
Forms of Chemisorption

A system of adsorbed particles is often treated as a two-dimensional gas that covers the surface of the adsorbent. Such an approach is valid and fruitful only as long as we are dealing with physisorption, where the influence of the adsorbent on the adsorbate may be regarded as a small perturbation. In the case of chemisorption, however, the picture of a two-dimensional gas becomes unacceptable. As we have already noted, the adsorbed particles and the adsorbent lattice form a single quantum-mechanical system and must be regarded as a whole. In such an approach the electrons of the crystal lattice are direct participants in the chemical processes that develop at the crystal surface and in some cases are even regulators of these processes.

We will start from a picture that in a certain sense is opposite to that of a two-dimensional gas. We will interpret the chemisorbed particles as "impurities" that have penetrated the crystal surface or, in other words, as structural defects of a kind that destroy the strict periodicity of the surface. In such an interpretation, first carried out in 1948 [16] by the present author, the chemisorbed particle and the lattice of the adsorbent emerge as a single quantum-mechanical system, with the participation of the chemisorbed particles in the electronic system of the lattice being automatically ensured.

Note that this in no way means that there is a strict localization of the adsorbed particles, since their ability to migrate over the surface to a certain extent is preserved. Of course, in such migration the particles encounter energy barriers. If the height of these barriers is less than the energy of bonding a particle with the lattice, the particle can wander over the surface without becoming detached. In this event the mobility of a particle, naturally, increases with temperature.

In interpreting chemisorbed particles as surface "impurities" the inherent difference between chemisorbed particles and defects that arise from the prehistory of the surface and are present on every real surface becomes blurred. The only difference is that chemisorbed particles are able to leave the surface for the gas or the gas for the surface, while the defects we have just mentioned can be regarded as firmly bound to the surface and not able to go over to the gaseous phase.

In a number of theoretical papers [17–19] it has been shown that a chemisorbed particle regarded as a sort of structural defect of the surface proves to be a localization center for a lattice free electron, acting as a trap for this electron and thus serving as an acceptor for it. Or it may serve (and this depends on the nature of the particle) as a localization center for a free hole, thus emerging as a donor.

These papers also show that in general a chemisorbed particle on an adsorbent can simultaneously be an acceptor and a donor, with an affinity for both a free electron and a free hole. Note that structural defects of this type, i.e., being simultaneously an acceptor and a donor, are well known in solid state theory. For example, an F-center, which, as is well known, can trap a free electron and thus become an F'-center (which leads to a change in the coloration of the crystal), can at the same time trap a free hole, which leads to the disappearance of the F-center (and is accompanied by loss of color by the crystal).

Finally, these papers show (and this is important) that localization of a free electron or hole on an adsorbed particle (or near it), forming a charged particle, changes the nature of its bond with the surface, and the bond will become stronger (see Section 2.6). In the process the electron or hole is involved in the bond.

Therefore, we must distinguish between two forms of chemisorption:

(1) "Weak" chemisorption, in which the chemisorbed particle C (considered together with its adsorption center) remains electrically neutral and in which the bond between the particle and the lattice is established without the participation of a free electron or hole from the crystal lattice. We will denote such a bond by CL, where L stands for "lattice."

(2) "Strong" chemisorption, in which the chemisorbed particle retains in its neighborhood a free electron or hole from the crystal lattice (and is thus an electrically charged compound) and in which the free electron or hole plays a direct part in the chemisorption bond.

Note that the terms "weak" and "strong" have a relative meaning. It is only a matter of more or less stable forms of bonding in chemisorption. As noted earlier, although "weak" chemisorption and physisorption differ drastically in their nature, experimentally it is extremely difficult to draw a distinct line between the two.

Since in "strong" chemisorption a free electron or hole is involved in the bond, we may distinguish two types of "strong" bonds and introduce the following terminology:

(a) A "strong" n-bond (or acceptor bond) is one in which a free electron captured by the adsorbed particle participates. We denote such a bond by CeL, where eL denotes a free electron in the lattice.

(b) A "strong" p-bond (or donor bond), denoted by CpL (where pL is the symbol for a free hole in the lattice), is one in which a hole captured by the adsorbed particle participates.

By its very nature an acceptor bond, just as a donor bond, may be a purely ionic or a purely homopolar bond or, in the general case, a bond of mixed type. As we will see below, this depends on how the electron or hole captured by the particle and participating in the bond is distributed between the adsorbed particle and the adsorption center. In other words, this depends on the nature of the localization of this electron or hole, which is determined by the nature of the adsorbate and adsorbent.

2.3.2. The Various Forms of Chemisorption on Ionic Crystals

Figure 2.4 shows the various forms of chemisorption for a particle C on a ionic crystal MR composed of singly charged M^+ and R^- ions. We recall that the presence of a free electron in such a crystal implies the existence of a neutral state M wandering about the M^+ ions of the lattice, while the presence of a free hole implies (although not always; see Section 1.3) the presence of a neutral state R wandering about the R^- ions from one ion to the neighboring one.

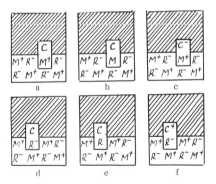

Fig. 2.4. Various forms of chemisorption on ionic crystals: a, d) "weak" bond; b, c) "strong" acceptor bond; e, f) "strong" donor bond.

Figures 2.4a and 2.4d correspond to a "weak" bond. Which of the two cases, a or d, is realized depends on the nature of the particle C and the lattice. Figures 2.4b and 2.4c demonstrate a "strong" acceptor bond and represent two limiting cases, the first corresponding to a purely homopolar bond (Fig. 2.4b) and the second to a purely ionic bond (Fig. 2.4c). As a rule, however, we have something intermediate between what is shown in Fig. 2.4b and 2.4c. Finally, Figs. 2.4e and 2.4f show a "strong" donor bond and represent two limiting cases, too. The real picture lies somewhere in the middle.

Note that with the bonds depicted in Figs. 2.4b and 2.4e, the M or R atom retaining the chemisorbed particle C is connected with the lattice more weakly than the normal M^+ or R^- ion, respectively. As a result of this we may expect volatilization of the CM or CR molecule in some cases; i.e., the particle C on desorbing may carry off an atom of the lattice, which would destroy the stoichiometric composition of the crystal. In all cases such adsorption would facilitate flat creep, which plays such an important role in the sintering, recrystallization, and disintegration of solids in a reaction. The well-known effect of adsorption on the surface mobility of adsorbent atoms may also be connected with this.

The reader must bear in mind that the electrons or holes drawn into a bond are not always taken from the supply of free electrons or holes in the crystal. They may also be taken from the atoms or ions of the lattice proper. For instance, an M^+ ion acting as an adsorption center (Fig. 2.4a) can obtain an electron by removing it from a neighboring R^- ion. As a result we have a hole that, having overcome its bond with the electron, can start to wander about the crystal and thus is included in the supply of free holes. In Fig. 2.4d an R ion, which is an adsorption center, can be neutralized; i.e., it acquires a hole and uses it to form a "strong" donor bond with the particle C chemisorbed on it, transferring its electron to a neighboring M^+ ion. The electron by moving through the lattice joins the supply of free electrons.

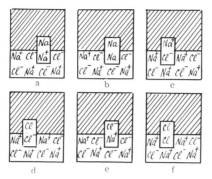

Fig. 2.5. Various forms of chemisorption on NaCl lattice: a) "weak" bond for Na atom; b) "strong" acceptor bond for Na atom; c) "strong" donor bond for Na atom; d) "weak" bond for Cl atom; e) "strong" acceptor bond for Cl atom; f) "strong" donor bond for Cl atom.

We see that "strong" forms of chemisorption are not necessarily accompanied by depletion of the electron or hole gas. On the contrary, chemisorption may be accompanied by enrichment of these gases. Therefore, the presence of an electron or hole gas is not necessary for the formation of "strong" bonds in chemisorption.

2.3.3. Examples

Here are some examples that illustrate the various types of bonds for the same particle on the same adsorbent.

Figure 2.5 shows the various forms of chemisorption of a Na atom and a Cl atom on the NaCl lattice [20]. Figure 2.5a corresponds to "weak" bonding of the Na atom with the lattice; this type of bond was studied by Bonch–Bruevich and the present author [21–23]. Grimley used the same method (the molecular orbital method) and obtained the same results [24] (this was noted by Koutecky [25]). We will examine the mechanism of formation of such a bond in Section 2.5 in greater detail.

Bonding in this case is implemented by the valence electron of the Na atom, the electron drawn to a certain degree away from the Na atom into the lattice. In other words, the electron cloud surrounding the positively charged core of the Na atom, and in the case of an isolated atom having spherical symmetry, proves to be deformed and to some degree drawn into the lattice. The wave function of the electron (the problem can be treated in the oneelectron approximation if the positive and negative ions of the lattice are regarded in the first approximation as point charges) falls off inside the crystal with distance from an adsorption center, which in this case is one of the atoms in the surface layer of the lattice. Here we have a one-electron bond of the same type as in the molecular ion Na_2^+ (or H_2^+). Of course, the strength of the bond differs from that in an isolated molecular ion Na_2^+ since in the case at hand the Na atom is bound not only to a Na^+ ion of the lattice, which is an adsorption center, but by way of this ion to the lattice as a whole. Note that the chemisorbed Na atom (considered together with its adsorption center) acquires a certain dipole moment, which has a purely quantum-mechanical origin and whose magnitude may exceed that of the dipole moment induced in physisorption by several orders of magnitude (see Section 2.5).

In the case of "weak" chemisorption of a Cl atom shown in Fig. 2.5d the bond is provided by the attraction of an electron from a Cl^- ion of the lattice, which serves as an adsorption center, to the adsorbed atom or, so to say, by the attraction of a hole from the Cl atom into the lattice. Here we are dealing with a bond of the same type as in the molecular Cl_2^- ion. The dipole moment that emerges in the process has a direction opposite to that in the previous case.

Figure 2.5b depicts a "strong" acceptor bond for a Na atom. It is formed from the weak bond shown in Fig. 2.5a, for instance, by trapping and localizing a free electron, i.e., as a result of converting a Na^+ ion in the lattice, which acts as an adsorption center, into a neutral Na atom. We obtain a bond of the same type as in a Na_2 or H_2 molecule. Of course, the strength of the bond differs from that in the case of an isolated molecule, which is clear if only from the fact that in breaking this bond (i.e., in desorption of the Na atom) the lattice electron does not remain on the Na^+ ion (the adsorption center) but is completely delocalized, returning to the supply of free electrons. This is a typical homopolar bond in which both the valence electron of the adsorbed Na atom and an electron from the crystal lattice proper, taken from the supply of free electrons, participate. A quantum-mechanical approach to this problem shows [17, 22] that these two electrons are bound in the process by exchange forces that are also the adsorption forces which (1) retain the adsorbed Na atom on the surface and (2) keep a free lattice electron near the adsorbed atom. We will examine this type of bond in more detail in Section 2.6.

Figure 2.5c corresponds to a "strong" donor bond of the Na atom formed from a "weak" bond (see Fig. 2.5a) as a result of ionizing the adsorbed Na atom, i.e., as a result of transition of the valence electron of the Na atom to a free state (the supply of free electrons in the crystal receives an additional electron) or, which is the same, as a result of capture of a free hole by the adsorbed Na atom [19, 23]. In this case the adsorption bond has a purely ionic nature; i.e., we arrive at a NaCl quasimolecule.

Figures 2.5f and 2.5e depict, respectively, a "strong" donor and a "strong" acceptor bond for a Cl atom, both bonds obtained from a "weak" bond (see Fig. 2.5d) by involving a free hole or a free electron in the bond. In the first instance (Fig. 2.5f) we arrive at a Cl_2 quasimolecule with a typical homopolar bond and in the second (Fig. 2.5e) at a NaCl quasimolecule with a typical ionic bond.

Here is another example. Figure 2.6 gives a rough sketch of two forms of chemisorption of an O_2 molecule on ZnO and Cu_2O crystals, both treated as purely ionic crystals (this is permissible in the first approximation). Note that the pres-

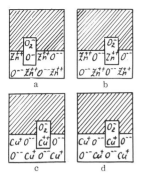

Fig. 2.6. Two forms of chemisorption of O_2 molecule on ZnO and Cu_2O crystals: a, c) "weak" bond for O_2 atom; b, d) "strong" acceptor bond for O_2 atom.

ence of a free electron in the ZnO crystal means that there is a Zn^+ ion among the normal Zn^{++} ions of the lattice, while the presence of a free hole means that there is a O^- ion among the O^{--} ions. In the Cu_2O crystal, made up of Cu^+ ions and O^{--} ions, a neutral Cu atom corresponds to a free electron and a doubly charged Cu^{++} ion (instead of a singly charge O^- ion) to a free hole (see Section 1.3).

Figures 2.6a and 2.6c depict a "weak" bonding of an O_2 molecule with the lattice. It is provided by the attraction of an electron from a lattice ion to the O_2 molecule. As a consequence of the great affinity of the O_2 molecule to an electron the electron can be considered as completely pulled from the lattice to the molecule, as a result of which the O_2 molecule is converted into a molecular ion O_2^- and a localized hole held by the O_2^- ion is produced in the lattice. The overall structure (an adsorbed O_2 molecule plus the adsorption center) acquires a considerable dipole moment (while remaining electrically neutral as a whole) directed with the negative pole outward. Here the bond does not include the participation of a free lattice electron. The transition to a "strong" acceptor bond means the localization of an electron or, which is the same, the delocalization of a hole. Such a "strong" acceptor bond is shown in Figs. 2.5b and 2.5d.

As a further example we will examine the adsorption of an H_2 molecule on an ionic crystal of the MR type. A "weak" bond with the lattice (see Fig. 2.4a, where the symbol C stands for an H_2 molecule) can be realized by the two electrons of the H_2 molecule which, while remaining paired, are to a certain extent drawn into the lattice, thus forming the quasimolecule (MH_2^+). We arrive at a two-electron bond of the same type as in the molecular ion H_3^+. The problem treated in the two-electron approximation was examined by Nagaev [26]. Here there cannot be any transition to a "strong" bond since involving a free electron or hole leads, as can be demonstrated (see Section 2.4.4), to the disruption of the bond between the two H atoms in the H_2 molecule.

The idea of different forms of chemisorption that differ in the nature of the bonding of the adsorbed particle to the adsorbent lattice plays an important role in the physical chemistry of the surface. The possibility of different types of bonds existing in chemisorption is due to the ability of the chemisorbed particle to draw a free electron or hole from the crystal lattice to participate in the bond or, which is the same, the ability of the chemisorbed particle to produce a free electron or hole and donate either to the lattice.

2.4. RADICAL AND VALENCE-SATURATED FORMS
OF CHEMISORPTION

2.4.1. Free Valences of a Surface

We see that the free electrons and holes in a crystal lattice are important participants in chemical processes on surfaces. Their role results from the fact that in such processes, as has been shown in a number of papers [17, 27–29] and will be evident from the following discussion, they act as free valences capable of breaking valence bonds within chemisorbed particles and becoming saturated because of these bonds. In fact, because of this important chemical role of the electrons and holes, we will introduce the term "free valences" and, as follows, identify free valences with free electrons or free holes. These functions of free electrons and holes follow from the very concept of "free electron" or "free hole." We will clarify this using two limiting cases, the purely homopolar crystal and the purely ionic crystal.

As an example of a homopolar crystal we will consider the germanium crystal. In such a crystal each Ge atom, being tetravalent, is surrounded by the four nearest neighbors to which it is bound by valence bonds. Two electrons participate in each such bond, an electron from the atom in question and an electron from its neighbor. Thus, all four valence electrons of each atom in the germanium lattice are used to form bonds and cannot take part in conduction. The presence of a free electron or hole in such a crystal implies the presence of a Ge^+ or Ge^- ion, respectively, among the Ge atoms. Such ionic states are able to migrate through the lattice by jumping from one Ge atom to another. The Ge^- ion is pentavalent and, since it is surrounded by four Ge atoms, keeps its fifth valence unsaturated. Consequently, a free electron or hole in the germanium lattice can be regarded as a free (unsaturated) valence migrating through the crystal.

As a typical example of an ionic crystal let us examine a NaCl crystal. The Na^+ and Cl^- ions have closed electron shells and are in this sense similar to atoms of the noble gases in group VIIIA. The presence of a free electron in a NaCl crystal implies the presence of an "excess" electron placed on the Na^+ ion above the closed shell (see Section 1.3). Such an electron can be interpreted as a free positive valence. On the other hand, the presence of a hole implies that an electron has been removed from the closed shell of one of the Cl^- ions. Such a hole can be interpreted as a free negative valence.

As a further example we will examine a Cu_2O crystal, which we will treat as an ionic crystal and in which, as we have already remarked (see Sections 1.3 and 2.3), the Cu state corresponds to a free electron and the Cu^{++} state to a free hole, both states migrating among the normal Cu^+ ions of the lattice. In the Cu atom and the Cu^+ and Cu^{++} ions the distribution of the electrons in groups and shells is as follows:

$$Cu \quad (1s)^2(2s)^2(2p)^6(3s)^2(3p)^6(3d)^{10}(4s)^1,$$
$$Cu^+ \quad (1s)^2(2s)^2(2p)^6(3s)^2(3p)^6(3d)^{10}, \qquad (2.27)$$
$$Cu^{++} (1s)^2(2s)^2(2p)^6(3s)^2(3p)^6(3d)^9.$$

The Cu^+ ion has a closed electronic shell (valence 0), the Cu atom has one electron above the closed shell (valence +1), and the Cu^{++} ion is characterized by the absence of one electron from the closed shell (valence −1). We see that in this case too a free electron is equivalent to an unsaturated positive valence and a free hole to an unsaturated negative valence. The use of the expression "free valence" to emphasize the chemical role of the electrons and holes is not to be confused with the term "valence state of the ions," where, in the example above, Cu^{+1} would be considered to be in the +1 valence state.

We note in passing that, apart from free electrons and holes, Frenkel excitons can emerge as free valences in a crystal (a Frenkel exciton is a tightly bound exciton in which the electron and the hole are usually on the same atom, although the pair can travel anywhere in the crystal; see Section 1.3). An example is again provided by the Cu_2O lattice in which the Frenkel exciton is an excited Cu^+ ion characterized by the following electronic structure:

$$Cu^+ \quad (1s)^2 (2s)^2 (2p)^6 (3s)^2 (3p)^6 (3d)^9 (4s)^1 ,$$

which differs from the normal structure in (2.27) by the displacement of an electron in the Cu^+ ion from the $3d$ shell to the $4s$ shell. In such excitation the Cu^+ ion, of course, preserves its charge but acquires a free valence.

Free valences having this excitonic nature can play a part in semiconductors containing a transition metal, which has an unfilled or easily vacated inner electronic shell, as one of its components. Certain catalytic properties of such semiconductors may be due to this. However, the role of Frenkel excitons in chemisorption phenomena and catalysis has hardly been investigated and, therefore, we will not study free valences having an excitonic nature.

2.4.2. The Reactivity of Chemisorbed Particles

The interpretation of free electrons and holes as free valences is convenient in describing the chemical processes that take place on the surface of a semiconductor, but it requires attributing the following properties to the free valences of a semiconductor surface [28, 29]:

(1) Each free valence has a certain mean lifetime; i.e., the valences are capable of appearing and disappearing, and the crystal continuously generates and absorbs valences.

(2) Free valences are not localized in the lattice but are capable of wandering about the crystal. In other words, as long as we are dealing with an ideal lattice, a free valence may be encountered with equal probability at any place in the crystal.

(3) The equilibrium concentration of free valences in a crystal and on its surface depends not only on the nature of the crystal but also on the conditions; for instance, it increases with temperature and can be artificially increased or decreased as a result of external influences on the crystal (illumination, introduction of impurities, and others).

(4) There is a permanent exchange of valences between the bulk and the surface of the crystal; i.e., valences leave the surface for the bulk and arrive at the surface from the bulk, so that the bulk of the crystal plays the role of a reservoir, absorbing free valences from the surfaces and supplying them to it.

(5) Free valences in a crystal can form pairs, each of which can wander about the crystal as a whole until it is dissociated. Such structures are well known in the theory of solids. A pair of valences of different signs in an ionic crystal (an electron and a hole bound together by the Coulomb interaction) constitutes a Mott exciton (see Section 1.3), while a pair of valences of the same sign (an electron and an electron or a hole and a hole, bound together by exchange interaction) constitutes a so-called doublon. Such structures have been investigated by Bonch–Bruevich and the present author [30, 31].

Since free electrons and holes in a crystal perform the functions of free positive and negative valences, respectively (we are speaking of crystals with more or less pronounced ionic bonds), a "weak" form of chemisorption is one that takes place without the participation of free surface valences, while a "strong" form indicates that there is a free surface valence involved in the bond. In the latter case the valence becomes localized and bound to the valence of the adsorbed particle. An acceptor or donor form of "strong" chemisorption will be formed depending on what kind of surface valence (positive or negative) comes into the picture.

The fact that a free surface valence is involved in the chemisorption bond leads to a valence-saturated particle being converted into an ion-radical, while a radical is converted into a valence-saturated (and electrically charged) structure. Therefore, among the various forms of chemisorption we must distinguish between those in which the chemisorbed particle sits on the surface in the form of a radical or ion-radical and those in which the same particle forms a valence-saturated structure with the surface. Naturally, in the radical forms of chemisorption the chemisorbed particles have an enhanced reactivity, i.e., they possess an increased reactivity to enter into chemical combinations with each other or with particles from the gaseous phase. We can therefore conclude that the various forms of chemisorption are distinguished not only by the character of the bonds and their strength but also by the reactivity of the chemisorbed particles.

2.4.3. Examples of Radical and Valence-Saturated Forms of Chemisorption

Figure 2.7 shows the various forms of chemisorption for the Na atom with the help of symbolic valence lines. In "weak" bonding the valence electron of the Na atom remains unpaired (see Fig. 2.5a and Section 2.5) and in this sense the free valence of the Na atom can be considered unsaturated. This form of bond, therefore, constitutes a radical form of chemisorption, shown symbolically in Fig. 2.7a. When we proceed to a "strong" n- or p-bond, a free lattice electron localized and paired with the valence electron of the Na atom (see Fig. 2.5b and Section 2.6) or a free lattice hole that recombines with the valence electron of the Na atom (see

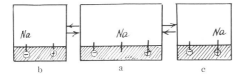

Fig. 2.7. Various forms of chemisorption of Na atom: a) radical form; b) "strong" *n*-bond; c) "strong" *p*-bond.

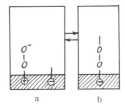

Fig. 2.8. Acceptor bonds for O_2 molecule: a) "weak" form; b) "strong" form.

Fig. 2.5c) is drawn into the bond. In both cases we can assume that the free valence of the Na atom is saturated by a surface valence (positive or negative, respectively). Mutual saturation of two like valences (the positive valence of the Na atom and a free positive surface valence) leads to the formation of a homopolar bond (see Fig. 2.5b), while mutual saturation of two unlike valences (the positive valence of the Na atom and a free negative surface valence) leads to the formation of an ionic bond (Fig. 2.5c). Both the "strong" *n*-bond and the "strong" *p*-bond constitute in this case a valence-saturated form of chemisorption. They are represented in Figs. 2.7b and 2.7c, respectively.

In Fig. 2.8 "weak" and "strong" acceptor bonds are represented for an O_2 molecule. In the "weak" bond (Fig. 2.8a) all the valences are closed, i.e., the O_2 molecule forms a valence-saturated structure with the surface (cf. Figs. 2.6a and 2.6c). In the "strong" acceptor bond (Fig. 2.8b) the chemisorbed O_2 molecule forms an ion-radical (cf. Fig. 2.6b and 2.6d). The chemisorption in this case is radical chemisorption. The same reasoning can be repeated for an O atom.

The various forms of chemisorption of the molecule of water are of interest. Figure 2.9a depicts the "weak" form of chemisorption of H_2O. Note that while the O atom has two vacancies in the closed electronic shell, i.e., is bivalent, the O^- ion has one vacancy and is thus univalent (see Fig. 2.8a), and the O^+ ion has three vacancies and is trivalent (Fig. 2.9a). In the "weak" bond the chemisorbed molecule of water, as seen from Fig. 2.9a, forms a valence-saturated (and electrically neutral) structure. As a result of ionization such a molecule becomes an ion-radical, as shown in Fig. 2.9b, and the "weak" bond becomes "strong."

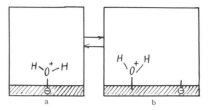

Fig. 2.9. Various forms of chemisorption of H_2O: a) "weak" form; b) "strong" acceptor form.

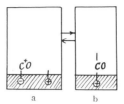

Fig. 2.10. Various forms of chemisorption of CO molecule: a) "weak" form; b) "strong" form.

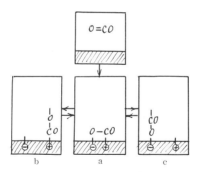

Fig. 2.11. Various forms of chemisorption of CO_2 molecule: a) "weak" donor form; b) "strong" donor form; c) "strong" acceptor form.

Figure 2.10 depicts "weak" (Fig. 2.10a) and "strong" bonds for a CO molecule, while Fig. 2.11 depicts the various forms of chemisorption for a CO_2 molecule. In "weak" chemisorption the CO_2 molecule is bound to the surface by two valence bonds (Fig. 2.11a). Here we have an example of adsorption on a virtual

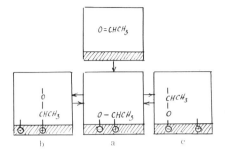

Fig. 2.12. Various forms of chemisorption of CH_3CHO molecule: a)
"weak" form; b) "strong" donor
form; c) "strong" acceptor form.

Mott exciton, i.e., not on an available Mott exciton, which is, as we have seen, a
pair of unlike free valences (a pair consisting of an electron and a hole) wandering
about the crystal as a whole, but on an exciton generated in the very act of adsorption [32]. The possibility of such a process was demonstrated by E. L. Nagaev for
the simple example of a one-electron atom.* The "weak" form of chemisorption
in the case of the CO_2 molecule is, as seen from Fig. 2.11a, valence-saturated and
electrically neutral. As a result of delocalization of an electron this form becomes
a "strong" donor form (Fig. 2.11b), while as a result of delocalization of a hole it
becomes a "strong" acceptor form (Fig. 2.11c). Both are ion-radical forms. Note,
however, that the ion-radicals obtained in these two forms differ considerably (cf.
Figs. 2.11b and 2.11c) and on entering into a reaction may take it in different
directions.

In a similar way we can imagine the adsorption of an acetaldehyde molecule
CH_3CHO (when this adsorption takes place without dissociation). The donor
(Fig. 2.12b) or acceptor (Fig. 2.12c) character of the chemisorption process is
determined by what bond in the molecules is saturated by the surface valence. In
the case shown in Fig. 2.12b the acetaldehyde molecule assumes the role of a
donor, while in the case shown in Fig. 2.12c it assumes the role of an acceptor.

Thus, the fact that the lattice electrons and holes participate in chemisorption
can be described in terms of valence lines, which constitutes only the chemical
aspect of the electronic mechanism.

Note that valence lines are often used in papers on catalysis (e.g., in describing
radical and chain mechanisms in catalysis). The physical meaning of these lines,
however, remains obscure and their properties, which determine the feasibility of
this or that valence scheme, are completely ignored. We have seen, and this is im-

---

*This problem is an example of the many-electron approach in the theory of
chemisorption. Bonch–Bruevich and Glasko [33] have examined the problem of
chemisorption on a metal by applying this method.

portant, that when speaking of the valences of a catalyst we must distinguish between valences of two types (positive and negative), which perform different functions. It is also important to note that like valences on the surface of a catalyst repel one another and, consequently, keep away from each other.* This fact compels one to consider many valence schemes used in theoretical papers on catalysis as physically impossible.

Finally, it must be emphasized that the formation of valence bonds between chemisorbed particles and the surface (i.e., "strong" forms of chemisorption) does not require, as we have seen in Section 2.3, that free surface valences be available beforehand; i.e., they may be generated in the process of chemisorption, and they are always generated in pairs (a positive valence and a negative valence).

2.4.4. The Dissociation of Molecules in Adsorption and the Recombination of Chemisorbed Atoms

The role of a free lattice electron as a free valence manifests itself most vividly in the dissociation of molecules, a process that often accompanies adsorption. Indeed, particles in a chemisorbed state may differ in nature from the corresponding molecules in the gaseous phase: they are parts of these molecules leading an independent existence on the surface. In other words, the act of adsorption may in some cases be accompanied by dissociation of the molecules, and this can be assumed to be experimentally established. Such adsorption requires a certain activation energy, as shown by Lennard-Jones [12] in the case of the H_2 molecule (see Section 2.24). There may be different mechanisms for such dissociation, but we will consider only one mechanism in which a free lattice electron plays the major part.

Consider the following problem. A molecule AB consisting of two univalent electropositive atoms A and B (H_2 may serve as an example) approaches the surface of a semiconductor as shown in Fig. 2.13a. We will study the behavior of a free electron and the energy of the system as a function of the distance b, which enters into the formulas as a parameter. We will treat the problem as a three-electron one (one electron from each of the atoms A and B and the free lattice electron) and restrict our discussion to qualitative reasoning (the interested reader can refer to the original paper [27]).

As the molecule AB approaches the surface of the crystal, the free lattice electron, as was shown, becomes more and more localized at the point on the surface that AB approaches (point M in Fig. 2.13a); and a bond arises between the atom B and the surface. This bond is provided by the localized electron and becomes stronger as the molecule AB approaches the surface, while the bond between the atoms A and B, which form the molecule, becomes gradually weaker. As b gets smaller, the distance a increases, so that as a result the atom B proves to

---

*Formation of complexes of like valences is possible on surface structural defects (See Section 6.3.1).

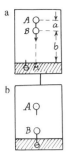

Fig. 2.13. Dissociation of a molecule and the formation of a "strong" n-bond: a) process of motion of molecule AB toward surface; b) dissociation completed and a "strong" n-bond is formed.

be bound to the surface by a "strong" n-bond, while the atom A becomes free and remains in the gaseous phase (as depicted in Fig. 2.13b) or is bound to the surface by a "weak" bond. The reaction proceeds according to the equation

$$AB + eL \rightarrow ABeL \rightarrow A + BeL \tag{2.28}$$

and, as has been shown, requires that a certain "activation" barrier be overcome. The unstable transient state ABeL, in which there is a bond of the same type as in the H_3 molecule, corresponds to the top of the barrier.

Along with the dissociation reaction (2.28) the reverse reaction is possible:

$$A + BeL \rightarrow AB + eL, \tag{2.29}$$

which is the reaction of recombination of the atom A arriving from the gaseous phase with the chemisorbed atom B, the latter being bound to the surface by a "strong" acceptor bond. As a result the molecule AB returns to the gaseous phase, while the electron is delocalized and returns to the supply of free lattice electrons. Denoting the energy of the "strong" bond between B and the lattice by q and the energy of dissociation of AB in the gaseous phase by D, we obtain

$$Q = q - D, \tag{2.30}$$

where Q is the heat for the reaction (2.28). If the dissociation reaction (2.28) is exothermic ($Q > 0$), the recombination reaction (2.29) is endothermic ($Q < 0$). This feature of the recombination reaction involving a chemisorbed atom makes it quite different from the recombination reaction with two free atoms, which, if favorable, is exothermic. We see that the transition of the reaction from the gaseous phase to the crystal surface may convert an exothermic reaction into an endothermic reaction and vice versa [34]. In Section 8.4 we will consider the recombination reaction (2.29) and some phenomena associated with it in greater detail.

The above problem of dissociation in adsorption is similar to the well known Slater problem [35], in which there were three univalent atoms A, B, and C lying on a straight line. Slater examined the substitution reaction

$$AB + C \rightarrow A + BC.$$

In our problem the lattice of the crystal regarded as a whole performs the role of the atom C.

The free lattice electron once more assumes the role of the free valence. This free valence, which wanders about the crystal, ruptures the valence bond inside the AB molecule and becomes saturated at the expense of the radical A thus released. The crystal in this case plays the role of a free radical, and we can write our reaction as a normal reaction with a free radical:

$$AB + \dot{L} \rightarrow \dot{A} + BL,$$

where L is the symbol of the lattice, and the dot above a letter denotes a free valence.

In general, if the chemisorbed molecule AB consisting of two atoms or two atomic groups A and B joined by a single bond is in the state of "weak" bonding with the surface, then involvement of a free surface valence leads to rupture of the valence bond inside the molecule, i.e., the chemisorbed molecule is dissociated into two radicals A and B, with the valence of one radical being free and the valence of the other saturated by the free surface valence. Thus, one of the products of dissociation is in the state of "weak" bonding with the surface, while the other is in the state of "strong" bonding.

In conclusion we must note that dissociation of a molecule does not require a free surface valence. The rupture of a molecule in adsorption can also occur without the free surface valence. As an example let us take the H_2O molecule, in which the H atom and the OH group are coupled by a single bond but which is to a certain extent polarized (the positive pole on the H atom and the negative on the OH group). Let us imagine that such a molecule approaches the surface of an ionic crystal as shown in Fig. 2.14a. Given a suitable crystallographic structure of the adsorbent, the bond that couples the H atom and the OH group will get more and more polarized as the molecule moves closer to the surface, and as a result the molecule may be broken into two ions, H^+ and $(OH)^-$, by the lattice field (see Fig. 2.14a). Each of the dissociation products is joined to the surface by a "strong" bond (donor and acceptor, respectively) and these bonds do not have to be purely ionic. The extent to which the bond is ionic is determined by the localization of the electron (belonging to the OH group) and the hole (belonging to the H atom) between the adsorbed particle and the corresponding adsorption center (see Section 2.3.2 and Fig. 2.4). Figure 2.14b depicts this mechanism of dissociation by means of valence lines. No free surface valence participates in this mechanism, while the rupture of the valence bond in the molecule occurs on account of surface valences generated in the adsorption act.

Finally, the dissociation of a molecule in adsorption may also occur in such a way that as a result of dissociation both products will form a "weak" form of chemisorption rather than a "strong." This can be illustrated by the example of the O_2 molecule, in which the double valence bond joining the oxygen atoms may be ruptured as a result of the transfer of two electrons from two negative ions in the lattice to the O_2 molecule and the formation of two localized holes, as shown symbolically via valence lines in Fig. 2.15. The oxygen atoms resulting from the

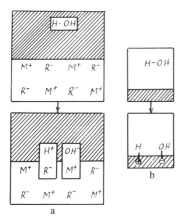

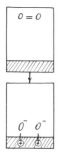

Fig. 2.14. Dissociation of H_2O molecule not involving a free surface valence: a) process of motion of H_2O molecule toward surface; b) mechanism of dissociation.

Fig. 2.15. Dissociation of a molecule and the formation of a "weak" bond.

disintegration of the O_2 molecule are joined to the surface by "weak" bonds and may be considered as unreactive or weakly reactive (the valence-saturated form of chemisorption). Subsequent delocalization of a hole or, vice versa, localization of a free electron that recombines with a hole, transfers the oxygen atom to a reactive state (the radical form of chemisorption). Whether the O_2 molecule dissociates in adsorption according to the above-mentioned mechanism (Fig. 2.15) or combines with the surface and forms a unit without disintegrating depends, of course, on the nature and crystallographic structure of the adsorbent. The role of geometric factors in chemisorption is most pronounced in this case. These factors were analyzed in detail in the works of Balandin and his collaborators on the multiplet theory of catalysis (e.g., see [36]).

2.5. THE ONE-ELECTRON BOND IN CHEMISORPTION

2.5.1. Statement of the Problem

In Section 2.3 we introduced the concepts of "weak" and "strong" forms of chemisorption, which differ in the nature of the bond joining the chemisorbed particle with the adsorbent lattice. Here we will study the mechanism of formation of the "weak" bond in greater detail using a simple example [21, 22].

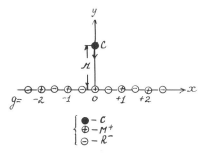

Fig. 2.16. Adsorption of a univalent electropositive atom on a one-dimensional ionic crystal.

Let us examine the adsorption of a univalent electropositive atom, i.e., an atom with one electron above the closed shell (we will denote such an atom by C; an example is the Na atom), on a one-dimensional ionic crystal MR composed of singly charged ions M^+ and R^-, both being treated as point charges in the first approximation. Such a one-dimensional model of a crystal, which is a chain of alternating M^+ and R^- ions, is shown in Fig. 2.16. We number all the metal ions in the lattice as is done in Fig. 2.16 (let g be the number of the ion, with $g = 0, \pm 1, \pm 2, ...$) and assume that the atom C approaches the lattice from above, with r the distance between the lattice and the atom.

Here we have a one-electron problem. The single electron is the valence electron of C. This electron is in the field of ion C^+ and the field generated by all the positive and negative ions M^+ and R^- of the lattice. We are primarily interested in its behavior as r varies. It is this electron that, as we have already noted and as we will prove below, is responsible for the formation of the bonding of the C atom with the lattice.

Let

$$\psi(x, y, z; r) \tag{2.31}$$

be the wave function that describes the behavior of our electron with the distance r entering into this function as a parameter. The function ψ can be found from the Schrödinger equation

$$\hat{H}\psi = E\psi, \tag{2.32}$$

where the Hamiltonian $\hat{H}$ is

$$\hat{H} = -\frac{\hbar^2}{2m}\Delta + \{V(x, y, z) + U(r)\}.$$

Here V is the potential energy of the electron in the field generated by the ion C^+

and all the ions M^+ and R^- of the lattice, and U is the energy of the interaction of C^+ with all the M^+ and R^-. The eigenvalue E of $\hat{H}$, which is the total energy of the system in state ψ, is obviously a function of parameter r:

$$E = E(r). \tag{2.33}$$

The problem consists in determining the function (2.31) and the respective eigen-value (2.33).

We will look for ψ in the following form:

$$\psi = a_C \varphi_C + \Sigma_g a_g \varphi_g, \tag{2.34}$$

where φ_C and φ_g are the atomic wave functions that describe the behavior of the electron in the field of the isolated C^+ ion and in the field of the gth M^+ ion (with the C^+ ion and the gth M^+ ion, respectively, being fixed and all the other ions being removed to infinity; all the wave functions are assumed to be s functions). Since the atomic wave functions in (2.34) can be considered known, the problem is reduced to finding the expansion coefficients a_C and a_g, where $g = 0, \pm 1, \pm 2, ...$. This can be done by applying a variational method, which ensures a set of coefficients a_C and a_g with which the wave function ψ given by (2.34), which is only an approximation of the solution of Eq. (2.32), satisfies this equation in the best possible way. The a_C and a_g, which characterize the relative weights of the wave functions φ_C and φ_g in the expansion (2.34), have a simple physical meaning, namely, $|a_g|^2$ is the probability of finding the electron on the gth metal ion M^+ of the lattice and $|a_C|^2$ is the probability of finding the electron on the C atom.

Just as in Sections 1.4.1 and 1.4.5, we arrive at the following system of equations for the a_g:

$$\Sigma_{g'} \{(E_g + U - E)S_{gg'} + P_{gg'}\}a_{g'} = 0, \tag{2.35}$$

where we have introduced the following notations:

$$S_{gg'} = \int \varphi_g \varphi_{g'} d\tau,$$

$$P_{gg'} = \int \varphi_g (V - U_{g'}) \varphi_{g'} d\tau,$$

and

$$g, g' = C, 0, \pm 1, \pm 2, \pm 3, ...$$

Note that

$$E_g = \begin{cases} E_M & \text{for} \quad g = 0, \pm 1, \pm 2, ..., \\ E_C & \text{for} \quad g = C, \end{cases}$$

$$U_g(x, y, z) = \begin{cases} U_M(x - ga, y, z) & \text{for} \quad g = 0, \pm 1, \pm 2, ..., \\ U_C(x, y, z) & \text{for} \quad g = C, \end{cases} \tag{2.36}$$

where E_M and E_C are the energies of the electron in the isolated atoms M and C, respectively, and $U_M(x - ga, y, z)$ and $U_C(x, y, z)$ are the potential energies of the electron in the field of the gth metal ion M^+ of the lattice and the ion C^+, respectively.

Assuming that the φ_g of neighboring atoms do not overlap too strongly and that the perturbation introduced by the atom C influences only the lattice atom with $g = 0$, we may put

$$S_{gg'} = \begin{cases} 1, & \text{if} \quad g' = g, \\ 0, & \text{if} \quad g' \neq g, \end{cases}$$

$$P_{gg'} = \begin{cases} \alpha_C, & \text{if} \quad g' = g =, C, \\ \alpha_M, & \text{if} \quad g' = g = 0, \\ \alpha, & \text{if} \quad g' = g \neq C, 0, \\ \beta_C, & \text{if} \quad g' = 0, \ g = C, \\ \beta_M, & \text{if} \quad g' = C, \ g = 0, \\ \beta, & \text{if} \quad g' = g \pm 1, \ g \neq C, \\ 0 & \text{otherwise.} \end{cases} \tag{2.37}$$

With (2.37) and (2.36) the system of equations (2.35) takes the following form:

$$\begin{cases} (E_C + U - E + \alpha_C)a_C + \beta_C a_0 = 0, \\ (E_M + U - E + \alpha_M)a_0 + \beta_M a_C + \beta(a_{-1} + a_{+1}) = 0, \\ (E_M + U - E + \alpha)a_g + \beta(a_{g-1} + a_{g+1}) = 0, \end{cases} \tag{2.38}$$

where $g = \pm 1, \pm 2, \ldots$. The problem is reduced to solving this system of equations.

2.5.2. Eigenfunctions and Eigenvalues

Before we begin to solve the system (2.38) for the general case, let us examine the limiting case with $r = \infty$. Then

$$\alpha_C = 0, \quad \alpha_M = \alpha, \quad \beta_C = \beta_M = 0$$

and the system (2.35) has two solutions,

$$\text{a)} \quad a_C = 1, \quad a_g = 0,$$

$$\text{b)} \quad a_C = 0, \quad a_g = a_0 \exp(i \lambda g), \tag{2.39}$$

where $g \neq C$ and λ is an arbitrary but real parameter. The first solution corresponds to a state of the system in which the electron is entirely in the atom C (the state L + C, where L is the symbol for the lattice). As shown by (2.38), in this case

$$\psi = \varphi_C, \quad E = E_C.$$

The second solution corresponds to states in which the electron is entirely transferred to the lattice and "smeared" over the crystal lattice (so that it may be encountered with equal probability on any M^+ ion of the lattice), while the C is left completely bare, i.e., is converted into a C^+ ion (the state $L^- + C^+$). In this case

$$\psi = a_0 \sum_g \exp(i\lambda g)\varphi_g, \quad E = E_M + \alpha + 2\beta \cos\beta,$$

where $g = 0, \pm 1, \pm 2, \dots$. We have arrived at the problem considered in Section 1.4 [see formulas (1.20) and (1.21)].

The energy spectrum of the system when r is infinite is depicted in the right-hand part of Fig. 2.17; we have a discrete level E_C corresponding to the solution (2.39a) and an energy band of width 4β (forming the so-called conduction band of the crystal) corresponding to the solution (2.39b). Figure 2.17 was constructed on the assumption that the energy band lies above the level E_C; i.e., the ionization potential of atom C is higher than the energy due to the affinity of the lattice for an electron (this is generally the case).

Let us now turn to the system (2.38) for the general case of a finite r. The symmetry of the problem implies

$$|a_g|^2 = |a_{-g}|^2. \tag{2.40}$$

Bearing this in mind, we will look for the expansion coefficients a_g in the form

$$a_g = \begin{cases} A \exp(i\lambda|g|) + B \exp(-i\lambda|g|), & \text{if} \quad g = 0, \pm 1, \pm 2, \\ C, & \text{if} \quad g = C, \end{cases} \tag{2.41}$$

where A, B, and C are arbitrary coefficients, and λ is an arbitrary (and generally complex-valued) parameter,

$$\lambda = \lambda' + i\lambda''. \tag{2.42}$$

with λ' and λ'' weak. There are two cases here:

(a) Both coefficients A and B in (2.41) are nonzero.
(b) One of the coefficients, A or B, is zero.

Let us study these cases separately.

(a) Assume that $A \neq 0$ and $B \neq 0$. The requirement that the wave function must be finite at infinity implies that λ is real, or $\lambda'' = 0$. Substituting (2.41) into the third equation in (2.38), we arrive at

$$E = E_M + U + \alpha + 2\beta \cos \lambda. \tag{2.43}$$

From the first two equations in (2.38) we obtain $a_C = a_0 = 0$ and, hence, $A = -B$; i.e., the coefficients A and B in (2.41) prove to be coupled. The states correspond to an electron in the conduction band of the crystal.

(b) We will now assume that one of the coefficients in (2.41), A or B, is zero. For the sake of definiteness we assume that $B = 0$. The requirement that the wave function be finite at infinity now implies that $\lambda'' \geq 0$. In what follows we will assume that

$$\exp(\lambda'') \gg 1. \tag{2.44}$$

From the third equation in (2.38) we again obtain (2.43) or, taking (2.42) into account,

$$E = E_M + U + \alpha + 2\beta(\cos \lambda' \cosh \lambda'' - i \sin \lambda' \sinh \lambda''). \tag{2.45}$$

Since E must be real,

$$\lambda' = n\pi, \tag{2.46}$$

where n is an integer. After we substitute

$$a_{-1} + a_{+1} = 2a_0 \exp(-\lambda'')$$

into (2.38) and employ (2.44), the condition that the first two equations in (2.38) must be compatible yields

$$(-1)^n \exp(\lambda'') = \frac{E_i - E_0}{\beta}, \tag{2.47}$$

where $i = 1, 2$ and

$$E_0 = E_M + \alpha + U,$$

$$E_1 = \frac{1}{2}\{(E_C + \alpha_C) + (E_M + \alpha_M) - \sqrt{[(E_C + \alpha_C) - (E_M + \alpha_M)]^2 + 4\beta_C\beta_M}\} + U, \tag{2.48}$$

$$E_2 = \frac{1}{2}\{(E_C + \alpha_C) + (E_M + \alpha_M) + \sqrt{[(E_C + \alpha_C) + (E_M + \alpha_M)]^2 + 4\beta_C\beta_M}\} + U.$$

The energy spectrum of the system is shown in Fig. 2.17. As the atom C approaches the lattice, i.e., as r decreases, the energy spectrum, as can be shown, changes as shown in the left-hand part of Fig. 2.17, namely, the energy band, whose middle is denoted by E_0, rises steadily (at the same time remaining undistorted), while the level $E_C = E_1$ moves downward, reaches its lowest position at a certain $r = r_0$, and then, as r decreases still further, moves upward. As we can see, adsorption occurs only in the ground (i.e., the lowest from the standpoint of energy) state of the system, while excited states do not lead to adsorption.

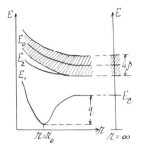

Fig. 2.17. Energy spectrum of the system of a univalent electropositive atom and a one-dimensional ionic crystal.

In the ground state of the system (curve E_1 in Fig. 2.17) with $r \neq 0$ (i.e., distinct from the case where $r = \infty$), all the expansion coefficients a_C and a_g in (2.34) prove to be nonzero. This means that the electron is now shared between the atom C and the lattice. Moreover, according to (2.41),

$$|a_0|^2 > |a_{\pm 1}|^2 > |a_{\pm 2}|^2 > \dots ,$$

i.e., the electron is distributed among the M^+ ions of the lattice symmetrically in relation to the $g = 0$ ion, and the farther the given M^+ ion is from the $g = 0$ ion (i.e., the greater the number g in absolute value) the smaller the probability of encountering the electron on this ion.

2.5.3. The Polarization of a Chemisorbed Atom

For the ground state ($E = E_1$) the condition (2.44), according to (2.47), takes the form

$$\left| \frac{E_1 - E_2}{\beta} \right| \gg 1.$$

Thus, the condition (2.44) implies that the distance between the middle of the energy band, E_0, and the local level E_1 is large compared with the width of the band 4β. In this case, according to (2.41) and (2.47), for all $g \neq 0$ we have

$$a_g = \left| \frac{E_1 - E_0}{\beta} \right|^{-|g|} ,$$

which means that among the a_g only a_C and a_0 are practically different from zero. Then, according to (2.34), the wave function is

$$\psi = a_C \varphi_C + a_0 \varphi_0 ,$$

where obviously

$$|a_C|^2 + |a_0|^2 = 1. \tag{2.49}$$

The electron is then shared between the atom C and the $g = 0$ atom in the lattice, i.e., spends time in both atom C and the $g = 0$ atom, the latter thus being an adsorption center.

We see that the valence electron of the atom C is drawn to a certain extent into the lattice, and the extent can be characterized by the quantity

$$\mu = 1 - |a_C|^2 = |a_0|^2, \tag{2.50}$$

which shows what fraction of the electron is transferred from atom C into the lattice or, in other words, the probability of finding the electron in the lattice. According to (2.50) and (2.49) we can write

$$\mu = \frac{1}{1 + \left|\dfrac{a_C}{a_0}\right|^2}, \tag{2.51}$$

and from the first two equations in (2.38) with $E = E_1$ we find that

$$\left|\frac{a_C}{a_0}\right|^2 = \left|\frac{\beta_C}{E_C + \alpha - E_1}\right|^2 = \left|\frac{E_M + \alpha_M - E_1}{\beta_M}\right|^2. \tag{2.52}$$

Obviously, μ varies with r, i.e., $\mu = \mu(r)$, and in the ground state (which is the only one we are considering here) we have

$$\alpha_C \to 0, \quad \alpha_M \to \alpha, \quad \beta_C \to 0, \quad \beta_M \to 0, \quad E_1 \to E_C$$

as $r \to \infty$ and, hence,

$$\left|\frac{a_C}{a_0}\right|^2 \to \infty, \qquad \mu \to 0,$$

i.e., the atom C in moving away from the lattice carries the electron with it. At $r = r_0$, where r_0 is the equilibrium distance between the adsorbed atom and the adsorbent surface, i.e., the bottom of the energy well in Fig. 2.17 we have

$$E_C - E_1 = q,$$

with q the depth of the well (the adsorption energy). We see that the strength of the bond, q, and μ are directly related at $r = r_0$, namely, the larger the value of μ the greater the value of q. Thus, the strength of the bonding between the atom C and the lattice depends on how strongly the valence electron of C is drawn into the lattice. The greater the pulling the stronger the bond.

The value of μ at $r = r_0$ is determined by the nature of the lattice and the atom C. Substituting (2.48) into (2.52) and then (2.52) into (2.51), we see that μ depends on

$$E_C - E_M = J_M - J_C = \Delta J,$$

where $J_M = -E_M$ and $J_C = -E_C$ are the ionization potentials of the atoms M and C, respectively (note that here we are dealing with the ionization potentials of free atoms). The greater the value of ΔJ (taking into account its sign) the greater the value of μ. Thus, the smaller the value of J_C as compared to that of J_M, the stronger the pulling of the electron from C into the lattice.

The quantity μ can be expressed by a simple formula in the limiting case where the exchange integrals β_C and β_M for a given value of r are large in absolute value. For instance, if

$$4\beta_C\beta_M \gg \{(E_C + \alpha_C) - (E_M + \alpha_M)\}^2,$$

then, according to (2.52) and (2.48), we have

$$\mu = \frac{1}{1 + \left|\dfrac{\beta_C}{\beta_M}\right|^2}.$$

Thus, to a certain extent a chemisorbed atom is polarized. The degree of this polarization is given by μ. In the process there emerges an electric field around the atom, and the greater the value of μ the stronger the field. At great distances from the atom (where the wave function is for all practical purposes zero) this field is equivalent to that of a dipole, with its moment M being normal to the surface and equal to $e\mu r$. This dipole moment, which is induced during chemisorption, has a purely quantum-mechanical nature and may, as shown in [21], exceed by several orders of magnitude the dipole moment induced during physisorption.

We note in conclusion that not only an M^+ ion in the lattice but also, as calculations show [26], an R^- ion (see Fig. 2.18b) may serve as an adsorption center for the C atom (see Fig. 2.16 or 2.18a). In the latter case the electron cloud is drawn into the lattice from the C atom symmetrically with respect to the R^- ion, which serves as an adsorption center. Which of two models depicted in Fig. 2.18 (a or b) is preferable from the standpoint of energy depends on the nature of the atom C and the atoms M and R of which the lattice is composed (see Figs. 2.4a and 2.4d).

2.6. THE TWO-ELECTRON BOND IN CHEMISORPTION

2.6.1. Statement of the Problem

Let us now examine in more detail the mechanism of formation of a "strong" acceptor bond, in which a free lattice electron is involved [17]. Let us limit ourselves to the simple model we used in Section 2.5 (a univalent electropositive atom C is adsorbed on a one-dimensional lattice composed of M^+ and R^-; see

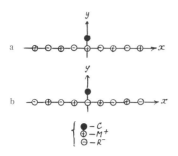

Fig. 2.18. Adsorption of a univalent electropositive atom by various surfaces: a) above a positive ion; b) above a negative ion.

Fig. 2.16). Let us assume that there is a free electron in the crystal, which we will call electron No. 1, while by electron No. 2 we denote the valence electron in C. The setting is essentially two-dimensional. The problem of formation of the bond between atom C and the lattice is to a certain extent similar to the problem of formation of the H_2 molecule from two hydrogen atoms, namely the atom C with its valence electron No. 2 plays the role of one of the hydrogen atoms, and the lattice (taken as a whole) with the free electron No. 1 that belongs to it plays the role of the other hydrogen atom.

To simplify the problem we will neglect the fact that the lattice draws electron No. 2 from the atom C and assume that this electron (for any distance r between C and lattice) belongs completely to C. In other words, we neglect the "weak" bonding of C with the lattice (later we will drop this assumption). Suppose that C approaches the lattice as shown in Fig. 2.16. We wish to find the energy of the system E as a function of r and the behavior of the free lattice electron (electron No. 1) in the process.

The Schrödinger equation for the two electrons is

$$\hat{H}\psi = E\psi, \tag{2.53}$$

where now

$$\hat{H} = -\frac{\hbar^2}{2m}(\Delta_1 + \Delta_2) + \{V(1) + V(2) + V(1, 2) + U\}.$$

Here the numbers 1 and 2, which figure as arguments, are abbreviated symbols for the triplets of coordinates x_1, y_1, z_1 and x_2, y_2, z_2 of the first and second electrons, respectively, $V(i)$, with $i = 1, 2$, and U have the same meaning as in Section 2.5, and $V(1, 2)$ is the energy of interaction of the two electrons with one another.

We will look for the solution of Eq. (2.53) in the form

$$\psi(1, 2) = \sum_g a_g \{\varphi_g(1)\varphi_C(2) \pm \varphi_g(2)\varphi_C(1)\}, \tag{2.54}$$

where φ_g and φ_C are the atomic wave functions, which we assume to be s functions. Here g is the number of the metal ion M^+ in the lattice ($g = 0, \pm 1, \pm 2, ...$). The upper sign in (2.54) corresponds to the case where the electron spins are antiparallel and the lower sign to the case where they are parallel. In what follows we will limit ourselves to symmetrical solutions (with respect to the interchange of the electrons); i.e., we will keep only the upper sign in (2.54). (Antisymmetrical solutions correspond to excited states of the system and will not interest us here.)

The function (2.54) is a wave function that describes the behavior of the two electrons for a given distance r [which enters (2.54) as a parameter]. We must therefore find the expansion coefficients a_g. Obviously, $|a_g|^2$ is the probability of finding one of the electrons on the gth M^+ ion and the other on the C atom. Here we can employ a variational method. The variational integral has the form

$$J_e = \int \psi^* (\hat{H} - E) \psi \, d\tau_1 \, d\tau_2,$$

where $d\tau_1 = dx_1 \, dy_1 \, dz_1$ and $d\tau_2 = dx_2 \, dy_2 \, dz_2$. The minimization condition for J is given by the system of equations

$$\partial J_E / \partial a_g^+ = 0, \tag{2.55}$$

which makes it possible to find the coefficients a_g we are interested in. The compatibility condition for these equations enables finding the eigenvalues E.

We label the atom C by $g = C$ and introduce the notations:

$$S_{gg'} = \int \varphi_g(i)\varphi_{g'}(i)d\tau_i,$$
$$P_{gg'} = \int \varphi_g(i) \{V(i) - U_{g'}(i)\} \, \varphi_{g'}(i)d\tau_i,$$
$$Q_{gg'} = \int \varphi_g(1) \, \varphi_C(2) \, V(1,2) \, \varphi_{g'}(1) \, \varphi_C(2)d\tau_1 d\tau_2,$$
$$R_{gg'} = \int \varphi_g(1) \, \varphi_C(2) \, V(1,2) \, \varphi_{g'}(2) \, \varphi_C(1)d\tau_1 d\tau_2,$$

We can write the system of equations (2.55) thus:

$$\sum_{g'} a_{g'} \{(E_C + E_M + U - E)(S_{gg'} + S_{gC} S_{Cg'})$$
$$+ P_{CC} S_{gg'} + P_{gC} S_{Cg'} + P_{Cg'} S_{gC} + P_{gg'} + Q_{gg'} + R_{gg'}\} = 0, \tag{2.56}$$

where g and g' assume the following values:

$$g, g' = 0, \pm 1, \pm 2, \pm 3, \ldots$$

Let us assume that the wave functions of neighboring atoms overlap only weakly and that only the atom in the lattice that is closest to C, i.e., the atom with $g = 0$, is perturbed by C. These approximations enable us to write

$$S_{gg'} = \begin{cases} 1, & \text{if } g' = g, \\ S, & \text{if } g' = C, g = 0 \quad \text{or} \quad g' = 0, g = C, \\ 0 & \text{otherwise;} \end{cases}$$

$$\begin{cases} \alpha_C, & \text{if } g' = g = C, \\ \alpha_M, & \text{if } g' = g = 0, \\ \alpha, & \text{if } g' = g \neq C, 0, \end{cases}$$

$$P_{gg'} = \begin{cases} \beta_C, & \text{if} \quad g' \neq 0, \ g = C, \\ \beta_M, & \text{if} \quad g' = C, \ g = 0, \\ \beta, & \text{if} \quad g' = g \pm 1, \ g \neq C, \\ 0 & \text{otherwise;} \end{cases}$$

$$Q_{gg'} = \begin{cases} \eta_0, & \text{if} \quad g' = g = 0, \\ 0 & \text{otherwise;} \end{cases} \tag{2.57}$$

$$R_{gg'} = \begin{cases} \xi_0, & \text{if} \quad g' = g = 0, \\ 0 & \text{otherwise.} \end{cases}$$

Introducing the abbreviated notations

$$\eta = \alpha_C, \ \alpha_M + \eta_0, \quad \zeta = (\beta_C + \beta_M)S + \xi_0,$$
$$E_0 = (E_C + E_M + U) + (\alpha_C + \alpha), \quad E_1 = (E_C + E_M + U) + (\eta + \zeta) \tag{2.58}$$

where η and ζ are the Coulomb and exchange integrals, respectively, and assuming that $S^2 \ll 1$, we can write the system of equations (2.56) and (2.57) and (2.58) thus:

$$\text{for} \quad g = 0 \quad (E_1 - E)a_0 + \beta(a_{-1} + a_{+1}) = 0,$$
$$\text{for} \quad g \neq 0 \quad (E_0 - E)a_g + \beta(a_{g-1} + a_{g+1}) = 0. \tag{2.59}$$

The problem is therefore reduced to solving these equations.*

2.6.2. Eigenfunctions and Eigenvalues

Before solving the system of equations (2.59) for the general case, we will consider the limiting case with $r = \infty$, i.e., when the atom C and the crystal MR are isolated from each other. Then

$$\alpha_M = \alpha, \quad \beta_C = \beta_M = 0, \quad \eta_0 = \zeta_0 = 0$$

and, hence, according to (2.58),

$$E_1 = E_0.$$

The two equations in (2.59) merge, and we return to the problem discussed in Section 1.4 [see Eqs. (1.18)], namely, one of the electrons (electron No. 1) is in the conduction band of the crystal and the other (electron No. 2) is localized on the atom C. The energy spectrum of this system is shown in the right-hand part of Fig. 2.19.

---

*The condition $S^2 \ll 1$ is introduced only for the sake of simplifying the calculations. The subsequent calculations can be done without this condition; in the final formulas we will have $(1 + S^2)$.

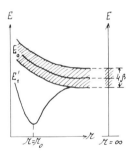

Fig. 2.19. Energy spectrum of a two-dimensional system formed at the surface of a one-dimensional lattice.

Now let us turn to the general case where $r \neq \infty$. We will look for the solutions of Eqs. (2.59) in the form

$$a_g = \begin{cases} A \exp(i\lambda|g|) + B \exp(-i\lambda|g|) & \text{for} \quad g = \pm 1, \pm 2, \ldots, \\ C & \text{for} \quad g = 0, \end{cases} \tag{2.60}$$

where λ is an arbitrary complex-valued parameter:

$$\lambda = \lambda' + i\lambda''. \tag{2.61}$$

Substituting (2.60) into (2.59), we find that

$$[A \exp(i\lambda) + B \exp(-i\lambda)] \, 2\beta + C(E_1 - E) = 0,$$

$$A \exp(i\lambda)[E_0 - E + \beta \exp(i\lambda)] + B \exp(-i\lambda) [E_0 - E + \beta \exp(-i\lambda)] + C\beta = 0, \tag{2.62}$$

$$[A \exp(i\lambda|g|) + B \exp(-i\lambda|g|)] \, (E_0 - E + 2\beta\cos\lambda) = 0.$$

The unknown coefficients A, B, and C can be found from this system of equations. There are two typical cases here:

(a) Both coefficients A and B are nonzero.
(b) One of the coefficients, A or B, is zero.

We will study these cases separately.

(a) $A \neq 0$ and $B \neq 0$. To ensure that the wave function is finite at infinity we must put $\lambda'' = 0$, i.e., λ is real. Equations (2.62) yield

$$E = E_0 + 2 \beta \cos \lambda, \tag{2.63a}$$

$$A + B = C, \tag{2.63b}$$

$$[A \exp(i\lambda) + B \exp(-i\lambda)] \, 2\beta = C(E_0 - E_1 + 2\beta \cos \lambda). \tag{2.63c}$$

With the notations

$$D = 2 \sqrt{AB}, \quad \exp(i\delta) = \sqrt{\frac{A}{B}}, \quad \gamma = \frac{E_1 - E_0}{2\beta}, \tag{2.64}$$

we can write Eqs. (2.63b) and (2.63c) in the following form:

$$D \cos \delta = C, \quad D \cos (\lambda + \delta) = C(\cos \lambda - \gamma),$$

where

$$\cos \delta = \frac{\sin \lambda}{\sqrt{\sin^2 \lambda + \gamma^2}} \tag{2.65}$$

Combining (2.65) with (2.60) and (2.64), we find that

$$a_g = D \cos(\lambda |g| + \delta),$$

where the phase δ is expressed by (2.65), and the coefficient D can be found from the normalization condition. This case corresponds to an electron in the conduction band of the crystal (2.63a).

(b) $A \neq 0$ but $B = 0$. Note that we will arrive at the same results of $A = 0$ and $B \neq 0$. Equations (2.62) yield

$$E = E_0 + 2\beta \cos \lambda, \tag{2.66a}$$

$$A = C, \tag{2.66b}$$

$$\sin \lambda = i\gamma. \tag{2.66c}$$

Substituting (2.61) into (2.66c), we obtain

$$\sin \lambda' \cosh \lambda'' + i \cos \lambda' \sinh \lambda'' = i\gamma. \tag{2.67}$$

Since γ is necessarily real, Eq. (2.67) implies that $\lambda' = n\pi$, where n is an integer. Note that λ'' must be nonnegative for the finiteness of the wave function (2.60). This together with (2.67) implies that

$$\begin{aligned} n & \quad \text{is even, if} \quad \gamma > 0, \\ n & \quad \text{is odd, if} \quad \gamma < 0. \end{aligned} \tag{2.68}$$

Returning to (2.66), where we put $\lambda = n\pi + i\lambda''$, we find that

$$E = E_0 + (-1)^n 2\beta \sqrt{1 + \gamma^2}, \tag{2.69a}$$

$$(-1)^n \exp(\lambda'') = \gamma + (-1)^n \sqrt{1 + \gamma^2} \tag{2.69b}$$

and, according to (2.60),

$$a_g = A \left[\gamma + (-1)^n \sqrt{1 + \gamma^2} \right]^{-|g|}. \tag{2.70}$$

This case corresponds to an electron on the local level (2.69a).

The energy spectrum of the system as a function of r and for $\gamma < 0$ is depicted in the left-hand part of Fig. 2.19 [it is simply a graphic representation of (2.63a) and (2.69a)]. As r decreases, the energy band, whose middle we denote by E_0, rises steadily (without being distorted) and then a discrete level, which we denote by E_1, splits off from the lower edge of the band; as r decreases still further, the level passes through a minimum (at $r = r_0$).* The states corresponding to the energy band (the lattice electron remains free) do not lead to adsorption (see Fig. 2.19), while the state corresponding to the level E_1' does lead to adsorption. According to (2.70), for this state we have

$$|a_0|^2 > |a_{\pm 1}|^2 > |a_{\pm 2}|^2 > \dots, \tag{2.71}$$

i.e., the lattice electron is to a certain extent localized at the $g = 0$ ion. The stronger the inequalities in (2.71), the sharper the localization.

We see that a foreign atom C, when brought closer to the crystal surface, creates a local level for a lattice electron, a level that lies under the conduction band. In this sense we can say that C is a trap for a conduction electron. The free electron drops out of the conduction band and is localized on the crystal surface in the neighborhood of C.

2.6.3. Free Lattice Electrons as Adsorption Centers

Let us examine the case where

$$\gamma < 0, \quad \gamma^2 = \left(\frac{E_1 - E_0}{2\beta} \right)^2 \gg 1. \tag{2.72}$$

According to (2.68) and (2.70), for all values of g we have

$$a_g = a_0 \left| \frac{E_1 - E_0}{\beta} \right|^{-|g|} \tag{2.73}$$

Under these conditions only one term in expansion (2.54) is retained, and instead of (2.54) we have

$$\psi(1, 2) = a_0 \{ \varphi_0(1) \varphi_C(2) + \varphi_0(2) \varphi_C(1) \}. \tag{2.74}$$

This means that the lattice electron is localized at the $g = 0$ ion. From (2.69a) we obtain

$$E = E_1 = E_C + E_M + U + \eta + \zeta. \tag{2.75}$$

---

*Note that Fig. 2.19 has a somewhat different meaning than Fig. 2.17, since Fig. 2.19 depicts the energy levels for a lattice electron, while Fig. 2.17 depicts the energy levels of a foreign electron, introduced into the system by the atom C.

From (2.72) it follows that the local level $E_1' = E_1$ (see Fig. 2.19) lies deep under the conduction band, so that the distance between this level and the middle of the conduction band E_0 is large compared with the width of the band, which is possible, as seen from Fig. 2.19, for not very small values of r.

The expressions (2.74) and (2.75) coincide with the corresponding expressions in the problem of the H_2 molecule (if in the latter we put $S^2 \ll 1$), the only difference being that now the Coulomb and exchange integrals, η and ζ, take into account the potentials of all of the surrounding ions in the lattice.

We see that the atom C and the lattice are coupled by exchange forces. When the adsorbed atom C moves away from the lattice, we move up the curve E_1' in Fig. 2.19 from left to right from the minimum point. In the process the exchange coupling of atom C with the lattice becomes weaker, the valence electron of C is carried off together with C, and the electron remaining in the lattice becomes gradually delocalized. In the limit, when $r = \infty$, the lattice electron is raised into the conduction band, i.e., becomes completely delocalized. If we now move in the opposite direction (from right to left in Fig. 2.19), i.e., bring the atom C from infinity to the lattice, we can either move upward, remaining in the energy band, or descend along the E_1' vs. r curve. In the first case the lattice electron remains free, i.e., does not enter into a bond with the atom C. In the process the atom C, as is evident, is repelled from the surface and adsorption proves impossible. In the second case (the E_1' vs. r curve) the free lattice electron and the atom C is linked by exchange coupling, which becomes stronger as r gets smaller. This leads to adsorption of the atom C and localization of the free lattice electron in the neighborhood of C, i.e., it drops out of the conduction band.

Thus, we can say that the adsorption of the atom C in the second case takes place "on a lattice electron," and the free lattice electron acts as an adsorption center. Since the presence of a free electron in the lattice means that there is a neutral state M that wanders among the M^+ ions of the lattice, the above result can be formulated in the following manner: a neutral atom M among the ions M^+ of the lattice can serve as an adsorption center. The adsorption of the atom C on such a center leads to the localization of this wandering center and signifies the formation of a quasimolecule CM with a characteristic two-electron bond. Two electrons participate in this bond, the valence electron of C and a lattice electron. We have arrived at a homopolar bond of the same type as in the H_2 molecule.

2.6.4. Allowing for a "Weak" Bonding

Until now we have neglected "weak" bonding of the atom C to the lattice. This simplification is contained in the very form of the wave function (2.54). If we wish to take into account this bond, we must look for the wave function $\psi(1, 2)$ in the form

$$\psi(1, 2) = \sum_g \sum_h a_{gh} \{ \varphi_g(1) \varphi_h(2) + \varphi_g(2) \varphi_h(1) \}, \qquad (2.76)$$

where each label g and h runs through the values

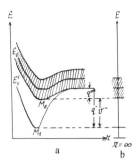

Fig. 2.20. Energy spectrum of a two-electron system formed at the surface of a one-dimensional lattice with "weak" bonding taken into account.

$$g, h = C, \quad 0, \pm 1, \pm 2, \ldots$$

Just as before, our problem is similar to the problem of the H_2 molecule. However, while in the original formulation (no "weak" bond) it was similar to the problem of the H_2 molecule in the Heitler–London setting, now (i.e., with the "weak" bond) it is similar to the same problem in the Mulliken-Hund setting, in which, as is well known, the possibility of ionic (polar) states is taken into account. The terms in (2.76) with $g = h$, i.e., terms that indicate the presence of both electrons simultaneously on the same ion M^+ or C^+, correspond to such ionic (polar) states.

When "weak" bonding is taken into account, the energy spectrum of the system has the form shown in Fig. 2.20a. This diagram differs from that in Fig. 2.19 by the presence of a shallow well in the conduction band. The minimum M_0 (its depth is denoted by q^0) represents the "weak" bonding of the atom C with the lattice, while the minimum M_1 (its depth is denoted by q^-) represents the "strong" bond. A transition of the system from point M_0 to point M_1 means that the free lattice electron is localized and the bond between C and the lattice is strengthened. The chemisorbed atom C, coupled with the lattice by a "weak" bond, thus emerges as a localization center for the free lattice electron and can be represented in the energy spectrum of this electron by a local acceptor level situation below the conduction band at a depth of

$$v^- = q^- - q^0,$$

as shown in Fig. 2.20b.

2.7. QUANTUM-MECHANICAL CALCULATIONS IN ADSORPTION THEORY*

2.7.1. The Cluster Approximation

One of the limitations of the band approximation in describing chemisorption is the fact that in this approximation it becomes difficult to account for fine details

---

*This section was written by G. M. Zhidomirov.

of the local interactions of a chemisorbed particle with the surrounding atoms on the surface. However, in many cases of practical interest there are grounds to assume that these aspects of chemisorption interactions play an important role, for instance, in the interaction of adsorbed molecules with Brönsted and Lewis acid sites on the surface and with the atoms and ions of transition elements on the surface [37]. The simplest approach to a description of local interactions in chemisorption is to take into account only a relatively small fragment of the solid's lattice (a cluster) built from a finite (and not very great) number of atoms from the surface layer and, possibly, from a small number of lattice atoms from the layers near the surface. This opens up the possibility of using directly the methods of quantum chemistry dealing with limited molecular systems to calculate various effects.

This "cluster" approach to chemisorption calculations has lately received wide recognition. Both nonempirical and semiempirical MO LCAO methods (EHT, CNDO/2, INDO, CNDO/BW) were used as well as X_α methods (X_α-SW, X_α-DW). The spectrum of the systems considered and the problems solved in the various works are broad, starting from traditional problems of chemisorption (geometrical and energy aspects of chemisorption) to calculations of such physical characteristics as electronic and EPR spectra. It would be difficult to include an extensive bibliography of the various works in this field, and so we limit ourselves to review articles [38–44]. Great attention has been paid to justifying the cluster approach. In this respect various authors studied the convergence of the characteristics as the size of the cluster increases and compared theoretical calculations with experimental results as well as with the results of calculations in which the periodic structure of the surface was taken into account. Especially important are the calculations carried out by applying nonempirical methods of quantum chemistry (e.g., see [45]), since in semiempirical calculations the errors in the calculation scheme are added to the errors of the cluster approach, which occasionally considerably decreases the value of the conclusions. As a somewhat intermediate result, we can say that the cluster approximation proves sufficiently effective in calculating such chemisorption characteristics as the energy and geometry of a process, but is less satisfactory in describing the properties of solids, which depend on the periodicity of the extended lattice, such as ionization potentials, electron affinity, and the mean binding energy per atom. A pronounced and often nonmonotonic dependence on the size and shape of the cluster is often found.

The cluster approximation is best suited for heteropolar lattices with a noticeable fraction of the ionic bond and is less effective in the case of atomic lattices bound by the covalent bond. Apparently, one of the least suitable systems for the cluster approach is graphite; calculations show that in this case even the energy aspects of chemisorption are influenced by the cluster approximation, although the geometry is still reflected correctly [46].

2.7.2. "Covalent" Clusters for Oxide Lattices

One of the main deficiencies of the cluster model is the unnatural rupture of the valence bonds of the boundary cluster atoms with the remaining part of the lattice. This rupture leads to "surface" levels in the forbidden band and nonuniform distribution of electric charge even in a monatomic cluster. These "surface" levels of a cluster distort the energy pattern at the top of the valence band and the bottom of the conduction band, and it is these regions that are essential in chemisorption and catalysis problems. There are various ways of counteracting this deficiency of the cluster approximation, but actually there exists a whole hierarchy of more and more complicated approaches, which enables finding results that are ever more exact and thus verifying the correctness of variants of the cluster approach. Here is a simple procedure that is useful in practical calculations. We introduce model univalent atoms A (pseudoatoms) on the boundary of the cluster to saturate the broken valence bonds. The quantum parameters of these atoms can be varied, which enables choosing them in such a way as to remain within the framework of the cluster model and yet reproduce in an optimal manner the characteristics of a solid that are most important for the chemisorption problem, namely, the charge distribution, the nature of the top of the valence band and the bottom of the conduction band, the width of the forbidden band, and the various properties (spectroscopic and others) of surface groups. The discussed cluster approach was realized by the semiempirical CNDO/BW and MINDO/3 methods [47]. The choice of these methods was justified by the fact that on the semiempirical level they are the most reliable and widely used calculation schemes aimed at calculating the total energy of the system, which is of special interest to chemistry. The cluster scheme we have just discussed has been widely tested, both in terms of reproduction of known experimental data on silica gel and zeolites [47–49] and in terms of logical consistency, i.e., preservation of the principal characteristics of the electronic structure during the successive expansion (within the framework of the scheme) of the cluster [50]. It has been applied to a study of the structure of Brönsted [47–49] and Lewis [51] acid sites in aluminosilicates, to a discussion of the various factors that influence their acidity, and to several catalytic reactions with these aluminosilicates. Cluster calculations vividly expose the strong dependence of the strength of Lewis acid sites in aluminosilicates on the screening of these sites by the neighboring atomic groups in the lattice and the lability of their structure, i.e., a marked geometric rearrangement during chemisorption [51].

2.7.3. "Ionic" Clusters for Oxide Lattices

The cluster scheme with pseudoatoms encounters certain difficulties when we are dealing with lattices with high coordination, e.g., cubic lattices. In this case a convenient scheme is that of a charged cluster, constructed on the principle of the purely ionic lattice; the "excess" electrons in such a cluster saturate the abovementioned "surface" states of the cluster. To compensate somewhat for the serious deficiencies of the ionic cluster scheme, we must superimpose a field equivalent to the crystalline field of the neglected part of the lattice. Korsunov et al. [52] developed, within the CNDO/2 method, a computer program for cluster calculations of chemisorption on oxide surfaces, including oxides of 3d transition elements, which takes into account the special features of calculating transition element compounds. Recently the program was augmented to enable calculating the parameters of electronic and EPR spectra of surface centers.

Although the ionic cluster scheme has shown its effectiveness in calculating spectroscopic properties, the question of how justified this scheme is in calculations of energy characteristics of chemisorption remains open because of the anomalously great electronic charge of the boundary atoms in the cluster. The first comparative calculations of the electronic structure and the protonization energy of surface hydroxyl groups [53] point to the usefulness of the ionic cluster model in describing the chemical properties of surface centers, but a much broader body of data is needed to draw more definite conclusions. At present such work is under way.

Chapter 3

ELECTRON TRANSITIONS IN CHEMISORPTION

3.1. TRANSITIONS BETWEEN VARIOUS FORMS OF CHEMISORPTION

3.1.1. Transitions between Energy Levels

As we have seen, various forms of chemisorption can transform into one another. In other words, a chemisorbed particle, while remaining in the adsorbed state, may change the character of its bond with the surface, i.e., may change from a state with one type of bond to a state with another. These transitions indicate that there is localization or delocalization of a free electron or hole at the adsorbed particle or in its neighborhood (see Figs. 2.7–2.12).

Such transitions may conveniently be described by means of the energy band scheme of a semiconductor depicted in Fig. 3.1. The y axis is parallel to the adsorbing surface of the semiconductor, the surface being flat. Here we have two energy bands (the valence band and the conduction band, both shaded in the diagram) separated by a forbidden region (band) of width $E_C - E_V$. As shown by the present author [1–3], a foreign particle C chemisorbed on the surface and bound by a "weak" bond is taken into account by the structure of the crystal's energy band scheme. A particle with an affinity for a free electron is represented by a local acceptor level (level A in Fig. 3.1), while a particle with an affinity for

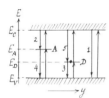

Fig. 3.1. Energy band scheme of a semiconductor.

83

a free hole is represented by a local donor level (level D in Fig. 3.1). In general, when an adsorbed particle is in a state of "weak" bonding with the surface and has affinity for both a free electron and a free hole, it is represented by two levels, an acceptor level and a donor level. The location of levels A and D in the forbidden band depends on the nature of the lattice and the adsorbed particle C [3–5].

Between the valence band and the conduction band and between the energy bands and the local levels depicted in Fig. 3.1 there may be electron transitions, which in the case of semiconductors and at not too low temperatures have a thermal origin. As a result of such transitions an electron may be ejected into the conduction band, trapped on the acceptor level A, or removed from the donor level D.

The fact that there is an electron on the local level A indicates that the chemisorbed particle C has moved from a state of "weak" bonding with the surface to a state of "strong" acceptor bonding. As is evident from Fig. 3.1, this can happen in two ways: a free electron may fall onto the level A from the conduction band or an electron may be ejected form the valence band onto the same level. The removal of an electron from the local level D indicates that the chemisorbed particle C has moved from the state of a "weak" bonding to the state of a "strong" donor bonding with the surface. Here there are also two ways in which this may happen: by the recombination of the electron on level D with a free hole wandering about the valence band or by the ejection of the electron on the D level into the conduction band.

These electron transitions, shown in Fig. 3.1 by thick vertical arrows (transitions 1, 2, 3, 4, and 5) can be written using the notation of Section 2.3.1 as follows:

$$
\begin{aligned}
&1)\, eL + pL \rightleftarrows L && E_C - E_V, \\
&2)\, CL + eL \rightleftarrows CeL && E_C - E_A, \\
&3)\, CL + pL \rightleftarrows CpL && E_D - E_V, \\
&4)\, CeL + pL \rightleftarrows CL && E_A - E_V, \\
&5)\, CpL + eL \rightleftarrows CL && E_C - E_D.
\end{aligned}
\tag{3.1}
$$

where the arrows directed from left to right correspond to exothermic transitions and those directed from right to left to endothermic transitions, i.e., transitions depicted in Fig. 3.1 by arrows directed downward and upward, respectively; the energies released or consumed in the transitions are given on the right-hand side in (3.1). In a particular case we may have $E_A > E_C$ or $E_D < E_V$, which means that the acceptor or donor type of bond is not realized for the given particle on the given adsorbent.

Figure 3.2 gives the electron transitions 1–5 by means of valence lines, with downward transitions in Fig. 3.1 corresponding to transitions from left to right in Fig. 3.2. Transition 1 corresponds to the recombination of a free electron with a free hole, i.e., the annihilation of two unlike free valences, while the reverse transition corresponds to the creation of a pair in the crystal consisting of a free electron and a free hole, which is equivalent to the creation of two unlike free valences. Transitions 2, 3, 4, and 5 correspond to transitions from one form of

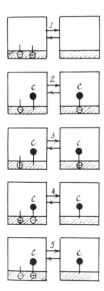

Fig. 3.2. Electron transitions expressed by valence lines corresponding to the transitions in Fig. 3.1.

chemisorption to another, i.e, transitions of the chemisorbed particle from a state of "weak" bonding with the surface to a state of "strong" bonding and vice versa. We see that the transition from a "weak" type of bonding to a "strong" type may be accompanied by the disappearance of a free valence on the surface (transitions 2 and 3), but may also occur without the participation of a free valence, in which case a free valence is created on the surface instead of disappearing on it (transitions 4 and 5). In the first case (transitions 2 and 3) the "strengthening" of the bond is accompanied by a lowering of the energy of the system (which was initially in an excited state), while in the second case (transitions 4 and 5) it is accompanied by excitation of the system.

We see that participation of the electrons and holes of a semiconductor can be described in terms of the energy band representation, which therefore constitutes yet another (energy) aspect of the electronic mechanism of chemisorption.

Note that the local surface levels generated by adsorbed particles occur not only during chemisorption, as is usually assumed, but in some cases in physisorption, when the wave functions of the adsorbed particle and the adsorbent lattice do not overlap. Indeed, if a physisorbed particle is polarized, as is usually the case, a free electron or hole in the lattice moves in the field of this dipole. The dipole creates a potential well (trap) for the electron or the hole, depending on which of its poles (positive or negative) is directed into the lattice. This results in a local level in the energy spectrum (situated not very deeply as a rule) of an acceptor or donor nature, respectively.

3.1.2. Transitions between Adsorption Curves

The electron transitions shown in Fig. 3.1 correspond to transitions of the system between states characterized by different adsorption curves. Such adsorp-

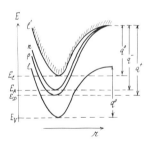

Fig. 3.3. Energy curves (or adsorption curves) corresponding to the transition between states shown in Fig. 3.1.

tion curves, which represent the energy of the system E as a function of the distance r between the particle C and the adsorbent surface when the particle C is a univalent atom, are shown in Fig. 3.3 (see Section 2.6.4 and [3, 4]). The curve l represents adsorption on an unexcited crystal, i.e., a crystal that has no free electrons or holes. The curve l' is simply curve l shifted upward parallel to itself through a distance $E_C - E_V$, which corresponds to adsorption on an excited crystal with a free electron in the conduction band or a free hole in the valence band. Curves p and n represent adsorption curves from "strong" donor and "strong" acceptor types of chemisorption (curve n may lie either lower or higher than curve p). States CL, CeL + pL, CpL + eL, and CL + eL + pL correspond to the minima on the curves l, n, p, and l', respectively.

Rising from left to right from the minima on the curves in Fig. 3.3, which correspond to various states of the system, implies desorption of the atom C. For the states l, n, p, and l' this process can be described by the following equations:

$$
\begin{aligned}
&l)\ \mathrm{CL} \to \mathrm{C} + \mathrm{L} && q^0, \\
&n)\ \mathrm{CeL} + \mathrm{pL} \to \mathrm{C} + \mathrm{eL} + \mathrm{pL} && q^-, \\
&p)\ \mathrm{CpL} + \mathrm{eL} \to \mathrm{C} + \mathrm{eL} + \mathrm{pL} && q^+, \\
&l')\ \mathrm{CL} + \mathrm{pL} + \mathrm{eL} \to \mathrm{C} + \mathrm{eL} + \mathrm{pL} && q^0.
\end{aligned}
\tag{3.2}
$$

In these equations the energies released in the processes are written on the right, and obviously (see Figs. 3.1 and 3.3)

$$
q^+ = q^0 + (E_C - E_D), \quad q^- = q^0 + (E_C - E_A). \tag{3.3}
$$

In all the cases in (3.2) the desorption product is the neutral particle C. Indeed, as r increases, i.e., as particle C moves away from the surface, the level A in Fig. 3.1, as was shown earlier (see Section 2.6 and [2]), moves up to the conduction band and in the limit, where $r = \infty$, is drawn into it, while the level D in Fig. 3.1 moves down to the valence band and at $r = \infty$ is drawn into it. In other words, an electron localized on the acceptor level A (producing an n-bond) or a hole localized on the donor level D (producing a p-bond) gets delocalized as r increases and in the limit, where $r = \infty$, returns to the conduction or valence band, respectively, i.e., increases the supply of free electrons or holes.

Note that the atom C may, in general, be desorbed from the states CeL or CpL in the form of a C^- or C^+ ion, respectively. Then instead of Eqs. (3.2n) and (3.2p)

we have

$$CeL + pL \rightarrow C^- + pL \qquad Q^-,$$
$$CpL + eL \rightarrow C^+ + eL \qquad Q^+,$$

where, as is easily shown,

$$Q^- = q^0 + (P_A - A),$$
$$Q^+ = q^- + (J - P_D).$$

Here A is the energy of affinity of C for an electron, J the ionization potential of atom C, and P_A and P_D the photoelectric work functions for an electron on the levels A and D, respectively.

The following transitions between states l, n, p, and l' in Fig. 3.3 are possible: $l' \rightleftarrows l$, $l' \rightleftarrows n$, $l' \rightleftarrows p$, and $p \rightleftarrows l$. These are, in fact, the transitions that in Fig. 3.1 are depicted by thick vertical arrows, 1, 2, 3, 4, and 5 [see also Eqs. (3.1)]. Note that these electronic transition reactions [the reactions in (3.1)] are characterized by heats of the same order of magnitude as the heats of adsorption [the reactions in (3.2)]; in the case of semiconductors these are tenths of electron volts. Consequently, in studying chemisorption processes, we cannot ignore the electron transitions (3.1) that occur parallel with adsorption and desorption reactions.

3.1.3. Equilibrium of Various Forms of Chemisorption

Let us consider the case where the forward and reverse transitions in (3.1) are in equilibrium (steady-state electronic equilibrium on the surface). We will assume that the same chemisorbed particle possesses affinity both for an electron and for a hole. Then a certain fraction of the total number of acceptor levels A will be occupied by electrons and a certain fraction of the total number of donor levels D will be freed from electrons; i.e., of the total number of particles N of a given type chemisorbed on a unit surface area a certain fraction will be in a state of "weak," "strong" acceptor, or "strong" donor bonding with the surface. Let us denote the number of particles adsorbed on the surface by N^0, N^+, N^-, each corresponding to a definite bonding state, and introduce the notations

$$\eta^0 = N^0/N, \quad \eta^+ = N^+/N, \quad \eta^- = N^-/N, \tag{3.4}$$

where obviously

$$\eta^0 + \eta^+ + \eta^- = 1.$$

The quantities η^0, η^+, and η^- characterize the relative fractions content of the various forms of chemisorption at equilibrium or, in other words, the probabilities of finding a chemisorbed particle in a particular state (characterized by a particular type of bond with the surface) or, to put it still another way, the average relative lifetimes of a chemisorbed particle in the corresponding states.

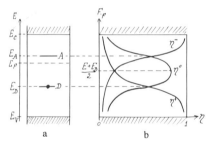

Fig. 3.4. Acceptor and donor levels vs. the relative content of various forms of chemisorption: a) position of acceptor and donor levels; b) relative content curves.

In accordance with Fermi statistics we will have [6]

$$\frac{N^+}{N^0 + N^+} = \frac{1}{1 + \exp\dfrac{E_F - E_D}{kT}},$$

$$\frac{N^-}{N^0 + N^-} = \frac{1}{1 + \exp\dfrac{E_A - E_F}{kT}},$$

where E_F is the Fermi level (see Fig. 3.4a).* From this we obtain

$$\eta^0 = \frac{1}{1 + \exp\left(-\dfrac{E_F - E_D}{kT}\right) + \exp\left(-\dfrac{E_A - E_F}{kT}\right)},$$

$$\eta^+ = \frac{\exp\left(-\dfrac{E_F - E_D}{kT}\right)}{1 + \exp\left(-\dfrac{E_F - E_D}{kT}\right) + \exp\left(-\dfrac{E_A - E_F}{kT}\right)},$$

$$\eta^- = \frac{\exp\left(-\dfrac{E_A - E_F}{kT}\right)}{1 + \exp\left(-\dfrac{E_F - E_D}{kT}\right) + \exp\left(-\dfrac{E_A - E_F}{kT}\right)}.$$

(3.5)

---

*Strictly speaking, as a consequence of degeneracy there should be factors connected with the ratios of the statistical weights of the above-mentioned states in these formulas before the exponentials. However, these factors are unimportant for our discussion below (see Section 1.6.3 and [7]).

The meaning of the notations adopted is clear from Fig. 3.4a. Note that the formulas in (3.5) can also be obtained from the mass action law for the reactions in (3.1) [4, 6].

The variation of η^0, η^+, and η^- with E_F is shown schematically in Fig. 3.4b. We see that as the Fermi level moves upward in Fig. 3.4b (i.e., as it moves away from the valence band and closer to the conduction band) the value of η^+ steadily decreases and that of η^- steadily increases, i.e., the number of particles bound to the surface by an acceptor bond increases and the number bound to the surface by a donor bond decreases. As for the value of η^0, which gives the relative fraction of the "weak" form of chemisorption, it is evident that it passes through a maximum as the Fermi level moves steadily upward.

Equations (3.5) refer to the general case where an adsorbed particle acts both as an acceptor and as a donor. If the particles act only as donors, then

$$\eta^0 = \frac{1}{1 + \exp\left(-\dfrac{E_F - E_D}{kT}\right)} \ , \quad \eta^+ = \frac{1}{1 + \exp\left(\dfrac{E_F - E_D}{kT}\right)} \ . \tag{3.6a}$$

If the chemisorbed particles act as acceptors,

$$\eta^0 = \frac{1}{1 + \exp\left(-\dfrac{E_A - E_F}{kT}\right)} \ , \quad \eta^- = \frac{1}{1 + \exp\left(\dfrac{E_A - E_F}{kT}\right)} \ . \tag{3.6b}$$

Equations (3.6a) and (3.6b) can be obtained from the general equations (3.5) as limiting cases. Indeed, if the level E_D is situated far enough below the Fermi level E_F, so that (see Fig. 3.4a)

$$\exp\left(-\frac{E_A - E_F}{kT}\right) \gg \exp\left(-\frac{E_F - E_D}{kT}\right),$$

then (3.6b) follows from (3.5). In this case, as (3.5) shows, $\eta^- \gg \eta^+$, i.e., almost all the chemisorbed particles are acceptors. On the other hand, if the level E_A is far enough above the Fermi level E_F, so that (see Fig. 3.4a)

$$\exp\left(-\frac{E_A - E_F}{kT}\right) \ll \exp\left(-\frac{E_F - E_D}{kT}\right),$$

then (3.5) yields (3.6a). In this case $\eta^+ \gg \eta^-$, i.e., almost all the chemisorbed particles act as donors.

Note that both η^+ and η^- in (3.6a) and (3.6b) differ from zero or unity only in narrow energy intervals, i.e., when the Fermi level lies sufficiently close to the level E_A or, respectively, to the level E_D (at a distance of the order of or less than kT). If the Fermi level coincides with E_D, we have $\eta^+ = 0.5$, and if E_F coincides with E_A, we have $\eta^- = 0.5$. For $T \neq 0$ the variation of η^- and η^+ with E_F is depicted in Fig. 3.5a; the case where $T = 0$ is depicted in Fig. 3.5b.

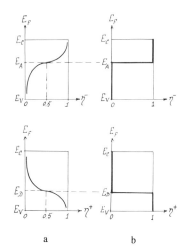

Fig. 3.5. Variation of η^- and η^+ with E_F: a) $T \neq 0$; b) $T = 0$.

On the basis of (3.5), (3.6a), and (3.6b) we can say that when electronic equilibrium has been established, the relative fractions of various forms of chemisorption on a surface and, therefore, the reactivity of the chemisorbed particles are uniquely determined by the position of the Fermi level. Thus, by moving the Fermi level we can change the ratio of concentrations of the neutral form of chemisorption to the charged form.

In experiments the separation of the different forms of chemisorption from the data obtained in electrophysical measurements presents serious difficulties and, as shown by Zarif'yants [8], requires in each case a detailed study of the electronic parameters of the surface before and after adsorption and the nature of the adsorption bonds that emerge and a direct estimate of the number of molecules adsorbed.

3.1.4. The Notion of Electron Transitions in Chemisorption Theories

In conclusion of this section let us examine a notion that is widely current among physical chemists, namely, the notion of electron transitions in chemisorption.

The notion of an electron transition in works on chemisorption theory and heterogeneous catalysis is usually understood literally, namely, an electron transition is the transition of an electron from a chemisorbed particle to the adsorbent or in the reverse direction, from the adsorbent to a chemisorbed particle. In other words, an electron belonging to a chemisorbed particle becomes the property of the crystal lattice or, in the reverse direction, an electron belonging to the lattice becomes the property of a chemisorbed particle.

When we say that an electron belongs to an adsorbed particle, we mean that the electron wave function has a maximum at the adsorbed particle and falls off as the distance from the particle increases. When we say that an electron belongs to a lattice, we mean that the electron wave function is nonzero inside the lattice and vanishes outside. From this viewpoint let us consider the act of chemisorption on the hydrogen atom, which is a simple example. This atom, as we saw in Section 2.3, may form three types of bonding with the lattice: the "weak" bond, the "strong" donor bond (in which case the adsorbed atom is a donor), and the "strong" acceptor bond (in which case the adsorbed atom is an acceptor). Let us see to what extent the idea of an electron transition can be applied to these three forms of chemisorption.

When the "weak" form of chemisorption operates, the electron of the chemisorbed atom, as shown in Section 2.5, proves to be drawn to a certain extent into the adsorbent lattice (sometimes almost completely). The extent is determined by the nature of the lattice. It is doubtful whether we can speak of the electron being transferred to the adsorbent since it remains bound to the chemisorbed particle and is localized at it.

Under a "strong" donor bond the electron of the atom proves to be absorbed by the valence band of the crystal or transferred to the conduction band. In the latter case its wave function is smeared over the entire crystal and vanishes outside the crystal. We can then speak of the electron being transferred from the atom to the crystal in the real sense of the word.

Finally, if the "strong" acceptor bond is acting, then it is commonly said that the electron is transferred from the lattice to the adsorbed particle. In actual fact, however, we saw that in this case a free lattice electron is localized on the crystal surface in the neighborhood of the chemisorbed particle and proves to be drawn into this particle to a certain extent. In some cases, e.g., for the hydrogen atom, this effect is negligible (see Section 2.6). But is there any sense in speaking of the electron being transferred to the chemisorbed particle if the electron remains with the crystal although it is drawn to the particle?

We see that the term "electron transition" is rather fuzzy if we take its primitive geometric meaning, which is the meaning most commonly used. Most authors use the term in this meaning (e.g., see [9]). This is an example of classical terminology and notions being employed to describe quantum-mechanical effects.

In quantum-mechanics (and in chemisorption we are concerned with a quantum-mechanical problem) the term "electron transition" has an energy meaning rather than a geometric one. It means that an electron goes over from one energy level to another, i.e., from one state to another state, which generally differs from the initial state by the electron cloud in space, i.e., by the wave function.

Electron transitions occurring in chemisorption constitute transitions of electrons from the conduction or valence band to a local level of the chemisorbed particle or back. As we have seen, these transitions are responsible for the transfer of one form of chemisorption into another. Precisely, these are transitions of an electron or hole from the free state in which the electron or hole wanders through the crystal to a localized state in which the electron or hole is tied to the chemisorbed particle, or transitions in the opposite direction. While in a free state an electron or hole belongs to the adsorbent as a whole, in a localized state the electron or hole belongs to a certain extent to both the adsorbent and the adsorbate, being distributed between the two.

The idea of electron transitions in chemisorption as transitions of electrons from the adsorbate to the adsorbent or back must be banned from works on chemisorption and catalysis, just as the idea of electron orbits was banned from atomic theory. At best these ideas have no clear and unambiguous physical meaning, and at worst they lead to an incorrect understanding of the actual processes.

3.2. ADSORPTION EQUILIBRIUM

3.2.1. Adsorptivity of a Surface

Let us now assume that along with electronic equilibrium established on the surface there is also adsorption equilibrium between the surface and the gaseous phase [10]. The condition for adsorption equilibrium (for the case where adsorption is not accompanied by dissociation) has the form

$$\alpha P(N^* - N) = \nu^0 N^0 \exp\left(-\frac{q^0}{kT}\right) + \nu^+ N^+ \exp\left(-\frac{q^+}{kT}\right) + \nu^- N^- \exp\left(-\frac{q^-}{kT}\right), \quad (3.7)$$

where P is the pressure, the factor α has the form (2.3), and N^* is the surface concentration of the adsorption centers, i.e., the maximum number of particles that a unit area of the surface can adsorb. When the surface is ideal (i.e., contains no defects), we can assume that $N = 1/s$, where s is the effective area of a chemisorbed particle. Note that by s we mean an area over which the wave functions of two neighboring atoms do not overlap. With this in mind we can interpret chemisorbed particles as surface impurities for any (within a monolayer) surface coverage.*

On the basis of (3.4) and (3.3) we can rewrite condition (3.7) as follows:

$$\alpha P(N^* - N) = \left[1 + \frac{\nu^+}{\nu^0} \cdot \frac{\eta^+}{\eta^0} \exp\left(-\frac{E_D - E_V}{kT}\right)\right.$$
$$\left. + \frac{\nu^-}{\nu^0} \cdot \frac{\eta^-}{\eta^0} \exp\left(-\frac{E_C - E_A}{kT}\right)\right] \nu^0 \eta^0 N \exp\left(-\frac{q^0}{kT}\right),$$

or, according to (3.5),

$$\alpha P(N^* - N) = \left[1 + \frac{\nu^+}{\nu^0} \exp\left(-\frac{E_F - E_V}{kT}\right)\right.$$
$$\left. + \frac{\nu^-}{\nu^0} \exp\left(-\frac{E_C - E_F}{kT}\right)\right] \nu^0 \eta^0 N \exp\left(-\frac{q^0}{kT}\right).$$

If the electron and hole gases on the semiconductor surface are nondegenerate, then, according to the definition in Section 1.6,

---

*Strictly speaking, the effective area s of a chemisorbed particle may be different for different types of bonding with the surface.

$$\exp\left(-\frac{E_C - E_F}{kT}\right) \ll 1 \text{ and } \exp\left(-\frac{E_F - E_V}{kT}\right) \ll 1,$$

and we have, assuming that the frequency factors ν^0, ν^-, and ν^+ are of the same order of magnitude,

$$\alpha P(N^* - N) = \nu^0 \eta^0 N \exp\left(-\frac{q^0}{kT}\right). \tag{3.8}$$

We see that here equilibrium between the surface and the gaseous phase is maintained exclusively by the "weak" form of chemisorption. In other words, the only particles that are desorbed are those in a state of "weak" bonding with the surface, while the particles that are "strongly" bound to the surface hardly participate in the exchange with the gaseous phase.

If we introduce the notation $\theta = N/N^* = sN$ (where, obviously, θ is the degree of surface coverage, i.e., the fraction of the surface occupied by the adsorbed particles), we can obtain from (3.8) the following:

$$\theta = \frac{1}{1 + b^*/P}, \tag{3.9}$$

or in the low coverage region (with $P \ll b^*$)

$$\theta = P/b^*, \tag{3.10}$$

where

$$b^* = b\eta^0 = \frac{\nu^0}{\alpha} N^* \eta^0 \exp\left(-\frac{q^0}{kT}\right) = b_0^* \exp\left(-\frac{q^*}{kT}\right). \tag{3.11}$$

Here, according to (3.5) and (3.3) (see also Fig. 3.4a),

$$q^* = q^- - (E_C - E_F), \quad \text{if} \quad \exp\left(-\frac{E_A - E_F}{kT}\right) \gg 1, \tag{3.12a}$$

$$q^* = q^0, \quad \text{if} \quad \exp\left(-\frac{E_A - E_F}{kT}\right) \ll 1, \quad \exp\left(-\frac{E_F - E_D}{kT}\right) \ll 1, \tag{3.12b}$$

$$q^* = q^+ - (E_F - E_V), \quad \text{if} \quad \exp\left(-\frac{E_F - E_D}{kT}\right) \gg 1. \tag{3.12c}$$

In the case (3.12a) practically all the chemisorbed particles are negatively charged ("strong" acceptor bonding), in the case (3.12c) they are positively charged ("strong" donor bonding), and in the case (3.12b) practically all the particles are in the neutral state ("weak" bonding). Neglecting electron transitions, i.e., putting $\eta^0 = 1$ and, via (3.11), $b^* = b$, we return to the ordinary Langmuir theory.

To avoid misunderstandings we must note that Eq. (3.9) is not a Langmuir isotherm just as Eq. (3.10) is not a Henry isotherm, as might appear at first glance, since η^0 in (3.11) is generally a function of θ, as will be clear from what follows.

By moving the Fermi level (other conditions remaining unchanged) we can change the adsorptivity of a surface. In relation to acceptor particles the adsorptivity of a surface steadily decreases as the Fermi level E_F lowers, while in relation to donor particles it steadily increases, as seen from (3.6a) and (3.6b) and Fig. 3.5. In relation to particles that are both acceptors and donors the adsorptivity of a surface, as seen from (3.5) and Fig. 3.4b, passes through a minimum for a certain position of the Fermi level in the forbidden band between the conduction and valence bands and decreases if the Fermi level is moved upward or downward from this position. In what follows (see Chapter 4) we will examine the factors that influence the position of the Fermi level and, therefore, the adsorptivity of the surface.

3.2.2. Surface Charging in Adsorption

The presence of a "strong" form of chemisorption, i.e., a form in which the chemisorbed particle retains on itself (or in its neighborhood) a free electron or hole from the crystal lattice, results, among other things, in the semiconductor surface being charged in chemisorption. If we denote the density of the electric charge concentrated on the surface and brought on by chemisorption by σ, then under conditions of adsorption equilibrium and low surface coverage we have, according to (3.10),

$$\sigma = e(N^+ - N^-) = e(\eta^+ - \eta^-)N = \frac{e}{bs}\left(\frac{\eta^+}{\eta^0} - \frac{\eta^-}{\eta^0}\right)P,$$

or, according to (3.5),

$$\sigma = \frac{e}{bs}\left[\exp\left(-\frac{E_F - E_D}{kT}\right) - \exp\left(-\frac{E_A - E_F}{kT}\right)\right]P$$

$$= \frac{2e}{bs}\exp\left(-\frac{E_A + E_D}{2kT}\right)\sinh\left[\frac{\frac{1}{2}(E_A + E_D) - E_F}{kT}\right]P, \qquad (3.13)$$

where e is the absolute value of the electron charge.

We see that the degree and nature of the surface charging, i.e., the magnitude and sign of the surface charge that occurs in chemisorption, depend not only on the nature of the chemisorbed particles and the surface coverage but also on the position of the Fermi level, i.e., the state of the system as a whole. The dependence of σ on E_F given by (3.13) (there is an implicit dependence through b^*) is depicted in Fig. 3.6, which shows that the surface is positively charged if $E_F < (E_A + E_D)/2$, negatively charged if $E_F > (E_A + E_D)/2$, and electrically neutral, in spite of the presence of chemisorbed particles on it, if $E_F = (E_A + E_D)/2$, i.e., if the acceptor and donor types of bonding are present to the same degree. Therefore,

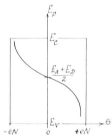

Fig. 3.6. The σ vs. E_F curve expressed by (3.13).

by moving the Fermi level (with other conditions remaining unchanged, i.e., with P and T constant) or, in other words, by subjecting the sample to one or another treatment, we can control the charge on the surface.

Consequently, while for one sample we might have $\eta^+ \gg \eta^-$, for another sample, treated differently, we might have $\eta^- \ll \eta^+$. This means that the same adsorbed species may be either a donor or an acceptor depending not only on the chemical composition of the semiconductor but on the prehistory of the sample serving as the adsorbent.

In this connection let us take the case where simultaneously with adsorption there is a certain amount of the adsorbed substance in solution in the bulk of the crystal. Suppose that the atoms chemisorbed on the semiconductor surface are those of hydrogen and that $\eta^+ \gg \eta^-$; that is, the chemisorbed hydrogen atoms act as donors and charge the surface positively. In addition, let us assume that some of the hydrogen atoms penetrate the crystal lattice in the process and dissolve into the crystal. The hydrogen atoms introduced into the interstices of the lattice are typical donors. The semiconductor, therefore, is enriched by donor impurities, which leads, as usual, to the Fermi level being shifted upward; i.e., the Fermi level moves up to the conduction band. If the concentration of these impurities is high, i.e., if the amount of dissolved hydrogen is great, the condition $\eta^+ \gg \eta^-$ may change, as shown in Fig. 3.4b, to $\eta^- \gg \eta^+$. The hydrogen atoms on the surface remain chemisorbed but lose their donor properties and become acceptors. Thus, while for small quantities of dissolved hydrogen the adsorbent surface is positively charged, for large quantities the sign of the surface charge may change.

Note that if besides surface states caused by chemisorbed particles the semiconductor possesses surface states of an intrinsic nature (surface energy bands, structural surface defects, surface impurities), then the surface may prove to be charged even in the absence of chemisorbed particles. In this case the appearance of chemisorbed particles leads to a change in magnitude (and also in sign) of the surface charge. Therefore, within the total surface charge we must distinguish between the adsorption and the intrinsic fraction. Note that since (see Chapter 4) the changes due to the adsorption and intrinsic surface states are not additive, chemisorbed particles not only introduce a new charge but change the magnitude of the intrinsic surface state charge.

Note further that in some cases adsorption may have little or no influence on the magnitude of the surface charge. This does not mean, however, that here we

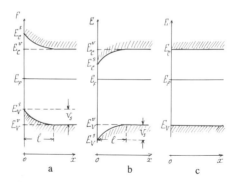

Fig. 3.7. Bending of bands: a) negatively charged surface; b) positively charged surface; c) electrically neutral surface.

are dealing with physisorption, as some authors suggest [11]. This effect (the constancy of the surface charge) may manifest itself, as we see, in chemisorption, too, namely, in the case of the "weak" form of chemisorption, in the case where the "strong" donor and "strong" acceptor forms of chemisorption are present on the surface more or less to the same extent, and finally in the case where the adsorption charge is small compared with the intrinsic charge.

Here is an important corollary that follows from the very fact of surface charging. Let us assume, following other authors, that the charge concentrated on the surface is uniformly distributed over the surface, so that its density is the same at all points of the surface. As a result of the surface charge there is a space in the near-the-surface layer of the semiconductor, the sign of the latter being opposite to that of the former and the charges compensating each other. This makes the energy bands bend near the surface of the semiconductor.

The bending of the band is shown in Fig. 3.7. Figure 3.7a corresponds to a negatively charged surface, Fig. 3.7b to a positively charged, and Fig. 3.7c to an electrically neutral. The x axis is directed into the crystal at right angles to the adsorbing surface, which is taken to coincide with the $x = 0$ plane. The local levels (surface and bulk levels) are not shown in Fig. 3.7. The distance denoted in Fig. 3.7a and 3.7b by l, over which the bending of the bands is perceptible (compared with kT), is called the screening length (see Section 4.2.2).

Note that the bending of the bands must not be understood as the dependence of the electron total energy E on the x coordinate, as is sometimes erroneously thought. The shaded regions in Fig. 3.7 (the energy bands) are those regions of values of E and x for which the square of the modulus of the wave function describing the behavior of the electron in the crystal remains periodic (with a period equal to the lattice constant). In other words, these are the regions of E and x within which the quasimomentum of the electron is a real (rather than complex) quantity. Thus, bending of the bands only indicates that for an electron with a given "allowed" (i.e., lying within a band) value of E not all the regions in the crystals are accessible (e.g., see [12]).

When electronic equilibrium is established, the surface and bulk (volume) of the semiconductor have a common Fermi level (depicted in Fig. 3.7 by a horizontal straight line E_F), i.e., a common electrochemical potential. However, in view of the bending of the bands the position of the Fermi level in the crystal energy spectrum (its position relative to the energy bands) will generally be different for different distances from the surface; i.e., $E_C = E_C(x)$ and $E_V = E_V(x)$. We adopt the following notations:

$$E_C^s = E_C(0), \qquad E_C^v = E_C(\infty),$$
$$E_V^s = E_V(0), \qquad E_V^v = E_V(\infty).$$

The Fermi level at the surface of the crystal proves to be shifted in relation to the Fermi level in the bulk by the amount (see Fig. 3.7)

$$V_s = E_C^s - E_C^v = E_V^s - E_V^v. \tag{3.14}$$

Note that the conditions $E_C = E_C^v$ and $E_V = E_V^v$ are met already at $x \geq l$. The quantity V_s characterizes the degree of bending of the bands. The greater the value of V_s (in absolute value), the greater the value of σ (in absolute value). We will find the precise dependence of V_s on σ in Section 4.1.2.

3.3. THE KINETICS OF ADSORPTION

3.3.1. Statement of the Problem

Let us now investigate the kinetics of chemisorption [13, 14]. This is an interesting problem because the experimenter studying chemisorption on semiconductors deals in the majority of cases with the kinetics rather than with adsorption equilibrium. The problem is aggravated by the fact the electronic equilibrium, which is certain to be achieved in adsorption equilibrium, is far from being achieved in the adsorption process and is never achieved at the beginning of such a process.

For the sake of definiteness we will study the adsorption of acceptors; donor particles and, in general, the case where the adsorbed particles are both acceptors and donors can be studied along similar lines. We must find N^0 and N^- separately as functions of t, i.e.,

$$N^0 = N^0(t), \quad N^- = N^-(t).$$

In other words, we will investigate the kinetics of the neutral and charged forms separately.

Let us denote by a_1 and a_2 the number of particles of a given kind that adsorb and desorb per unit time on and from a unit surface area of the semiconductor at a given constant pressure P and temperature T. By $b_1, b_2, b_3,$ and b_4 we denote the number of electron transitions per unit time per unit surface area; we denote these

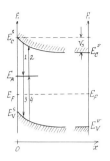

Fig. 3.8. Transitions of a chemisorbed particle from neutral to charged states and back.

by vertical arrows in Fig. 3.8, where the level E_A corresponds to the adsorbed acceptors. These are transitions that transfer a chemisorbed particle from a neutral state to a charged state (transitions 1 and 4 in Fig. 3.8) or back (transitions 2 and 3). If we assume that the only particles desorbed from the surface are those in the neutral state (while the particles in a charged state in no way participate in the exchange with the gaseous phase; see Section 3.2.1), we have (N^0 is constant)

$$\frac{dN}{dt} = (a_1 - a_2) - [(b_1 + b_4) - (b_3 + b_2)],$$
(3.15)

$$\frac{dN^-}{dt} = (b_1 + b_4) - (b_3 + b_2),$$

where

$$a_1 = \alpha P(N^* - N), \qquad a_2 = \nu^0 N^0 \exp\left(-\frac{q_0}{kT}\right),$$

$$b_1 = \beta_1 n_s N^0, \qquad b_2 = \beta_2 N^- \exp\left(-\frac{E_C - E_A}{kT}\right),$$
(3.16)

$$b_3 = \beta_3 p_s N^-, \qquad b_4 = \beta_4 N^0 \exp\left(-\frac{E_A - E_V}{kT}\right),$$

with [see (2.3)]

$$\alpha = \frac{\kappa}{N^* \sqrt{2\pi M k T}}, \qquad \kappa = \kappa_0 \exp\left(-\frac{E_0}{kT}\right);$$

here M is the adsorbed particle mass, κ the probability of a particle being captured by the surface, E_0 the activation energy, and n_s and p_s the free electron and hole concentrations on the plane of the surface (the $x = 0$ plane in Fig. 3.8).
    Introducing the notations

$$\frac{1}{\tau} = \nu^0 \exp\left(-\frac{q_0}{kT}\right),$$

$$\frac{1}{\tau^0} = \beta_1 n_s + \beta_4 \exp\left(-\frac{E_A - E_V}{kT}\right), \tag{3.17}$$

$$\frac{1}{\tau^-} = \beta_3 p_s + \beta_2 \exp\left(-\frac{E_C - E_A}{kT}\right),$$

keeping for the sake of simplicity to the region of low surface coverage (the Henry region $N \ll N^*$), and noting that when adsorption equilibrium is attained (i.e., at $t = \infty$)

$$a_1 = a_2 \quad \text{and, consequently,} \quad \alpha P N^* = N_\infty^0 / \tau \tag{3.18}$$

(where N_∞^0 is the value of N^0 at $t = \infty$), we can rewrite Eqs. (3.15) as follows:

$$\frac{dN^0}{dt} = \frac{N_\infty^0 - N^0}{\tau} - \left(\frac{N^0}{\tau^0} - \frac{N^-}{\tau^-}\right),$$

$$\frac{dN^-}{dt} = \frac{N^0}{\tau^0} - \frac{N^-}{\tau^-}. \tag{3.19}$$

This is the starting system of equations. We must solve it for the initial conditions

$$N^0 = N^- = 0 \quad \text{at} \quad t = 0. \tag{3.20}$$

Note [see (3.17)] that $1/\tau$ is the desorption probability, and $1/\tau^0$ and $1/\tau^-$ are, respectively, the probability of charging and neutralizing the chemisorbed particle (i.e., probability of the local level E_A in Fig. 3.8 acquiring and losing an electron), related to a unit of time. At the same time τ is the mean lifetime of the particle in the chemisorbed state, and τ^0 and τ^- are the mean lifetimes of the chemisorbed particle in the electrically neutral and charged states, respectively.

For the concentrations n_s and p_s we will take the usual Boltzmann expressions:

$$n_s = C_n \exp\left(-\frac{E_C^s - E_F}{kT}\right), \quad p_s = C_p \exp\left(-\frac{E_F - E_V^s}{kT}\right), \tag{3.21}$$

thus assuming that inside the semiconductor there is electronic equilibrium and that the electron and hole gases are nondegenerate.

Note that the Boltzmann formulas (3.21) must be taken as approximations. In reality chemisorption is generally accompanied by variations of the surface charge and, hence, by a current in the semiconductor; i.e., strictly speaking it leads to violations of the electronic equilibrium in the bulk of the semiconductor. However, we will keep to the approximation (3.21) (just as is done in the theory of electrical conductivity or the theory of thermoelectricity), bearing in mind, when dealing with (3.21), the variation in the position of the Fermi level with respect to the energy bands that occurs in adsorption:

$$E_C^s - E_F = (E_C^v + V_s) - E_F, \quad E_F - E_V^s = E_F - (E_V^v + V_s). \tag{3.22}$$

In the course of this the parameters τ^0 and τ^- from (3.19), according to (3.17) and (3.21), must also be considered varying, i.e.,

$$\tau^0 = \tau^0(N^-), \qquad \tau^- = \tau^-(N^-).$$

Note that in the case of acceptors we are considering here,

$$V_s(0) \leqslant V_s(N^-) \leqslant V_s(N^-_\infty)$$

and, according to (3.21) and (3.22),

$$n_s(0) \geqslant n_s(N^-) \leqslant n_s(N^-_\infty), \qquad p_s(0) \leqslant p_s(N^-) \leqslant p_s(N^-_\infty), \tag{3.23}$$

i.e., in the course of adsorption the free electron concentration falls and the free hole concentration increases.

At $t = \infty$, i.e., when adsorption and electronic equilibria on the semiconductor surface are achieved, we have $dN^-/dt = 0$ and, hence, according to (3.19) and (3.6a),

$$\frac{\tau^-}{\tau^0} = \frac{N^-_\infty}{N^0_\infty} = \frac{\eta^-}{\eta^0} = \exp\left(-\frac{E_A - E_F}{kT}\right). \tag{3.24}$$

If, in addition, we assume that the Fermi level at the crystal surface lies very high or very low, the formulas (3.17) for $1/\tau^0$ and $1/\tau^-$ simplify considerably. Indeed, at $t = \infty$ we have $b_1 = b_2$ and $b_3 = b_4$, where, according to (3.16) and (3.24),

$$\frac{\beta_1 n_s N^0_\infty}{\beta_4 N^0_\infty \exp\left(-\dfrac{E_A - E_V}{kT}\right)} = \frac{\beta_2 N^-_\infty \exp\left(-\dfrac{E_C - E_A}{kT}\right)}{\beta_4 N^0_\infty \exp\left(-\dfrac{E_A - E_V}{kT}\right)}$$

$$= \frac{\beta_2}{\beta_4} \exp\left[-\frac{(E_C - E_A) - (E_F - E_V)}{kT}\right],$$

$$\frac{\beta_3 p_s N^-_\infty}{\beta_2 N^-_\infty \exp\left(-\dfrac{E_C - E_A}{kT}\right)} = \frac{\beta_4 N^0_\infty \exp\left(-\dfrac{E_A - E_V}{kT}\right)}{\beta_2 N^-_\infty \exp\left(-\dfrac{E_C - E_A}{kT}\right)}$$

$$= \frac{\beta_4}{\beta_2} \exp\left[\frac{(E_C - E_A) - (E_F - E_V)}{kT}\right],$$

and, consequently, on the basis of (3.17) we can put (assuming that the coefficients β_2 and β_4 are of the same order of magnitude)

$$\frac{1}{\tau^0} = \beta_1 n_s,$$

$$\frac{1}{\tau} = \beta_2 \exp\left(-\frac{E_C - E_A}{kT}\right),$$

$$\text{if} \quad \exp\left(\frac{E_C - E_A}{kT}\right) \ll \exp\left(\frac{E_F - E_V}{kT}\right), \tag{3.25a}$$

$$\frac{1}{\tau^-} = \beta_3 p_s,$$

$$\frac{1}{\tau^0} = \beta_4 \exp\left(-\frac{E_A - E_V}{kT}\right),$$

$$\text{if} \quad \exp\left(-\frac{E_C - E_A}{kT}\right) \ll \exp\left(\frac{E_F - E_V}{kT}\right). \tag{3.25b}$$

Note that if condition (3.25a) is met at $t = \infty$, i.e., when adsorption equilibrium is achieved (after the adsorption process has ceased), it is sure to be met, as follows from (3.23), when $0 \leq t \leq \infty$, i.e., prior to adsorption and in the process. In what follows we will use this remark.

When (3.25) is realized, the valence band does not act in electron transitions, while when (3.25b) is realized, the conduction band does not act in such transitions.

3.3.2. Adsorption at a Constant Surface Potential: The General Case

Let us study the system of equations (3.19). The solution process is complicated by the fact that generally τ^0 and τ^- are functions of N^-. But if we assume that τ^0 and τ^- are constant, i.e., $V_s = \text{const}$, then we can easily solve (3.19). We will start from this assumption [13]. It means that we can neglect the bending of the energy bands due to adsorption in comparison with initial bending (at $t = 0$) of the bands due to the surface charge of nonadsorption origin. Obviously, this can take place provided that the density of surface states on the initial surface is sufficiently high. The general case where the surface potential V_s changes in the process of adsorption will be considered in Section 3.3.4.

Now let us turn to the solution of (3.19) at τ^0 and τ^- constant. The first equation in (3.19) yields

$$N^- = \tau^- \left\{ \frac{dN^0}{dt} + \left(\frac{1}{\tau^0} + \frac{1}{\tau}\right) N^0 - \frac{N_\infty^0}{\tau} \right\}, \tag{3.26}$$

where

$$\frac{dN^-}{dt} = \tau^- \left\{ \frac{d^2 N^0}{dt^2} + \left(\frac{1}{\tau^0} + \frac{1}{\tau}\right) \frac{dN^0}{dt} \right\}. \tag{3.27}$$

The second equation in (3.19) yields

$$N^0 = \tau^0 \left(\frac{dN^-}{dt} + \frac{N^-}{\tau^-} \right),$$
(3.28)

where

$$\frac{dN^0}{dt} = \tau^0 \left(\frac{d^2 N^-}{dt^2} + \frac{1}{\tau^-} \cdot \frac{dN^-}{dt} \right).$$
(3.29)

Substituting (3.26) and (3.27) into (3.28), then (3.28) and (3.29) into (3.26), and noting that

$$\frac{N^0_\infty}{\tau^0} = \frac{N^-_\infty}{\tau^-},$$

we obtain

$$\frac{d^2 N^0}{dt^2} + \left(\frac{1}{\tau^0} + \frac{1}{\tau^-} + \frac{1}{\tau} \right) \frac{dN^0}{dt} - \frac{N^0_\infty - N_0}{\tau \tau^-} = 0,$$

$$\frac{d^2 N^-}{dt^2} + \left(\frac{1}{\tau^0} + \frac{1}{\tau^-} + \frac{1}{\tau} \right) \frac{dN^-}{dt} - \frac{N^-_\infty - N^-}{\tau \tau^-} = 0.$$
(3.30)

Thus, we have converted the coupled equations in (3.19) into two independent equations (3.30). The solutions of Eqs. (3.30) have the form

$$N^0(t) = N^0_\infty + C_1' \exp\left(-\frac{t}{\tau_1} \right) + C_2' \exp\left(-\frac{t}{\tau_2} \right),$$

$$N^-(t) = N^-_\infty + C_1'' \exp\left(-\frac{t}{\tau_1} \right) + C_2'' \exp\left(-\frac{t}{\tau_2} \right),$$
(3.31)

where

$$\frac{1}{\tau_1} = \lambda \left(1 + \sqrt{1 - \frac{\mu}{\lambda^2}} \right), \qquad \lambda = \frac{1}{2} \left(\frac{1}{\tau} + \frac{1}{\tau^0} + \frac{1}{\tau^-} \right),$$

$$\frac{1}{\tau_2} = \lambda \left(1 - \sqrt{1 - \frac{\mu}{\lambda^2}} \right), \qquad \mu = \frac{1}{\tau} \frac{1}{\tau^-}.$$
(3.32)

The constants of integration C_1', C_2', C_1'', and C_2'' in (3.31) can be found from the initial conditions [see (3.20) and (3.19)]

$$N^0 = 0, \quad N^- = 0, \quad \frac{dN^0}{dt} = \frac{N^0_\infty}{\tau}, \quad \frac{dN^-}{dt} = 0 \quad \text{at} \quad t = 0.$$

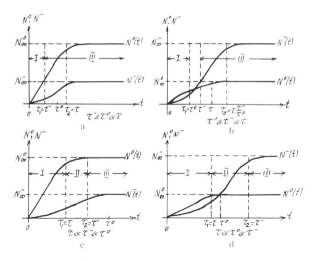

Fig. 3.9. N^0 and N^- as functions of time: a) $\tau^- \ll \tau^0 \ll \tau$; b) $\tau^0 \ll \tau^- \ll \tau$; c) $\tau \ll \tau^- \ll \tau^0$; d) $\tau \ll \tau^0 \ll \tau^-$.

We finally have

$$N^0(t) = \frac{N^0_\infty}{(\tau_2 - \tau_1)\tau}\left\{\tau_2(\tau - \tau_1)\left[1 - \exp\left(-\frac{t}{\tau_2}\right)\right] + \tau_1(\tau_2 - \tau)\left[1 - \exp\left(-\frac{t}{\tau_1}\right)\right]\right\},$$

(3.33)

$$N^-(t) = \frac{N^-_\infty}{\tau_2 - \tau_1}\left\{\tau_2\left[1 - \exp\left(-\frac{t}{\tau_2}\right)\right] - \tau_1\left[1 - \exp\left(-\frac{t}{\tau_1}\right)\right]\right\}.$$

Note that if τ^0 and τ^- are much smaller than τ or if τ is much smaller than τ^0 and τ^-, then $\mu/\lambda^2 \ll 1$, and the formulas (3.32) for $1/\tau_1$ and $1/\tau_2$ simplify considerably:

$$1/\tau_1 = 2\lambda = 1/\tau + 1/\tau^0 + 1/\tau^-,$$

(3.34)

$$1/\tau_2 = \mu/2\lambda = \frac{(1/\tau)(1/\tau^-)}{1/\tau + 1/\tau^0 + 1/\tau^-},$$

and, hence, in this case

$$\tau_1/\tau_2 = \mu/4\lambda^2 \ll 1.$$

(3.35)

The functions $N^0 = N^0(t)$ and $N^- = N^-(t)$ with $\tau^0, \tau^- \ll \tau$ and $\tau \ll \tau^0, \tau^-$ are depicted in Fig. 3.9a, b and Fig. 3.9c, d, respectively.* Here Figs. 3.9a and 3.9c

---

*We will estimate the values of τ, τ^0, and τ^- in Section 3.4.2.

correspond to the case where $\eta^- \ll \eta^0$ (or $\tau^- \ll \tau^0$), while Figs. 3.9b and 3.9d correspond to the case where $\eta^0 \gg \eta^-$ (or $\tau^0 \ll \tau^-$). We will restrict ourselves to these two limiting cases, which for established adsorption equilibrium correspond to one of the two coexisting forms of chemisorption (neutral or charged) being predominant. In reality either one of these cases is realized.

3.3.3. Adsorption at a Constant Surface Potential: Particular Cases

Let us study the kinetic law (3.33). We will take as particular cases those with low, high, and intermediate surface coverage, which correspond to regions I, II, and III in Fig. 3.9.

(a) We will start with the case of low surface coverage (region I), assuming that $t \ll \tau_1$, where τ_1 is the smallest of the numbers τ, τ^0, and τ^- [see (3.34) and Fig. 3.9]. If we take into account (3.35), we find that Eqs. (3.33) assume the form

$$N^0(t) = N^0_\infty \frac{t}{\tau} \ll N^0_\infty,$$

$$N^-(t) = N^-_\infty \frac{t^2}{2\tau_1\tau_2} = N^-_\infty \frac{t^2}{2\tau\tau^-} \ll N^-_\infty, \tag{3.36}$$

since, as follows from (3.34), $\tau_1\tau_2 = \tau\tau^-$. If we combine Eqs. (3.36) with (3.24), we obtain

$$\frac{N^-}{N^{0\cdot}} = \frac{\eta^-}{\eta^0} \frac{t}{2\tau^-} = \frac{t}{2\tau^0} \tag{3.37}$$

and, hence,

$$\frac{N^-}{N^0} \ll 1, \quad \text{since } t \ll \tau,$$

i.e., in the given region it is the neutral form of chemisorption that is dominant at the surface. Moreover,

$$\frac{N^-}{N^0} \ll \frac{\eta^-}{\eta^0}, \quad \text{since } t \ll \tau^-, \tag{3.38}$$

i.e., we are far from electronic equilibrium. In the same region, i.e., at $t \ll \tau_1$, we have, on the basis of (3.24) combined with (3.36),

$$\frac{dN^0}{dt} = \frac{N^0_\infty}{\tau}, \quad \frac{dN^-}{dt} = \frac{N^-_\infty}{\tau^-}\frac{t}{\tau} = \frac{N^0_\infty}{\tau^0}\frac{t}{\tau},$$

where, according to (3.18),

$$\frac{dN}{dt} = \frac{dN^0}{dt} + \frac{dN^-}{dt} = \frac{N^0_\infty}{\tau}\left(1 + \frac{t}{\tau^0}\right) = \frac{N^0_\infty}{\tau} = \alpha PN^*, \qquad (3.39)$$

where α is given in (3.16). We see that in the region under investigation adsorption proceeds with the activation energy E_0 being independent of electronic parameters.

(b) Now we will take the case of high surface coverage (region III in Fig. 3.9), assuming that $t \gg \tau^-$. According to (3.34), we have $t \gg \tau_1$ and, hence, Eqs. (3.33) with (3.35) taken into account assume the form

$$N^0(t) = N^0_\infty\left[1 - \frac{\tau - \tau_1}{\tau}\,\exp\left(-\frac{t}{\tau_2}\right)\right],$$

$$N^-(t) = N^-_\infty\left[1 - \exp\left(-\frac{t}{\tau_2}\right)\right]. \qquad (3.40)$$

When $\tau^0, \tau^- \ll \tau$ (Fig. 3.9a, b), we have $\tau_1 \ll \tau$ [see (3.34)] and, hence, Eqs. (3.40) yield

$$N^0(t) = N^0_\infty[1 - \exp(-t/\tau_2)], \qquad N^-(t) = N^-_\infty[1 - \exp(-t/\tau_2)]. \qquad (3.41)$$

while when $\tau \ll \tau^0, \tau^-$ (Fig. 3.9c, d), we have $\tau_1 = \tau$ and $\tau_2 = \tau^-$ [see (3.34)] and, hence, Eqs. (3.40) yield

$$N^0(t) = N^0_\infty, \qquad N^-(t) = N^-_\infty. \qquad (3.42)$$

Combining (3.41) and (3.42), we find that

$$\frac{N^-}{N^0} = \frac{N^-_\infty}{N^0_\infty} = \frac{\eta^-}{\eta^0},$$

i.e., in region II there is electronic equilibrium on the surface.

We see that when $\tau \ll \tau^0, \tau^-$, electronic equilibrium is established only when adsorption equilibrium is present, while when $\tau^0, \tau^- \ll \tau$, electronic equilibrium may be considered established at any moment of time t (for $t \gg \tau^-$) in the adsorption process. In this latter case we can find the adsorption rate from (3.41) combined with (3.24) and (3.34):

$$\frac{dN}{dt} = \frac{dN^0}{dt} + \frac{dN^-}{dt} = \frac{N^0_\infty}{\tau_2}\left(1 + \frac{\tau^-}{\tau^0}\right)\exp\left(-\frac{t}{\tau_2}\right) = \frac{N^0_\infty}{\tau}\exp\left(-\frac{t}{\tau_2}\right). \qquad (3.43)$$

Thus, at $t \ll \tau_2$ we again arrive at (3.39).

(c) We now turn to the intermediate region (region II in Fig. 3.9c, d; case $\tau \ll \tau^0, \tau^-$), in which $\tau_1 \ll t \ll \tau_2$. Here, according to (3.34), $\tau_1 = \tau$, $\tau_2 = \tau^-$, and, hence, $\tau \ll \tau_2$, according to (3.35). Equations (3.33) can be rewritten as follows:

$$N^{\upsilon}(t) = N_{\infty}^{\upsilon} \frac{(\tau_2 - \tau)\tau_1 + (\tau - \tau_1)t}{\tau_2 \tau} = N_{\infty}^0,$$

$$N^-(t) = N_{\infty}^- \frac{t}{\tau^-} \ll N_{\infty}^-.$$

(3.44)

We see that in the given region of values of t the fraction of the neutral form of chemisorption remains practically constant (after attaining the equilibrium value), while the fraction of the charged form increases with the passage of time but is far from equilibrium. Dividing the second equation in (3.44) by the first, we again arrive at the condition (3.38), which means that we are far from the equilibrium region.

Differentiating N^0 and N^- in (3.44) with respect to t, we obtain

$$\frac{dN^0}{dt} = N_{\infty}^0 \frac{\tau - \tau_1}{\tau \tau_2}, \qquad \frac{dN^-}{dt} = \frac{N_{\infty}^-}{\tau^-},$$

or, taking (3.34) and (3.24) into account, i.e.,

$$\frac{\tau - \tau_1}{\tau \tau_2} = \frac{\tau}{\tau^-} \left(\frac{1}{\tau^0} + \frac{1}{\tau^-} \right), \qquad \frac{N_{\infty}^-}{\tau^-} = \frac{N_{\infty}^0}{\tau^0},$$

we have

$$\frac{dN}{dt} = \frac{dN^0}{dt} + \frac{dN^-}{dt} = \frac{N_{\infty}^0}{\tau^0} \left(1 + \frac{\tau^0}{\tau^-} \cdot \frac{\tau}{\tau^-} \right)$$

or, if the Fermi level does not lie too low, when $E_F \geq E_A$ and, hence, $\tau^0 \leq \tau^-$ [see (3.24) and Fig. 3.9d], then

$$\frac{dN}{dt} = \frac{N_{\infty}^0}{\tau^0}.$$

(3.45)

If, in addition,

$$\exp\left(-\frac{E_C - E_F}{kT} \right) \ll \exp\left(\frac{E_A - E_V}{kT} \right),$$

(3.46)

then on the basis of (3.25a), (3.21), (3.18), (3.17), and (3.16) we can write Eq. (3.45) in the following form:

$$\frac{dN}{dt} = B \exp\left(-\frac{E}{kT} \right) P,$$

(3.47)

where

$$E = E_0 - q^0 + (E_C^s - E_F) = E_0 - q^0 + (E_C^{\upsilon} - E_F) + V_s.$$

We see that in the given region the adsorption activation energy proves to be dependent on the position of the Fermi level on the adsorbent surface. This offers the possibility of controlling the activation energy by acting externally on the sample.

3.3.4. Adsorption with a Varying Surface Potential

Equations (3.19) were solved in Section 3.3.2 on the assumption that $V_s =$ const, i.e., the bending of the energy bands did not change with surface coverage. This corresponds to the case where practically all of the surface charge is in intrinsic surface states, and the contribution due to adsorption is negligible.

Now let us examine Eq. (3.19) for V_s, a function of N^-, and, hence, $\tau^0 = \tau^0(N^-)$ and $\tau^- = \tau^-(N^-)$; i.e., we will allow for variations in the surface potential in the process of adsorption. Note that according to (3.23) and (3.17) we have

$$\tau_0^0 \leqslant \tau^0 \leqslant \tau_\infty^0, \qquad \tau_0^- \leqslant \tau^- \leqslant \tau_\infty^-. \tag{3.48}$$

where we have used the notations

$$\tau_0^0 = \tau^0(0), \qquad \tau_\infty^0 = \tau^0(N_\infty^-), \qquad \tau_0^- = \tau^-(0), \qquad \tau_\infty^- = \tau^-(N_\infty^-).$$

We will restrict ourselves to the case of greatest interest, namely, when

$$\tau \ll \tau_0^0 \ll \tau_\infty^- \tag{3.49}$$

(see Fig. 3.9d). In addition, we consider only the region where

$$\frac{N^-(t)}{N^0(t)} \ll \frac{N_\infty^-}{N_\infty^0}, \tag{3.50}$$

i.e., far from electronic and adsorption equilibriums. With (3.50) in mind and employing (3.48) and (3.24), we obtain

$$\frac{N^-}{N^0} \frac{\tau^0}{\tau^-} \leqslant \frac{N^-}{N^0} \frac{\tau_\infty^0}{\tau_\infty^-} = \frac{N^-}{N^0} \frac{N_\infty^0}{N_\infty^-} \ll 1,$$

and Eqs. (3.19) assume the form

$$\frac{dN^0}{dt} = \frac{N_\infty^0}{\tau} - \frac{N^0}{\tau^0} \left(\frac{\tau^0 + \tau}{\tau} \right), \tag{3.51a}$$

$$\frac{dN^-}{dt} = \frac{N^0}{\tau^0}. \tag{3.51b}$$

We will consider two regions, which we will call the regions of "low" and "high" coverage:

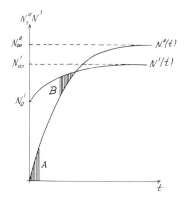

Fig. 3.10. "Low" coverage and "high" coverage regions in adsorption with varying surface potential.

(A) The region of "low" coverage, $N^0 = \delta N'$, (3.52a)

(B) The region of "high" coverage, $N^0 = (1 - \delta) N'$, (3.52b)

where

$$\delta \ll \frac{\tau}{\tau^0} \ll 1 \tag{3.53}$$

and

$$N^- = \frac{\tau^0}{\tau^0 + \tau} N_\infty^0. \tag{3.54}$$

The regions (A) and (B) are the shaded regions in Fig. 3.10, where the N^0 vs. t and N' vs. t curves are given and where the following notations are employed:

$$N_0' = \frac{\tau_0^0}{\tau_0^0 + \tau} N_\infty^0,$$

$$N_\infty' = \frac{\tau_\infty^0}{\tau_\infty^0 + \tau} N_\infty^0.$$

We will study each region separately.

(A) According to (3.52a) and (3.54),

$$N^0 = \delta \frac{\tau^0}{\tau^0 + \tau} N_\infty^0. \tag{3.55}$$

Substituting (3.55) into (3.51a) and (3.51b) and allowing for (3.49) and (3.53), we find that

$$\frac{dN^0}{dt} = \frac{N^0_\infty}{\tau},$$ (3.56a)

$$\frac{dN^-}{dt} = -\frac{N^0_\infty}{\tau^0}\,\delta,$$ (3.56b)

where, according to (3.49) and (3.53), we have

$$\frac{dN^0/dt}{dN^-/dt} = \frac{\tau^0}{\tau}\cdot\frac{1}{\delta} \gg 1$$

and, hence,

$$\frac{dN}{dt} = \frac{dN^0}{dt} = \frac{N^0_\infty}{\tau}.$$ (3.57)

We have arrived at formula (3.29). Combining (3.57) with (3.16) yields, finally,

$$\frac{dN}{dt} = A\,\exp\left(-\frac{E_0}{kT}\right)P.$$ (3.58)

We see that in the given region the adsorption activation energy is constant and does not depend on the electronic state of the system. It can be shown (see [15]) that this formula can be applied as long as

$$t \ll \tau.$$ (3.59)

Note that in the surface coverage region considered here

$$N^- \ll N^0,$$ (3.60)

i.e., the neutral form of chemisorption prevails over the charged form. Indeed, integrating Eq. (3.56a) yields

$$N^0 = N^0_\infty\,\frac{t}{\tau}.$$ (3.61)

Substituting (3.61) into (3.51b) and then integrating the result, we obtain [if we take (3.48) into account]

$$N^- = \frac{N^0_\infty}{\tau}\int_0^t \frac{t}{\tau^0}\,dt \leqslant \frac{N^0_\infty}{\tau_0^0}\,\frac{t^2}{2\tau}.$$ (3.62)

On the basis of (3.62) and (3.61) we obtain

$$\frac{N^-}{N^0} \leqslant \frac{t}{2\tau_0^0},$$

where, taking into account (3.59) and (3.49), we arrive at (3.60).

(B) This is the "high" surface coverage region. According to (3.52b) and (3.54), we have

$$N^0 = (1 - \delta) \frac{\tau^0}{\tau^0 + \tau} N_\infty^0. \tag{3.63}$$

Substituting (3.63) into (3.51a) and (3.51b), we obtain

$$\frac{dN^0}{dt} = \frac{N_\infty^0}{\tau} \delta, \tag{3.64a}$$

$$\frac{dN^-}{dt} = \frac{N_\infty^0}{\tau^0}, \tag{3.64b}$$

where, according to (3.53),

$$\frac{dN^0/dt}{dN^-/dt} = \frac{\tau^0}{\tau} \delta \ll 1$$

and, hence,

$$\frac{dN}{dt} = \frac{dN^-}{dt} = \frac{N_\infty^0}{\tau^0}. \tag{3.65}$$

We see that in the region considered here the neutral form of chemisorption practically attains equilibrium [see (3.63)] and adsorption proceeds exclusively through replenishment of the charged form. Expression (3.65) coincides with (3.45), the only difference being that τ^0 in (3.65) is not a constant but depends on N^-. If (3.46) holds, formula (3.65) takes the form of (3.47):

$$\frac{dN}{dt} = B \exp\left(-\frac{E}{kT}\right) P, \tag{3.66}$$

where, however,

$$E = E_0 - q^0 + (E_C^v - E_F) + V_s(N^-). \tag{3.67}$$

The activation energy proves to be a function of the position of the Fermi level. It can be shown (see [15]) that (3.66) holds if

$$\tau_0^0 \ll t \ll \tau_\infty^-.$$

Note that in the surface coverage region considered here

$$N^0 \ll N^-, \tag{3.68}$$

i.e., the charged form prevails considerably over the neutral form (practically all the adsorbed particles are in the charged state). Indeed, integrating Eqs. (3.64a) and (3.64b), we obtain via (3.53) the following:

$$N^0 = \frac{N^0_\infty}{\tau} \int_0^t \delta \, dt \ll N^0_\infty \int_0^t \frac{dt}{\tau^0}, \qquad N^- = N^0_\infty \int_0^t \frac{dt}{\tau^0},$$

which leads to (3.68).

In conclusion we must note that within the framework of these models it is still possible to analyze the kinetics of chemisorption when the intrinsic surface charge varies and when the kinetic curves reflecting the charging of the surface in experiments are not smooth [16].

3.4. THE KINETICS OF DESORPTION

3.4.1. Desorption with Electronic Equilibrium

Let us study the kinetics of desorption [13]. We assume that for given T and P there is adsorption equilibrium at the surface as well as electronic equilibrium. We also assume that at $t = 0$ the pressure suddenly drops and subsequently (at $t > 0$) is kept equal to zero. We are interested in the course of desorption with the passage of time; i.e., we wish to find $N^0 = N^0(t)$ and $N^- = N^-(t)$ for $t \geq 0$.

The problem consists in solving the system of equations (3.19) where, however, $N^0_\infty = 0$ [according to (3.18)]. The initial conditions have the form

$$N^0 = N^0_0, \quad N^- = N^-_0 \qquad \text{at} \quad t = 0, \tag{3.69}$$

where N^0_0 and N^-_0 satisfy the following condition [cf. (3.24)]:

$$\frac{N^-_0}{N^0_0} = \frac{\tau^-}{\tau^0} = \frac{\eta^-}{\eta^0}. \tag{3.70}$$

Just as in Section 3.3.2, we will assume that τ^0 and τ^- are constant; i.e., the bending of the energy bands does not change during desorption (the validity of this assumption will be considered below). In this case Eqs. (3.19) yield, just as in Section 3.3.2 [see (3.30)],

$$\frac{d^2 N^0}{dt^2} + \left(\frac{1}{\tau^0} + \frac{1}{\tau^-} + \frac{1}{\tau} \right) \frac{dN^0}{dt} + \frac{N^0}{\tau \tau^-} = 0,$$

$$\frac{d^2 N^-}{dt^2} + \left(\frac{1}{\tau^0} + \frac{1}{\tau^-} + \frac{1}{\tau} \right) \frac{dN^-}{dt} + \frac{N^-}{\tau \tau^-} = 0,$$

where instead of (3.31) we obtain

$$N^0(t) = C_1' \exp\left(-\frac{t}{\tau_1}\right) + C_2' \exp\left(-\frac{t}{\tau_2}\right),$$

$$N^-(t) = C_1'' \exp\left(-\frac{t}{\tau_1}\right) + C_2'' \exp\left(-\frac{t}{\tau_2}\right), \tag{3.71}$$

where τ_1 and τ_2 are defined in (3.32). The integration constants C_1', C_2', C_1'', and C_2'' in (3.71) can be found from the initial conditions, which according to (3.69), (3.19), and (3.71) have the following form:

$$
\begin{aligned}
N^0 &= N_0^0\,, & N^- &= N_0^- \\
\frac{dN^0}{dt} &= -\frac{N_0^0}{\tau}\,, & \frac{dN^-}{dt} &= 0
\end{aligned}
\quad\text{at}\quad t = 0.
$$

We finally obtain

$$N^0(t) = \frac{N_0^0}{(\tau_2 - \tau_1)\tau}\left[\tau_2(\tau - \tau_1)\exp\left(-\frac{t}{\tau_2}\right) + \tau_1(\tau_2 - \tau)\exp\left(-\frac{t}{\tau_1}\right)\right],$$

$$N^-(t) = \frac{N_0^-}{\tau_2 - \tau_1}\left[\tau_2\exp\left(-\frac{t}{\tau_2}\right) - \tau_1\exp\left(-\frac{t}{\tau_1}\right)\right]. \tag{3.72}$$

Adding the first equation in (3.72) to the second and putting $N_0^0 = \eta^0 N_0$ and $N_0^- = \eta^- N_0$, with $N_0 = N(0)$, we obtain

$$
\begin{aligned}
N(t) = \frac{N_0}{(\tau_2 - \tau_1)\tau}\Bigg\{ &\tau_2\left[\eta^0(\tau - \tau_1) + \eta^-\tau\right]\exp\left(-\frac{t}{\tau_2}\right) \\
&+ \tau_1\left[\eta^0(\tau_2 - \tau) - \eta^-\tau\right]\exp\left(-\frac{t}{\tau_1}\right)\Bigg\},
\end{aligned}
$$

where (since $\eta^0 + \eta^- = 1$)

$$N(t) = \frac{N_0}{(\tau_2 - \tau_1)\tau}\left[\tau_2(\tau - \eta^0\tau_1)\exp\left(-\frac{t}{\tau_2}\right) + \tau_1(\eta^0\tau_2 - \tau)\exp\left(-\frac{t}{\tau_1}\right)\right]. \tag{3.73}$$

We will restrict ourselves, just as we did in Section 3.3.2, to two limiting cases: $\tau^0, \tau^- \ll \tau$ and $\tau \ll \tau^0, \tau^-$. We will take τ_1 and τ_2 in the form (3.34), with $\tau_1 \ll \tau_2$, which means that for all values of t we have

$$\tau_2\exp\left(-\frac{t}{\tau_2}\right) \gg \tau_1\exp\left(-\frac{t}{\tau_1}\right),$$

so that the second equation in (3.72) takes the form

$$N^-(t) = N_0^-\exp\left(-\frac{t}{\tau_2}\right). \tag{3.74}$$

Dividing the first equation in (3.72) by (3.74), we find that

$$\frac{N^0}{N^-} = \frac{\eta^0}{\eta^-} \frac{1}{\tau} \left\{ (\tau - \tau_1) + (\tau_2 - \tau) \frac{\tau_1}{\tau_2} \exp\left[-\left(\frac{1}{\tau_1} - \frac{1}{\tau_2}\right)t\right] \right\}. \tag{3.75}$$

Let us take the case where $\tau^0, \tau^- \ll \tau$ (we will consider the case where $\tau \ll \tau^0, \tau^-$ in Section 3.4.2). If both τ^0 and τ^- are much smaller than τ, then, according to (3.34) and (3.24), we have

$$\tau_1 = \frac{\tau^0 \tau^-}{\tau_0 + \tau^-} = \tau^0 \eta^- = \tau^- \eta^0, \quad \tau - \tau_1 = \tau\left(1 - \eta^- \frac{\tau^0}{\tau}\right) = \tau,$$

$$\tau_2 = \frac{\tau}{\tau_0}(\tau^0 + \tau^-) = \frac{\tau}{\eta^0}, \quad \tau_2 - \tau = \tau\left(1 - \frac{1}{\eta^0}\right) = \tau^- \frac{\tau^-}{\tau_0}. \tag{3.76}$$

Substituting (3.76) into (3.73) and bearing in mind that $\tau_1 \ll \tau_2$, we find that

$$N(t) = N_0 \exp\left(-\frac{t}{\tau_2}\right). \tag{3.77}$$

If we substitute (3.76) into (3.75), we obtain

$$\frac{N^0}{N^-} = \frac{\eta^0}{\eta^-} \left\{ 1 + \frac{\tau^-}{\tau^0} \frac{\tau_1}{\tau_2} \exp\left[-\left(\frac{1}{\tau_1} - \frac{1}{\tau_2}\right)t\right] \right\},$$

where, noting that according to (3.76)

$$\frac{\tau^-}{\tau^0} \cdot \frac{\tau_1}{\tau_2} = \frac{\tau^-}{\tau} \eta^- \eta^0 \ll 1$$

and that

$$\exp\left[-\left(\frac{1}{\tau_1} - \frac{1}{\tau_2}\right)t\right] \ll 1,$$

we have

$$\frac{N^0}{N^-} = \frac{\eta^0}{\eta^-}. \tag{3.78}$$

We see that according to (3.77) and (3.76) desorption proceeds the faster the greater the value of η^0, i.e., the stronger the neutral form of chemisorption is prior to desorption than the charged form. Moreover, on the basis of (3.78) we can conclude that when $\tau \gg \tau_0$ or τ^- (viz., adsorption and desorption of the neutral species is rate limiting), electronic equilibrium on the surface is retained during desorption. The charged and neutral forms leave the surface simultaneously and in such a way that their fractions do not vary with time, thus retaining their value all through the desorption process.

The initial assumption that both τ^0 and τ^- are constants means (in the present case) that the adsorption charge on the surface is assumed to be small compared to the surface charge with no adsorption.

3.4.2. Violation of Electronic Equilibrium in Desorption

Now we will turn to the other limiting case, $\tau \ll \tau^0, \tau^-$. According to (3.34) and (3.24) we have

$$\tau_2 = \tau, \quad \tau - \tau_1 = \tau\left(\frac{\tau}{\tau^0} + \frac{\tau}{\tau^-}\right) = \frac{\tau^2}{\tau^- \eta^0} = \frac{\tau^2}{\tau^0 \eta^-},$$

$$\tau_2 = \tau^-, \quad \tau_2 - \tau = \tau^-.$$

(3.79)

Combining Eq. (3.73) with (3.79), we arrive at the following equation:

$$N(t) = N_0\left[\eta^- \exp\left(-\frac{t}{\tau^-}\right) + \left(\eta^0 - \frac{\tau}{\tau^-}\right)\exp\left(-\frac{t}{\tau}\right)\right].$$

(3.80)

Note that

$$\frac{\tau}{\tau^-}\frac{1}{\eta^0} = \frac{\tau}{\tau^-}\frac{\tau^0 + \tau^-}{\tau^0} = \frac{\tau}{\tau^-} + \frac{\tau}{\tau^0} \ll 1,$$

and therefore we can rewrite Eq. (3.80) thus:

$$N(t) = N_0\left[\eta^- \exp\left(-\frac{t}{\tau^-}\right) + \eta^0 \exp\left(-\frac{t}{\tau}\right)\right].$$

(3.81)

Equation (3.75) takes the form

$$\frac{N^0}{N^-} = \frac{\eta^0}{\eta^-}\left\{\frac{\tau}{\tau^0} + \frac{\tau}{\tau^-} + \exp\left[-\left(\frac{1}{\tau} - \frac{1}{\tau^-}\right)t\right]\right\}.$$

(3.82)

It then follows that

$$\frac{N^0}{N^-} = \begin{cases} \frac{\eta^0}{\eta^-} \exp\left[-\left(\frac{1}{\tau} - \frac{1}{\tau^-}\right)t\right] \leqslant \frac{\eta^0}{\eta^-} & \text{at} \quad t \ll t^*, \\[3mm] \frac{\eta^0}{\eta^-}\left(\frac{\tau}{\tau^0} + \frac{\tau}{\tau^-}\right) \ll \frac{\eta^0}{\eta^-} & \text{at} \quad t \gg t^*, \end{cases}$$

(3.83)

where

$$t^* = -\tau \ln\left(\frac{\tau}{\tau^0} + \frac{\tau}{\tau^-}\right).$$

(3.84)

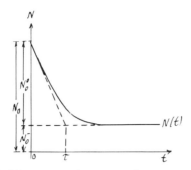

Fig. 3.11. Desorption curve for $t \ll \tau^-$.

Equation (3.83) shows that desorption leads to violation of electronic equilibrium on the surface. At $t \ll t^*$ we go farther and farther from equilibrium with the passage of time because the neutral fraction of chemisorption decreases faster than the charged fraction. On the other hand, at $t \gg t^*$ the fractions of the neutral and charged forms become constant (independent of time), with the charged form being predominant (in relation to the equilibrium value).

Mooveover, Eq. (3.81) shows that the N vs. t curve is a result of the superposition of two curves, a slowly decreasing curve and a rapidly decreasing curve [respectively, the first and second terms in (3.81)]. If desorption is observed over a period of time which is small compared to τ^- (i.e., if τ^- is greater than the duration of the experiment), then in (3.81) we can assume that $t \ll \tau^-$ and rewrite this equation as follows (see Fig. 3.11):

$$N(t) = N_0 \left[\eta^- + \eta^0 \, \exp\left(-\frac{t}{\tau} \right) \right],$$

i.e., only the fraction

$$N_r = N_0 - N_0'\eta^- = N_0 \eta^0 = N_0^0$$

of the total number of adsorbed molecules N_0 desorbs, while the remaining fraction

$$N_i = N_0 \eta^- = N_0^-$$

proves to be irreversibly bound to the surface. Hence, we are dealing here with a case of partially irreversible adsorption often met with in reality, and according to (3.24)

$$\frac{N_i}{N_r} = \frac{\eta^-}{\eta^0} = \exp\left(-\frac{E_A - E_F}{kT} \right), \tag{3.85}$$

i.e., the fraction of the irreversible and reversible forms of chemisorption on the surface is determined, all other conditions being equal, by the position of the Fermi level, or by the state of the system as a whole. Thus, by subjecting the adsorbent to one or another treatment we can control the ratio (3.85).

Note that as long as the first term in (3.81) remains constant, i.e., as long as $t \ll \tau^-$, the initial condition that τ^0 and τ^- are constant can be assumed to be automatically satisfied irrespective of the state of the surface, since during desorption only the electrically neutral form of chemisorption is removed and, hence, the surface charge and the near-the-surface bending of the energy bands remain the same in desorption.

Comparing the results obtained in this section we can conclude that it is important to discriminate between two limiting cases when studying the kinetics of desorption, namely,

$$\tau^0, \tau^- \ll \tau, \tag{3.86a}$$

$$\tau \ll \tau^0, \tau^-. \tag{3.86b}$$

In the case where (3.86a) is met, the electronic equilibrium in the desorption process is maintained, as we saw earlier, and desorption is complete, i.e., all the adsorbent is removed from the surface. Here the electronic equilibrium is maintained by two balancing mechanisms: the desorption of neutral particles, which violates the equilibrium, and the discharge of charged particles, which restores the equilibrium. The neutral and charged forms of chemisorption are removed from the surface simultaneously.

In the case where (3.86b) is met, the electronic equilibrium is violated during desorption; i.e., electronic equilibrium is restored more slowly than the neutral particles are desorbed. First the neutral particles leave the surface (over a time interval of the order of τ) and only then, as the discharge process proceeds and over a much longer time interval (of the order of τ), do the charged particles leave the surface. If this process is sufficiently slow, practically only the desorption of neutral particles will be observed, i.e., desorption will be incomplete (the case of partially irreversible adsorption).

In conclusion, let us estimate the values of the parameters τ, τ^0, and τ^-, which figure in the theory. We have (see [17–20])

$$\nu^0 = 10^{16} \text{ sec}^{-1}, \quad \beta_1, \beta_3 = 10^{-9} \text{ cm}^3 \text{ sec}^{-1}, \quad \beta_2, \beta_4 = 10^{10} \text{ sec}^{-1},$$

and, hence, according to (3.17), we have

$$\frac{1}{\tau} = 10^{16} \exp\left(-\frac{q_0}{kT}\right),$$

$$\frac{1}{\tau^0} = 10^{-9} n_s + 10^{10} \exp\left(-\frac{E_A - E_V}{kT}\right), \tag{3.87}$$

$$\frac{1}{\tau^-} = 10^{-9} p_s + 10^{10} \exp\left(-\frac{E_C - E_A}{kT}\right).$$

Putting $q^0 = 0.1$–1 eV, $E_C - E_A = 0.1$–1.5 eV, $E_A - E_V = 0.1$–1.5 eV, $n_s = 10^4$–10^{18} cm^{-3}, and $p_s = 10^4$–10^{18} cm^{-3} and employing (3.87), we find* that with $kT = 0.03$ eV we have

$$\tau = 10^{-14} \text{ to } 10^{-2} \text{ sec,}$$

$$\tau^0, \tau^- = 10^{-9} \text{ to } 10^5 \text{ sec.}$$

We see that the possible values of τ, τ^0, and τ^- lie in a wide range. Depending on the nature of the semiconductor, one of the limiting cases will be realized, i.e., either $\tau^0, \tau^- \ll \tau$ or $\tau \ll \tau^0, \tau^-$.

3.4.3. Incomplete Desorption

It is a well-known fact that it is often impossible to remove all the adsorbate from the adsorbent surface by evacuation of the space around the sample (at the same temperature at which the adsorption process took place). A fraction of the adsorbate remains on the surface; in other words, the desorption is incomplete. Complete desorption usually requires a significant rise in temperature. Apparently, this is the result of two forms of chemisorption coexisting on the surface: the reversible form, which leaves the surface in evacuation, and the irreversible form, which during the experiment practically remains where it is.

In Section 3.4.2 we discussed the possible nature of these two forms of chemisorption. The neutral form of chemisorption ("weak" bonding) can be taken as the reversible form, while in some cases the charged form ("strong" bonding) may serve as the irreversibly form. The irreversibility in this case is caused by the hampered desorption of the particles that are in the charged state. The desorption of such particles can be interpreted as an act in which an electron localized at the chemisorbed particle is delocalized and remains in the crystal lattice, while the particle becomes neutral and leaves the surface. The hampered nature of such electron delocalization, i.e., discharge of the charged particle, leads to the apparent incomplete desorption.

Here desorption passes through two stages: at first the neutral form of chemisorption is rapidly desorbed and then, slowly, the charged form disappears from the surface. The value of τ, which gives the order of the time interval necessary for completing the desorption process, may in some cases prove to be very large, of the order of 10^4–10^5 sec. If this is the case, then in a not too prolonged experiment we will observe only the neutral form, while the charged form will seem to be bound to the surface irreversibly.

---

*Note that we are dealing with values of the carrier concentration in the plane of the semiconductor surface, and these values may be small even for a good semiconductor. The value 10^4 cm^{-3} for n_s and p_s corresponds to a surface potential V_s of 1 eV.

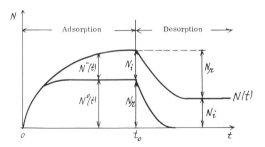

Fig. 3.12. Kinetics of adsorption and desorption.

This is illustrated by Fig. 3.12, where the kinetics of adsorption and desorption is depicted. At time $t = 0$ the gas is admitted into the adsorption volume (adsorption starts); by the time $t = t_0$ has elapsed, the adsorption (and electronic) equilibrium can be assumed to be established; at time t_0 the pressure is exhausted (desorption starts). We denote the surface concentrations of reversibly and irreversibly chemisorbed particles by N_r and N_i, respectively, and

$$N_r + N_i = N.$$

These two quantities obviously vary with the passage of time, i.e.,

$$N_r = N^0(t), \qquad N_i = N^-(t).$$

It should be noted that irreversible adsorption may have an entirely different origin. Partial irreversibility also occurs when on the surface of the semiconductor a secondary chemical process develops parallel to the chemisorption process in the proper sense of the word. The chemical process just mentioned is the creation of a new phase or the surface reaction of the chemisorbate with impurities that emerge at the semiconductor surface from the bulk. For instance, in adsorption of oxygen on zinc oxide the following occur:

$$O_{at}^- + Zn_{int}^+ \rightarrow Zn^{++} O^- \quad , \tag{3.88}$$

where O_{at}^- is a chemisorbed oxygen atom in the charged state, and Zn_{int}^+ is singly ionized interstitial zinc atom. Obviously, after reaction (3.88) has been completed, there is no way in which we can discriminate between the chemisorbed particle and impurity, on the one hand, and the lattice elements, on the other, because reaction (3.88) is actually the completion of the crystal lattice due to oxidation of the stoichiometric-in-excess zinc.

Note the difference between the above-mentioned mechanisms. The irreversibility due to the first mechanism is indeed only "apparent" since it is the result of the insufficient duration of the experiment in comparison with the time interval needed for the desorption of particles out of the charged state. The irreversibility brought on by the reaction of the chemisorbate with the impurity is independent of the duration of the experiment. This irreversibility can be said to be "true" irreversibility. The irreversible fraction N_i of chemisorption is determined in this case by the amount of gas that has reacted with the impurity up to the moment when desorption starts.

The mechanisms of "apparent" irreversibility can be effective only for moderate temperatures, since τ^- rapidly drops as the temperature rises. On the other hand, "true" irreversibility, which is due to the reaction between the chemisorbate and the impurity from the bulk and, therefore, requires a high mobility of the latter, must have a small effect at low temperatures but sharply increases its contribution at higher temperatures.

We note in conclusion that if electronic equilibrium is not maintained during desorption, "apparent" irreversibility may be superimposed on "true" irreversibility, and the irreversibility then has a mixed origin.

We will discuss "true" irreversibility in terms of a reaction of the adsorbate with the impurity dissolved in the adsorbent in greater detail in Section 4.5.4, which is devoted to the interaction of the surface with the bulk. This problem, which in the final analysis is the problem in the growth of the semiconductor lattice via its interaction with the gaseous phase, brings us to the problem of corrosion.

3.5. THE ROLE OF THE FERMI LEVEL IN CHEMISORPTION

3.5.1. The Fermi Level as Regulator of the Chemisorptive Properties of a Surface

In the course of this chapter we have seen that a variety of properties associated with chemisorption on a surface are determined by the position of the Fermi level on the adsorbent surface (we mean the position of the Fermi level relative to the energy bands in the surface plane). Let us name these properties.

(1) First, we saw (see Section 3.2.1) that, all other things being equal, the position of the Fermi level influences the chemisorptivity of the surface with respect to particles of a definite type, i.e., the total number of particles of this type that a surface retains under equilibrium with the gaseous phase for given pressure and temperature. Lowering the Fermi level lowers the adsorptivity with respect to acceptor molecules and increases the adsorptivity with respect to donor molecules.

(2) The position of the Fermi level influences the probability of finding a chemisorbd particle in a charged or neutral state, i.e., determines the surface charge for a given surface coverage by chemisorbed particles (see Section 3.2.2). Lowering the Fermi level makes the absolute value of the surface charge smaller if the surface is charged negatively (adsorption of acceptor particles) or greater if the surface is charged positively (adsorption of donor particles).

(3) The position of the Fermi level influences the probability of finding a chemisorbed particle in a "strong" or "weak" bonding with the surface, i.e., determines the relative fraction of various forms of chemisorption on the surface,

which differ in the nature of the bonding of the chemisorbed particle with the surface (see Section 3.1.3). Lowering the Fermi level decreases the fraction of the "strong" form of chemisorption of acceptor particles due to the increase in fraction of the "weak" form. When donor particles are involved in adsorption, lowering the Fermi level increases the "strong" fraction and decreases the "weak" fraction.

(4) The position of the Fermi level influences the reactivity of a chemisorbed particle, i.e., the probability of finding the particle in a radical or valence-saturated state (see Sections 3.1.3 and 2.4.2). This implies that the position of the Fermi level influences, as we will subsequently see (Section 5.2.2), the catalytic activity of the surface with respect to a given reaction and the selectivity of a catalyst with respect to two (or several) reactions occurring simultaneously.

(5) Finally, the position of the Fermi level under certain conditions influences the fraction of the reversible and irreversible forms of chemisorption on the surface or, in other words, the probability of a chemisorbed particle being adsorbed reversibly or irreversibly (see Section 3.4.3). Lowering the Fermi level decreases the irreversible fraction of chemisorbed acceptor particles, while in the case of donor particles this fraction grows. The reversible fraction does just the opposite.

The Fermi level, therefore, emerges as regulator of a variety of surface properties and proves to be, as we have seen, the key to controlling these properties. Note that the position of the Fermi level uniquely determines the concentrations of the electron and hole gases on the surface of the crystal. This reveals the physical meaning of the role that the Fermi level plays in chemisorption and, at the same time, establishes the characteristic correlation between the chemisorptive properties of the surface, on the one hand, and the concentrations of free electrons and holes on it, on the other.

What then determines the position of the Fermi level on the surface? Leaving a detailed analysis of this question to Section 4.1.3, we will note here that among other things the position is determined by the nature and amount of particles chemisorbed on the surface of the crystal. Indeed, as more and more acceptor particles become chemisorbed, and the surface acquires a negative charge, the energy bands bend upward and, hence, the Fermi level on the surface steadily moves downward (since the position of the Fermi level inside the crystal may be considered fixed; e.g., see Fig. 3.7). On the other hand, when donor particles are adsorbed, which leads to a positive charge on the surface, the energy bands bend downward and, hence, the Fermi level on the surface steadily moves upward.

Thus, the position of the Fermi level on the surface depends on the extent to which the surface is covered by chemisorbed particles. But, as we saw earlier, the surface coverage for given P and T (i.e., the adsorptivity of the surface) is itself dependent on the position of the Fermi level. We have thus arrived at a problem requiring a self-consistent solution.

We see that the probability of a chemisorbed particle of a given type being in a charged or neutral state, in a "strong" or "weak" state of bonding with the surface, or in a radical or valence-saturated state (i.e., an entire range of properties of a chemisorbed particle) proves to be dependent on the total number of particles of this type chemisorbed on a unit surface area (i.e., on the surface coverage). It is as if each particle "feels" the presence of other particles on the surface. Here we are dealing not with direct contact forces between the chemisorbed particles and not with the usual forces between them but with an action-at-a-distance of a very special type, namely, that the properties of each chemisorbed particle depend not only on the nature of this particle and that of the adsorbent but on the state of the system as a whole, which is a unique function of the position of the Fermi level on the surface of the crystal and which in turn depends on the concentration and nature of all the particles chemisorbed on the surface.

3.5.2. The Origin of Non-Langmuir Relations

The overall result of the dependence of the position of the Fermi level on the surface coverage is the violation of the Langmuir pattern of chemisorption. Specifically, this effect leads to a dependence of the differential adsorption heat on the surface coverage and to non-Langmuir equilibrium and kinetic isotherms.

(1) First let us turn to the phenomenon of adsorption equilibrium. The equation of state, i.e., the equation that connects P, T, and N, has the form (3.9):

$$N = \frac{N^*}{1 + \dfrac{b_0^*}{P} \exp\left(-\dfrac{q^*}{kT}\right)}, \tag{3.89}$$

where, as before, N^* is the adsorption center concentration, and q^* the differential adsorption heat. In Langmuir's theory (see Section 2.1.3) we have

$$q^* = \text{const.}$$

But if we take into account electronic processes that take place in chemisorption, then, as shown in Section 3.2.1, N^* proves to be a function of the position of the Fermi level E_F. If we are dealing with acceptor particles and assume that the Fermi level lies quite close to the conduction band, so that

$$\exp \frac{E_F - E_A}{kT} \gg 1,$$

or if we are dealing with donor particles and assume that the Fermi level lies quite close to the valence band, so that

$$\exp \frac{E_D - E_F}{kT} \gg 1,$$

then, according to (3.6a) and (3.6b), we have

$$N^- = N \quad \text{or} \quad N^+ = N$$

and, according to (3.12),

$$q^* = q^- - (E_C^s - E_F), \quad q^* = q^+ - (E_F - E_V^s), \tag{3.90}$$

where q^- and q^+ are the energies of the acceptor and donor "strong" bonding.

Combining (3.90) with (3.22), we obtain

$$q^* = q^- - (E_C^v - E_F) - V_s, \quad q^* = q^+ - (E_F - E_V^v) + V_s. \tag{3.91}$$

Since V_s increases in absolute value with surface coverage, i.e., with the increase in N, and since V_s is positive for acceptor particles (the bands are bent upward) and negative for donor particles (the bands are bent downward), we find in both cases, in accordance with (3.91), that the differential adsorption heat decreases as the surface coverage increases, which is often observed experimentally. The detailed dependence of q^* on N is determined by the dependence of V_s on N and is different in different approximations.

(2) Equation (3.89) at $q^* = $ const represents a Langmuir isotherm. With (3.90) this equation is not, however, the equation of an isotherm. But we can obtain such an equation if we substitute (3.91) into (3.90), then (3.90) into (3.89), and solve Eq. (3.89) for N. The shape of the isotherm is determined by (3.91). We hence arrive at an essentially non-Langmuir isotherm which in special cases can, as shown in [22], become a Langmuir isotherm or a Freundlich isotherm or an isotherm of the logarithmic type (see Section 2.1.3).

(3) Now let us turn to the kinetics of adsorption. For not too high surface coverages the kinetics equation has the following form:

$$\frac{dN}{dt} = C \exp\left(-\frac{E}{kT} \right) P, \tag{3.92}$$

where E is the adsorption activation energy, and P, the pressure.

In Langmuir's theory (see Section 2.2.3) we have

$$E = \text{const},$$

and the adsorption kinetics, i.e., the time dependence of N, is expressed by a linear law. With the electronic processes taken into account and under certain conditions, we have for acceptor and donor particles, respectively,

$$E = E^* + (E_C^s - E_F), \tag{3.93a}$$

$$E = E^* + (E_F - E_V^s), \tag{3.93b}$$

where $E_C^0 - E_F$ and $E_F - E_V^0$ are given in (3.22). The expression (3.93a) was obtained in Section 3.3.4, while (3.93b) can be obtained in a similar manner. In both cases the adsorption kinetics is obviously given by a more complex law than a linear law since E in (3.92) depends on N. The form of the kinetic dependence is determined by the function $V_s = V_s(N)$. We can obtain the kinetic law explicitly if we substitute (3.91) into (3.93a) and (3.93b), then (3.93a) and (3.93b) into (3.92), and then integrate Eq. (3.92). In particular cases we arrive, as can be shown, at a logarithmic law, Bangham's law, and other laws.

We see that deviations from the Langmuir pattern are caused by the fact that q^* and E depend on the position of the Fermi level and, in the final analysis, on N. Even for a homogeneous (and more than that, ideal) surface we find that these deviations are present. This result was studied in detail theoretically in a number of works [22–27]. In Sections 8.2 and 8.4 we will return to this question.

3.5.3. The Approximations of the "Boundary Layer Theory"

In conclusion a few remarks will be made about the many theoretical papers on the theory of chemisorption on semiconductors by non-Soviet authors and known under the general title "boundary layer theory" ("Randschichttheorie der Adsorption") (see, for instance, [24–26, 28–31]).

From the viewpoint of the "boundary layer theory" every chemisorbed particle is always charged and remains in the adsorbed state only as long as it is retained on the surface by the electron (or hole) localized at it; it leaves the surface as soon as the charge associated with it is neutralized.

In other words, the removal of an electron from a local acceptor level E_A or a hole from a local donor level E_D, both levels denoting a chemisorbed particle in the energy spectrum of the crystal (Fig. 3.5), is considered as desorption of this particle, i.e., the disappearance of the very local level. This notion of an acceptor level that exists only as long as it is occupied by an electron or of a donor level that is always deprived of its electron renders meaningless the very concept of a local level as a level capable of accepting or giving up an electron.

The removal of an electron from an acceptor level or a hole from a donor level means, as we have seen, not the desorption of the chemisorbed particle but only its transition from a state of "strong" bonding to a state of "weak" bonding with the surface. Neglecting this "weak" form of chemisorption (i.e., the electrically neutral form), which is characteristic of all the papers on "boundary layer theory," makes it impossible in general to represent the chemisorbed particle via a local level, i.e., to use the energy diagrams shown in Fig. 3.5 and figuring in these papers.

The "boundary layer theory" starts from the assumption that the η^- vs. E_F and η^+ vs. E_F curves depicted in Fig. 3.5a are approximated by the broken lines depicted in Fig. 3.5b, i.e., it is assumed that

$$\text{for acceptor particles: } \eta^- = \begin{cases} 1 & \text{for } E_F > E_A, \\ 0 & \text{for } E_F < E_A; \end{cases}$$

$$\text{for donor particles: } \eta^+ = \begin{cases} 1 & \text{for } E_F < E_D, \\ 0 & \text{for } E_F > E_D, \end{cases} \tag{3.94}$$

In reality this can never occur. From this, among other things, follows an incorrect conclusion characteristic of the "boundary layer theory," namely, that surface coverage cannot exceed a certain critical value (a very small value, as calculations show). In reality, as surface coverage increases, the Fermi level on the surface steadily moves downward in the case of acceptor particles or upward in the case of donor particles, so that

$$E_F \to E_A \quad \text{for} \quad E_F > E_A,$$
$$E_F \to E_D \quad \text{for} \quad E_F < E_D.$$

For sufficiently large coverage the condition $E_F = E_A$ or, respectively, $E_F = E_D$ is met, at which, according to (3.94), further adsorption ceases. This maximal attainable coverage, according to Weitz's [32] calculations, is of the order of 1%. Greater surface coverage is impossible within the framework of "boundary layer theory." The fact that we neglected "weak" bonding, which is a characteristic feature of this theory, leads to this limitation. Taking the "weak" (i.e., the electrically neutral) form of chemisorption into account lifts this restriction.

It must be noted, however, that the "boundary layer theory" has undergone a certain evolution (e.g., see [33]), with the neutral form of adsorption being introduced into its framework. The charged form is then related to chemisorption, while the neutral form is related to physisorption. This makes the approximation (3.94) invalid. According to this modified version, the position of the Fermi level at electronic equilibrium determines the fraction of the physical and chemical forms of adsorption.

The transition of an adsorbed particle from the neutral state to the charged state, i.e., the appearance of an electron on the acceptor level E_A or a hole on the donor level E_D indicates, from the viewpoint of modern "boundary layer theory," a transition from physisorption to chemisorption. We have seen, however, that such a transition constitutes a transition between two different forms of chemisorption. Physisorption is characterized by an entirely different local level (if any level at all corresponds to physisorption), which, as a rule, is much shallower than the level for chemisorption.

Chapter 4

THE INTERACTION OF THE SURFACE WITH THE BULK
IN A SEMICONDUCTOR

4.1. THE CONNECTION BETWEEN SURFACE AND BULK
PROPERTIES OF A SEMICONDUCTOR

4.1.1. The Connection between the Position of the Fermi Levels
at the Surface and in the Bulk of a Semiconductor

When we speak of the interaction of the surface of a crystal with its bulk, we mean the correlation between the properties of the bulk and surface. A number of surface properties, such as the chemisorptivity of the surface, the charge of the surface, and the reactivity of the chemisorbed particles, are determined, as we have seen, by the position of the Fermi level at the crystal surface. Here we will characterize the position of the Fermi level by the distance between the level and the bottom of the conduction band and denote this distance by ϵ_s. On the other hand, a number of bulk properties, such as the electrical conductivity of the crystal and the carrier recombination rate, are determined by the position of the Fermi level in the bulk, which we characterize by ϵ_v, where ϵ_v is the distance between the Fermi level and the bottom of the conduction band in the bulk of the crystal. There is a unique relationship between ϵ_s and ϵ_v.

This relationship can be obtained from the condition of electrical neutrality of the crystal as a whole:

$$\sigma + \int_0^\infty \rho(x)\,dx = 0, \tag{4.1}$$

where σ is the surface charge density, and $\rho(x)$ is the volume charge density in plane x (the semiconductor is assumed to occupy the half-space $x \geq 0$). Here we have

$$\sigma - \sigma\,(P,\,T;c_s)\,.\qquad(1.2)$$

If, in particular, all of the surface charge is due only to the chemisorbed particles (and particles of one particular type), then σ has the form (3.13). The second term on the left-hand side of (4.1) is also a function of ϵ_s and, in addition, a function of ϵ_v, i.e., we define R:

$$\int_0^\infty \rho\,(x)\,dx = R\,(T;\epsilon_s,\,\epsilon_v)\,.\qquad(4.3)$$

Let us show that

$$R\,(T;\epsilon_s,\,\epsilon_v) = \pm\sqrt{\frac{\chi}{2\pi e}\int_{\epsilon_v}^{\epsilon_s}\rho\,(\epsilon)\,d\epsilon},\qquad(4.4)$$

where, as usual, χ is the dielectric constant of the crystal, e is the absolute value of the electron charge, and $\epsilon = \epsilon(x)$ the distance between the Fermi level and the bottom of the conduction band in plane x [so that $\epsilon_s = \epsilon(0)$ and $\epsilon_v = \epsilon(\infty)$]; the upper sign in (4.4) corresponds to $\sigma < 0$ and the lower sign to $\sigma > 0$. The Poisson equation can be written thus:

$$\frac{d^2 V}{dx^2} = \frac{4\pi e}{\chi}\,\rho\,(\epsilon),\qquad(4.5)$$

where

$$V\,(x) = \epsilon\,(x) - \epsilon_v\qquad(4.6)$$

is the potential energy of the electron in the surface field. On the basis of (4.5) and (4.6) we have

$$\int_\epsilon^{\epsilon_v}\rho\,(\epsilon)\,d\epsilon = \frac{\chi}{2\pi e}\int_x^\infty\frac{d^2 V}{dx^2}\frac{d\epsilon}{dx}\,dx = \frac{\chi}{4\pi e}\int_x^\infty\frac{d^2 V}{dx^2}\frac{dV}{dx}\,dx$$

$$= \frac{\chi}{8\pi e}\int_x^\infty\frac{d}{dx}\left(\frac{dV}{dx}\right)^2 dx = -\frac{\chi}{8\pi e}\left(\frac{dV}{dx}\right)^2,\qquad(4.7)$$

since

$$\left(\frac{dV}{dx}\right)_{x=\infty} = 0.\qquad(4.8)$$

On the other hand, employing (4.8) once more, we obtain

$$\int_x^\infty\rho\,(x)\,dx = \frac{\chi}{4\pi e}\int_x^\infty\frac{d^2 V}{dx^2}\,dx = \frac{\chi}{4\pi e}\int_x^\infty\frac{d}{dx}\left(\frac{dV}{dx}\right)dx = -\frac{\chi}{4\pi e}\left(\frac{dV}{dx}\right).\qquad(4.9)$$

Taking $x = 0$ and $\epsilon = \epsilon_s$ as the lower limits in (4.7) and (4.9) and comparing (4.9) with (4.7), we arrive at (4.4). Note that the integral inside the parentheses in (4.4) is certain to be positive, as shown by (4.7).

On the basis of (4.2) and (4.3) we can rewrite Eq. (4.1) thus:

$$\sigma\,(P,\,T;\,\epsilon_s) + R\,(T;\,\epsilon_s,\,\epsilon_v) = 0. \tag{4.10}$$

From this we can find ϵ_s as an explicit function of ϵ_v, i.e.,

$$\epsilon_s = f\,(P,\,T;\,\epsilon_v). \tag{4.11}$$

Note, however, that the solution of Eq. (4.10) for ϵ_s encounters mathematical difficulties [Eq. (4.10) is transcendental in ϵ_s] and can be carried through to the end in practice only in certain special cases.

Equation (4.11) correlates the surface and bulk properties of the semiconductor. We will subsequently study some of the corollaries of this equation. The value of ϵ_s is fixed through Eq. (4.11) to a value of ϵ_v which, in turn, is fixed by the equation

$$\rho\,(\epsilon_v) = 0. \tag{4.12}$$

Equation (4.12) is the condition for electrical neutrality of the crystal in its bulk (at $x \geq l$). (We will assume here that the dimensions of the crystal are greater than the screening length l).

Finally, note that the following condition is valid:

$$\frac{d\epsilon_s}{d\epsilon_v} \geqslant 0, \tag{4.13}$$

i.e., when the Fermi level inside the crystal moves (or ϵ_v varies), so does the Fermi level at the surface (or so does ϵ_s) (this can be shown for the general case without solving Eq. (4.10) for ϵ_s, see [1]). However, condition (4.13) holds only if the factors that influence the Fermi level ϵ_v inside the crystal do not touch the system of surface levels.

4.1.2. The Surface Potential

The bending of the energy bands at the surface can be characterized by the quantity

$$V_s = V\,(0) = \epsilon_s - \epsilon_v,$$

which we will call the surface potential. This quantity plays an important role in surface physics. We will try to determine it. To this end we turn to Eq. (4.7). On the basis of (4.6) we can rewrite Eq. (4.7) as follows:

$$\left(\frac{dV}{dx}\right)^2 = \frac{8\pi e}{\chi} \int_0^V \rho\,(V)\,dV. \tag{4.14}$$

Let us evaluate the integral on the right-hand side of Eq. (4.14).

We will assume that the semiconductor contains donors D of one type and acceptors A of another type, the concentrations of the atoms of these impurities being X and Y. By X^+ and Y^- we denote the concentrations of the respective atoms in a charged state. Obviously,

$$\rho = e\,\{\,(p + X^+) - (n + Y^-)\,\}. \tag{4.15}$$

Assuming that the free electrons and holes obey the Boltzmann statistics (non-degenerate electron and hole gases) while the electrons and holes on local levels in general obey the Fermi statistics, we have the following relations for free electrons and holes, respectively,

$$n\,(x) = C_n \exp\left[-\frac{E_C\,(x) - E_F}{kT}\right] = n_v \exp\left[-\frac{V\,(x)}{kT}\right],$$

$$p\,(x) = C_p \exp\left[-\frac{E_F - E_V\,(x)}{kT}\right] = p_v \exp\left[\frac{V\,(x)}{kT}\right]. \tag{4.16}$$

The notations here are those used in the previous chapters, i.e., $E_C(x)$ and $E_V(x)$ are the bottom of the conduction band and the top of the valence band in the x plane, E_F the Fermi level, and

$$n_v = n\,(\infty) = C_n \exp\left(-\frac{E_C^v - E_F}{kT}\right),$$

$$p_v = p\,(\infty) = C_p \exp\left(-\frac{E_F - E_V^v}{kT}\right). \tag{4.17}$$

The meaning of the other notations is clear from Fig. 4.1, where the energy spectrum of a semiconductor is shown (for the sake of definiteness the surface is assumed to be positively charged; the surface levels are not shown in Fig. 4.1).

For electrons and holes on local levels we have

$$X^+(x) = \frac{X}{1 + \exp\dfrac{E_F - E_D\,(x)}{kT}},$$

$$Y^-(x) = \frac{Y}{1 + \exp\dfrac{E_A\,(x) - E_F}{kT}}, \tag{4.18}$$

where $E_D(x)$ and $E_A(x)$ give the positions of the donor and acceptor levels in plane x (see Fig. 4.1).

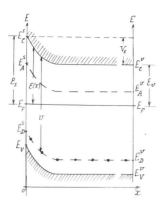

Fig. 4.1. Energy spectrum of a semiconductor.

Substituting (4.16) and (4.18) into (4.15) and then (4.15) into (4.14), we obtain after integration

$$\left(\frac{dV}{dx}\right)^2 = \frac{8\pi e^2 kT}{\chi}\left\{ p_v\left[\exp\left(\frac{V}{kT}\right) - 1\right] + n_v\left[\exp\left(-\frac{V}{kT}\right) - 1\right] + \right.$$

$$\left. + X \ln \frac{1 + \exp\left[-\dfrac{E_F - E_D^v - V}{kT}\right]}{1 + \exp\left[-\dfrac{E_F - E_D^v}{kT}\right]} + Y \ln \frac{1 + \exp\left[-\dfrac{E_A^v - E_F + V}{kT}\right]}{1 + \exp\left[-\dfrac{E_A^v - E_F}{kT}\right]} \right\} . \tag{4.19}$$

Here E_D^v and E_A^v are the positions of levels D and A in the bulk of the crystal.

There are two limiting cases that greatly simplify Eq. (4.19). We will use them often in our discussions. They are

$$\text{a)} \quad \text{for} \quad \exp\frac{V}{kT} \ll \exp\frac{E_F - E_D^v}{kT} ,$$

$$\exp\frac{V}{kT} \ll \exp\frac{E_A^v - E_F}{kT} ; \tag{4.20a}$$

$$\text{b)} \quad \text{for} \quad \exp\frac{V}{kT} \gg \exp\frac{E_F - E_D^v}{kT} ,$$

$$\exp\frac{V}{kT} \gg \exp\frac{E_A^v - E_F}{kT} . \tag{4.20b}$$

We will consider each case separately.

(a) In the case (4.20a) the electrons and holes on the local levels are described by the Boltzmann statistics (in Fig. 4.1 the donors lie above the acceptors). We assume that this is true for all values of V (in particular, for $V = 0$), and Eq. (4.19) takes the form

$$\left(\frac{dV}{dx}\right)^2 = \frac{8\pi e^2 kT}{\chi} \left\{ (p_v + X_v^+) \left[\exp\left(\frac{V}{kT}\right) - 1 \right] \right.$$

$$\left. + (n_v + Y_v^-) \left[\exp\left(-\frac{V}{kT}\right) - 1 \right] \right\}, \tag{4.21}$$

where

$$X_v^+ = X^+(\infty) = X \exp\left[-\frac{E_F - E_D^v}{kT} \right],$$

$$Y_v^- = Y^-(\infty) = Y \exp\left[-\frac{E_A^v - E_F}{kT} \right].$$

The condition of electrical neutrality in the bulk of the crystal (at $x = \infty$) yields

$$p_v + X_v^+ = n_v + Y_v^-,$$

and Eq. (4.21) assumes the following simple form*:

$$\frac{dV}{dx} = -\sqrt{\frac{8\pi e^2 kT n^*}{\chi}} \cdot 2 \sinh \frac{V}{2kT}. \tag{4.22}$$

where we used the notation

$$n^* = p_v + X_v^+ = n_v + Y_v^-.$$

Thus,

$$n^* = n_v \quad \text{for} \quad Y_v = 0, \quad \text{i.e., for } n\text{-type semiconductors,}$$
$$n^* = p_v \quad \text{for} \quad X_v = 0, \quad \text{i.e., for } p\text{-type semiconductors,}$$

which means that n^* is the concentration of the majority carriers in the bulk. Integrating (4.22), we obtain

$$\tanh \frac{V}{4kT} = \tanh \frac{V_s}{4kT} \exp\left(-\frac{x}{l}\right),$$

or

$$\exp\left(\frac{V}{2kT}\right) = \frac{\exp\left(\frac{x}{l}\right) + \tanh \dfrac{V_s}{4kT}}{\exp\left(\frac{x}{l}\right) - \tanh \dfrac{V_s}{4kT}}, \tag{4.23}$$

---

*We choose the minus sign in (4.22) because (see Fig. 4.1)

$$dV/dx \leq 0 \text{ for } V \geq 0 \text{ and } dV/dx \geq 0 \text{ for } V \leq 0.$$

where

$$l = \sqrt{\frac{\chi kT}{8\pi e^2 n^*}} \qquad (4.24)$$

has the dimensionality of length and is called the screening length or Debye length (see Section 3.2.2). The latter quantity is the distance from the surface over which $\tanh(V/4kT)$ diminishes (in absolute value) e times. Equation (4.23) gives us the behavior of the surface potential in the crystal.

(b) In the case (4.20b) all the impurity atoms are completely ionized (in Fig. 4.1 the donors lie above the acceptors). In this case, as shown by (4.17),

$$X^+ = X, \ Y^- = Y,$$

and the condition of electrical neutrality in the bulk of the crystal has the form

$$X - Y = n_v - p_v. \qquad (4.25)$$

Under condition (4.20b) (assuming that this condition is met for all values of V) and on the basis of (4.24) we can write Eq. (4.19) as follows:

$$\left(\frac{dV}{dx}\right)^2 = \frac{8\pi e^2 kT}{\chi} \left\{ p_v \left[\exp\left(\frac{V}{kT}\right) - 1 - \frac{V}{kT} \right] + n_v \left[\exp\left(-\frac{V}{kT}\right) - 1 + \frac{V}{kT} \right] \right\}. \qquad (4.26)$$

Note that in the case of an intrinsic semiconductor, where $n_v = p_v$, Eq. (4.26) transforms, as is readily seen, into Eq. (4.22).

Equation (4.19) can be used to find the surface potential V_s. Indeed, putting $x = 0$ in (4.19) and bearing in mind that, according to (4.9) and (4.1),

$$\left(\frac{dV}{dx}\right)_{x=0} = \frac{4\pi e}{\chi} \sigma,$$

we can rewrite Eq. (4.19) in the following form:

$$p_v \left[\exp\left(\frac{V_s}{kT}\right) - 1 \right] + n_v \left[\exp\left(-\frac{V_s}{kT}\right) - 1 \right]$$

$$+ X \ln \frac{1 + a \exp\left(\frac{V}{kT}\right)}{1 + a} + Y \ln \frac{1 + b \exp\left(-\frac{V}{kT}\right)}{1 + b} = n^* \left(\frac{\sigma}{\sigma^*}\right)^2, \qquad (4.27)$$

where

$$\sigma^* = \sqrt{\frac{\chi kT}{2\pi}} n^*. \qquad (4.28)$$

In the limiting cases (4.20a) and (4.20b) Eq. (4.27) takes the form

$$2 \sinh \frac{V_s}{2kT} = -\frac{\sigma}{\sigma^*}$$

(4.29)

when the impurity is nearly nonionized [case (4.20a); see (4.22)], and

$$p_v\left[\exp\left(\frac{V_s}{kT}\right) - 1 - \frac{V_s}{kT}\right] + n_v\left[\exp\left(-\frac{V_s}{kT}\right) - 1 + \frac{V_s}{kT}\right] = n^*\left(\frac{\sigma}{\sigma^*}\right)^2$$

(4.30)

when the impurity is almost completely ionized [case (4.20b); see (4.26)].

Note that the surface charge density σ in (4.27), (4.29), and (4.30) consists of two terms, the adsorption charge σ_A and the intrinsic charge σ_B, each of which is generally a function of V_s:

$$\sigma (V_s) = \sigma_A (V_s) + \sigma_B (V_s).$$

The value of the intrinsic charge for a given system of surface levels does not remain constant during adsorption but varies with the number of the adsorbed particles on the surface. Indeed, if the intrinsic charge on a clean surface, i.e., in the absence of adsorbed particles, is $\sigma_B(V_{s0})$, then when adsorbed particles are present the charge is $\sigma_B(V_s)$, where V_{s0} is a root of Eq. (4.27) at $\sigma_A = 0$, while V_s is a root of Eq. (4.27) at $\sigma_A \neq 0$.

Equation (4.27), or for particular cases Eqs. (4.29) and (4.30), serves as an equation for finding V_s.

4.1.3. The Dependence of the Surface Potential on Various Factors

We will consider only the case where the Fermi level on the surface is far (compared to kT) from the levels of the adsorbed particles as well as from the levels of the intrinsic defects. We put $\sigma = $ const and look at the various factors that influence V_s:

(1) the surface charge σ,
(2) the presence of impurities in the semiconductor,
(3) the temperature T,
(4) the surface concentration N of chemisorbed particles.

We will study only the limiting cases, i.e., weak ionization (4.29) and strong ionization (4.30) of the impurity.

(1) We study the V_s vs. σ dependence on the assumption that

the bands are bent only weakly, or $|V_s| \ll kT$,   (4.31a)

the bands are bent strongly, or $|V_s| \gg kT$.   (4.31b)

For weakly bent energy bands both Eq. (4.29) and Eq. (4.30) yield

$$V_s = -kT \frac{\sigma}{\sigma^*} \; , \tag{4.32}$$

where σ has the form (4.28). This case is realized, as we see, at $|\sigma| \ll \sigma^*$.

For strongly bent bands Eq. (4.29) yields

$$V_s = \pm kT \ln \left(\frac{\sigma}{\sigma^*} \right)^2 , \tag{4.33}$$

where we must choose the

$$\begin{aligned} &\text{upper sign if} \quad \sigma < 0, \\ &\text{lower sign if} \quad \sigma > 0. \end{aligned} \tag{4.34}$$

This case is realized at $|\sigma| \gg \sigma^*$.

But if we take Eq. (4.30) for the case of strongly bent bands, the result depends on whether the bending is "accumulation" or "depletion." If the bending is "accumulation," i.e., the bands are bent in such a way that the surface layer proves to be enriched by the majority carriers in comparison with the bulk [p-type semiconductors when the bands are bent upward ($V_s > 0$) and n-type semiconductors when the bands are bent downward ($V_s < 0$)], Eq. (4.30) again yields (4.33). If the bending is "depletion," i.e., the bands are bent in such a way that the surface layer proves to be depleted by the majority carrier in comparison with the bulk [p-type semiconductors when the bands are bent downward ($V_s < 0$) and n-type semiconductors when the bands are bent upwards ($V_s > 0$)], Eq. (4.30) yields

$$V_s = \pm kT \left(\frac{\sigma}{\sigma^*} \right)^2 , \tag{4.35}$$

where the signs are chosen according to the rule (4.34).*

We see that in the limiting cases we are considering here, $|V_s|$ increases with $|\sigma|$ by a linear, logarithmic, or quadratic law.

(2) Let us now see how impurities introduced into the crystal influence V_s. As shown by (4.32), (4.33), (4.35), and (4.28), the absolute value of V_s is the smaller the higher the concentration of the majority carriers n^* inside the crystal. On the other hand, all other conditions being equal, the concentration of the majority carriers is higher, the higher the concentration of impurities in the sample (donor or acceptor impurities, respectively; see Section 1.6.4). Thus, we can weaken the influence of the surface charge on the bending of the bands to some extent by introducing impurities into the crystal, and the greater the concentration of the im-

---

*Formula (4.35) is valid only if there is no conductance inversion; see Section 4.2.2.

purity, i.e., the greater the semiconductor differs from an intrinsic semiconductor, the smaller the absolute value of V_s for a fixed value of σ.

(3) Combining (4.32), (4.33), (4.35), and (4.28), we can also find the temperature dependence of V_s (assuming that the surface charge σ does not vary with temperature). Indeed, n^* steadily increases with temperature, while $|V_s|$ steadily decreases. This can be directly seen in the case (4.35): it is sufficient to substitute (4.28) into (4.35). In the cases (4.33) and (4.32) this can easily be seen if the temperature is not very low. Thus, for a given surface charge the bending of the bands decreases in absolute value as the temperature rises.

(4) Finally, let us consider the influence of adsorption on the surface potential. Suppose that there is only one type of particle adsorbed by the surface (we denote the surface concentration of these particles by N, as usual). Within the framework of the assumptions we made at the beginning of this section we have

$$\sigma_A = \mp eN.$$

Here and in what follows the upper sign corresponds to acceptor particles and the lower to donor particles. Let V_{s0} and V_s be the surface potentials without and with adsorption, so that

$$V_s = V_{s0} + \Delta V_s,$$

where ΔV_s is an additional bending of the bands caused by adsorption.

When the impurities are weakly ionized or when they are almost completely ionized and the bending of the bands is "enriching," we have [see (4.33)]

$$\Delta V_s = \pm 2kT \frac{eN}{|\sigma_B|} \quad \text{for} \quad eN \ll |\sigma_B|, \tag{4.36a}$$

$$\Delta V_s = \pm kT \ln\left(\frac{eN}{\sigma^*}\right)^2 \quad \text{for} \quad eN \gg |\sigma_B|. \tag{4.36b}$$

When the impurities are completely ionized and the bending of the bands is "depleting," we have [see (4.35)]

$$\Delta V_s = \pm kT \frac{|\sigma_B|}{(\sigma^*)^2} eN \quad \text{for} \quad eN \ll |\sigma_B|, \tag{4.37a}$$

$$\Delta V_s = \pm kT \left(\frac{eN}{\sigma^*}\right)^2 \quad \text{for} \quad eN \gg |\sigma_B|. \tag{4.37b}$$

Formulas (4.36a) and (4.37a) belong to the case of a small surface coverage or large intrinsic disorder, while formulas (4.36b) and (4.37b) belong to the case of a large surface coverage or small intrinsic disorder. Note that the above formulas remain valid only if the bending of the bands is considerable [condition (4.31b)].

As we see from (4.36a) and (4.37a), the same surface coverage N may lead to different values of ΔV_s, depending on the value of σ_B, i.e., depending on the treatment of the surface prior to adsorption and, in fact, depending on previously adsorbed gases.

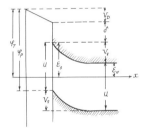

Fig. 4.2. Effect of adsorption on work function.

The formulas (4.36a), (4.36b), (4.37a), and (4.37b) lead to non-Langmuir patterns in adsorption. For instance, if we substitute (4.36a) into (3.89), we obtain for large values of N the logarithmic isotherm (2.20). Substitution of (4.36b) into (3.89) yields a Freundlich isotherm (2.19) with $n = 1/3$. Substitution of (4.37a) into (3.92) yields the Zel'dovich–Roginskii equation (2.7). Finally, substitution of (4.37b) leads to Thuillier kinetics (2.9) (see Sections 2.1.2 and 2.1.3).

4.2. EFFECTS DUE TO THE CHARGING OF THE SURFACE

4.2.1. Effect of Adsorption on Work Function

Generally speaking, the presence of surface levels leads to charging of the surface of the semiconductor. This, in turn, leads to certain observable effects as follows:

(1) The work function of the semiconductor depends on the surface charge. For this reason the work function proves to be dependent on the surface coverage by chemisorbed particles and their nature. We will study this effect in Section 4.2.1.

(2) The conductance of a semiconductor sample depends on the surface charge (for small samples). As a consequence, the conductance of a sample depends on the nature and concentration of the particles chemisorbed on the surface of the sample. Sections 4.2.2 and 4.2.3 are devoted to this effect.

(3) Finally, the distribution of impurities inside the semiconductor depends on the magnitude and sign of the surface charge. We devote Section 4.5 to this question.

How does adsorption affect the work function of a semiconductor? Note that here we must distinguish between the thermionic work function φ_T and the photoelectric work function φ_p. The definitions are (see Fig. 4.2, where the horizontal axis is assumed to coincide with the Fermi level)

$$\varphi_T = \epsilon_s + \delta + V_D, \quad \varphi_p = u + \delta + V_D, \tag{4.38}$$

where

$$\epsilon_s = \epsilon_v + V_s. \tag{4.39}$$

The quantities u and δ (where u is the width of the forbidden band, and δ the energy of affinity of the free electron for the lattice) are fixed by the type of semiconductor and are both independent of adsorption. The quantity ϵ_v is independent of adsorption and is determined by the nature and state of the semiconductor (its prehistory, temperature, nature, and the concentration of impurities which it contains). The quantities V_s and V_D depend on the nature and number of adsorbed particles; V_s characterizes the bending of the bands due to the charging of the surface. The quantity V_D is the dipole component of the work function; its cause is the potential drop across the electric double layer created by the adsorbed molecules. These may be physisorbed molecules with dipole molecules or molecules that in the free state have zero dipole moments but are polarized in the lattice field; finally, these may be chemisorbed particles in the state of "weak" bonding with the surface and possessing dipole moments of a quantum-mechanical origin (see Section 2.5.3).

The thermionic work function φ_T, in contrast to the photoelectric wave function φ_p, proves to be sensitive to impurities injected into the crystal (owing to the term ϵ_s in the expression for φ_T). Indeed, an acceptor impurity, which leads to an increase in ϵ_v and, hence, to an increase in ϵ_s [which follows from (4.13)], increases the work function φ_T, according to (4.38). The effect of a donor impurity on φ_T is just the opposite. Experimentally this effect has been observed by a number of authors. As an example we can cite the work of Enikeev, Margolis, and Roginskii [2], who studied the effect of various impurities (acceptor- and donor-type) on the work function of ZnO, CuO, and NiO. As the concentration of the impurity increased, the work function steadily grew or fell off (depending on the type of impurity).

Note that some authors (e.g., Bielanski and Deren [3], who studied the work function of NiO with a Li impurity) observed a maximum on the φ_T vs. Z curve, where Z is the impurity concentration. Apparently, the presence of such a maximum in the experiments of these authors suggests that for low concentrations of Li the latter forms a substitutional solution with the NiO lattice; i.e., Li acts as an acceptor, while at high concentrations it forms an interstitial solution with the NiO lattice, acting as a donor.

The changes that adsorption brings about in the thermionic and photoelectric work functions we denote by $\Delta\varphi_T$ and $\Delta\varphi_p$, respectively, while the corresponding changes in the surface potential and the dipole component of the work function we denote by ΔV_s and ΔV_D. According to (4.38) and (4.39), we have

$$\Delta\varphi_T = \Delta V_s + \Delta V_D, \qquad \Delta\varphi_p = \Delta V_D. \tag{4.40}$$

The value of $\Delta\varphi_T$ can be determined by measuring the change in the "contact potential difference" (cpd) between the sample and a reference electrode if the change in the work function of the reference electrode brought on by adsorption is known [4, 5]. The value of $\Delta\varphi_p$ can be found by measuring the shift of the surface photoelectric threshold brought on by adsorption [6]. Simultaneous measurement of $\Delta\varphi_T$ and $\Delta\varphi_p$ enables separating the two components, ΔV_s and ΔV_D, in the expression for $\Delta\varphi_T$. Both quantities, ΔV_s and ΔV_D, can be determined separately in cpd measurements (which yields $\Delta\varphi_T$) with simultaneous measurements of the field effect (which yields $\Delta\varphi_s$). According to the experimental data [7, 8], in the majority of cases we have

$$\Delta V_D \ll \Delta V_s.$$

In such cases the change in the thermionic work function can be used to find the change in the surface potential:

$$\Delta\varphi_T = \Delta V_s. \tag{4.41}$$

According to (4.36a) and (4.36b) or (4.37a) and (4.37b), ΔV_s is positive when acceptor molecules are adsorbed and negative when donor molecules are adsorbed. Thus, we can judge the acceptor or donor nature of the adsorbed molecules by the sign of $\Delta\varphi_T$. For instance, according to a vast body of experimental data (e.g., see [4, 5, 7]), adsorption of O_2, N_2O, and OH leads to an increase in the work function ($\Delta\varphi_T > 0$; acceptor molecules), while adsorption of H_2O, C_3H_6, and NH_3 on the same semiconductors (Cu_2O, NiO, ZnO, V_2O_5, MnO_2, PbS, MoS_2, Ge, and others) leads to a decrease in the work function ($\Delta\varphi_T < 0$; donor molecules).

In some cases the same adsorbate may, according to various sources, have opposite effects on the work function. According to Lyashenko and Stepko [7] and Keier and Kutseva [8], CO_2 molecules when adsorbed on CuO and NiO act as acceptors, while according to Lyashenko and Litovchenko [9] the same molecules adsorbed on Ge act as donors; according to Parravano and Dominicali [10], CO_2 molecules adsorbed on MnO_2 act as acceptors, while according to Elovich, Margolis, and Roginskii [11] they act as donors. This indicates that in different cases, corresponding to different positions of the Fermi level on the semiconductor surface, different forms of chemisorption bonding come into play (Figs. 2.11b and 2.11c show the donor and acceptor forms, respectively, of chemisorption of the CO_2 molecule).

We can easily obtain the dependence of $\Delta\varphi_T$ on N (where, as usual, N is the surface concentration of the chemisorbed molecules) by substituting (4.36a) and (4.36b) or (4.37a) and (4.37b) into (4.41). In particular, if we substitute (4.36b) and (4.37a) into (4.41), we have for "enriching" and "depleting" adsorptions, respectively,

$$\Delta\varphi_T = \pm\,(\alpha\ln N + \beta), \tag{4.42a}$$

$$\Delta\varphi_T = \pm\,\gamma N, \tag{4.42b}$$

where the upper sign corresponds to adsorption of acceptor particles and the lower to adsorption of donor particles.

Both (4.42a) and (4.42b) have been verified experimentally. For instance, Enikeev [12], who studied the adsorption of O_2 on NiO (acceptors on a p-type semiconductor) and C_3H_7OH on ZnO (donors on an n-type semiconductor), found a logarithmic dependence of $\Delta\varphi_T$ on N (enriching adsorption), while for adsorption of O_2 on ZnO (acceptors on an n-type semiconductor) and C_3H_6 on CuO (donors on a p-type semiconductor) the $\Delta\varphi_T$ vs. N dependence proved to be linear (depleting adsorption).

Finally let us discuss the kinetics of adsorption. Enikeev, Margolis, and Roginskii [13] studied the kinetics of chemisorption of O_2 on ZnO, using a sample of pure ZnO and samples doped with Li_2O (donor) and $ZnSO_2$ (acceptor), all of which had different work functions φ_T and adsorption activation energies ϵ. They found that the dependence of ϵ on the initial (i.e., $N = 0$) work function φ_T was linear:

$$\epsilon = \alpha + \varphi_T$$

This result should not be considered unexpected. We can arrive at such a dependence if we substitute (4.38) into (3.67), allow for (3.69), and assume that $V_D \ll V_s$. The same authors found that both φ_T and ϵ increase with N. The same result follows from (3.67).

4.2.2. Surface Conduction

If the surface of a semiconductor is electrically charged, then, due to the surface bending of the energy bands, the carrier concentration in the semiconductor will vary with the distance from the surface. Because of this, the apparent electrical conductivity κ of such a semiconductor will differ from the electrical conductivity κ_0 of the same semiconductor with horizontal energy bands, i.e., with an electrically neutral surface. The quantity $\kappa_s = \kappa - \kappa_0$ is commonly known as the surface conductivity. This is not a very suitable term since in the exact sense of the word we should call the conduction due to carrier transfer over Goodwin surface states (see Section 1.5.3) the surface conduction, while it would be more appropriate to call κ_s the conductivity of the near-the-surface layer, or the near-the-surface conductivity.

Let us find the dependence of κ_s and V_s. Assuming that we are dealing with a semi-infinite semiconductor occupying the space $x \geq 0$, and denoting the mobilities of the electrons and holes by μ_n and μ_p, respectively, and taking these mobilities to be constant (in the first approximation) over the entire bulk of the semiconductor, i.e., assuming that they are independent of x, we have

$$\kappa_s = e\mu_n \int_0^\infty [n(x) - n_v]\, dx + e\mu_p \int_0^\infty [p(x) - p_v]\, dx, \tag{4.43}$$

where the first term on the right-hand side corresponds to the electron component of the conductivity and the second to the hole component. If the electron and hole gases can be assumed to be nondegenerate (we will take this for granted), then $n(x)$ and $p(x)$ have the form (4.16) and n_v and p_v the form (4.18). Substituting (4.16) into (4.43), introducing the notations

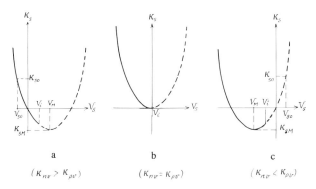

a
$(\kappa_{nv} > \kappa_{pv})$

b
$(\kappa_{nv} = \kappa_{pv})$

c
$(\kappa_{nv} < \kappa_{pv})$

Fig. 4.3. Surface conductivity vs. bending of bands: a) for n-type semiconductor; b) for intrinsic semiconductor; c) for p-type semiconductor.

$$\kappa_{nv} = e\mu_n n_v, \quad \kappa_{pv} = e\mu_p p_v,$$

$$\exp\left(\frac{V_M}{kT}\right) = \frac{\kappa_{nv}}{\kappa_{pv}}, \tag{4.44}$$

where κ_{nv} and κ_{pv} are the specified electron and hole conductivities in the bulk of the crystal, and going from integration with respect to x to integration with respect to V, we arrive at the following result:

$$\kappa_s = 4\sqrt{\kappa_{nv}\kappa_{pv}} \int_0^{V_s} \sinh\frac{V_M - V}{2kT} \sinh\frac{V}{2kT} \left(\frac{dV}{dx}\right)^{-1} dV. \tag{4.45}$$

This shows, for instance, that since the units of κ_{nv}, κ_{pv} are mhos per centimeter, the units of κ_s are mhos.

From (4.45) we can see that the κ_s vs. V_s curve has an extremum at $V_s = V_M$. Indeed, we have

$$\frac{d\kappa_s}{dV_s} = 4\sqrt{\kappa_{nv}\kappa_{pv}} \sinh\frac{V_M - V_s}{2kT} \sinh\frac{V_s}{2kT} \left(\frac{dV}{dx}\right)_{x=0},$$

which vanishes at $V_s = V_M$ provided that $(dV/dx)_{x=0} \neq 0$. It can also be proved that the extremum is a minimum. From (4.44) we can see that in the case of an n-type semiconductor ($\kappa_{nv} > \kappa_{pv}$) the minimum on the κ_s versus V_s curve lies in the region of positive values of V_s; in the case of a p-type semiconductor ($\kappa_{nv} < \kappa_{pv}$) it lies in the region of negative values of V_s, and in the case of an intrinsic semiconductor ($\kappa_{nv} = \kappa_{pv}$) it lies at $V_s = 0$ (see Figs. 4.3a, 4.3b, and 4.3c).

Note that if the depleting bending of the bands is so strong that

$$|V_s| \geqslant |V_i|,$$

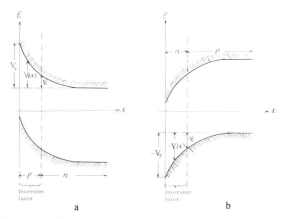

Fig. 4.4. Conduction inversion: a) for n-type semiconductor; b) for p-type semiconductor.

the conductivity at the surface may be of a different nature than that in the bulk of the semiconductor. For instance, in an n-type semiconductor the surface layer may have p-type conduction, while in a p-type semiconductor the surface layer may have n-type conduction. This phenomenon is known as conduction inversion, and V_i is called the inversion potential. The inversion potential can be found from the condition

$$e\mu_n n\,(V_i) = e\mu_p\,p\,(V_i),$$

which can be rewritten in the following form [using (4.16) and (4.44)]:

$$\kappa_{nv}\exp\left(-\frac{V_i}{kT}\right) = \kappa_{pv}\exp\left(\frac{V_i}{kT}\right). \tag{4.46}$$

Comparing (4.46) and (4.44), we find that $V_i = V_M/2$. In Figs. 4.3a, 4.3b, and 4.3c the left branch of the curve (at $V_s < V_i$) corresponds to n-type conduction and the right branch (at $V_s > V_i$) to p-type conduction. Obviously the layer where conduction inversion takes place (the inversion layer) involves the value of κ for which

$$|V_s| \geqslant |V(x)| \geqslant |V_i|. \tag{4.47}$$

Figure 4.4 demonstrates cases of conduction inversion for n- and p-type semiconductors (Figs. 4.4a and 4.4b, respectively; the horizontal axis in both figures coincides with the Fermi level).

Let us return to Eq. (4.45). If the impurity is completely ionized, then substituting (4.25) for the derivative dV/dx in (4.45), we obtain on the right-hand side of (4.45) an integral that can be evaluated only by numerical integration. After this we can construct the κ_s vs. V_s curve. But if the impurity is weakly ionized,

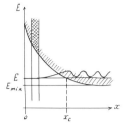

Fig. 4.5. An illustration of the concept of an extraband carrier.

when the electrons and holes on local levels can be described by the Boltzmann statistics, we can substitute (4.22) into (4.45) and, after integration, obtain

$$\kappa_s = 4l\sqrt{\kappa_{nv}\kappa_{pv}}\left(\cosh\frac{V_s - V_M}{2kT} - \cosh\frac{V_M}{2kT}\right), \qquad (4.48)$$

where l is the screening length [see (4.24)]. Substituting V_M for V_s in (4.48), we obtain the minimal value of κ_s (see Fig. 4.3):

$$\kappa_{sM} = -8l\sqrt{\kappa_{nv}\kappa_{pv}}\sinh^2\frac{V_M}{4kT}.$$

The value $\kappa_s = 0$ is obtained, as (4.48) shows, at $V_s = 0$ and $V_s = 2V_M$.

Note that formula (4.48) is valid not only for the Boltzmann statistics on local levels. It remains valid in the general case of the Fermi statistics if the semiconductor is of the i-type (as shown in Section 4.1.2), or if the impurity distribution is an equilibrium one, as we will show in Section 4.5.1 (in a semiconductor with a charged surface the equilibrium distribution of the impurity is not uniform; see Section 4.5.1).

There are two corrections that must be incorporated into formula (4.45), which expresses κ_s as a function of V_s. The first correction concerns the mobilities μ_n and μ_p, which we considered to be constant [the factors μ_n and μ_p in (4.43) were taken outside the integral sign]. In reality, however, the carrier mobility considerably decreases as we get closer to the surface, which is due to carrier scattering at the surface. This effect was studied in great detail by Schrieffer [14] (the Schrieffer correction).

The second correction concerns allowing for the so-called extraband carriers when calculating $n(x)$ and $p(x)$ for (4.43) [15, 16]. The concept of an extraband carrier, i.e., a carrier that is outside the allowed energy band, can be explained via Fig. 4.5, where a bent conduction band is depicted. Take the energy level E. It is allowed at $x \geq x_0$ and forbidden at $x < x_0$. The wave function corresponding to such a level is a plane harmonic wave with an amplitude modulated by the lattice period when $x \geq x_0$ and exponentially damped when $x < x_0$, as shown in Fig. 4.5 (see Section 3.2.2). The tail of this function propagates fairly far from the band edge. Thus, there is always a nonzero probability of finding an electron under the band (an extraband electron). When calculating $n(x)$ for each given x, the extraband electrons are usually ignored, and n is calculated via integration over all the levels of the band, starting from its lower edge (the crosshatched region in Fig. 4.5). The result is formula (4.16) for $n(x)$. If we wish to allow for the extra-band electrons, we must integrate (for each value of x) starting at $E = E_{min}$ (see Fig. 4.5). Calculations [17] show that the correction term in κ_s is 5 to 20%. A similar correction term must be introduced for extraband holes. We see that the correction for extraband carriers is of the same origin as the Franz–Keldysh effect.

4.2.3. Effect of External Field and Adsorption on Conduction

Surface (or near-the-surface) conduction contributes to the total conduction of the sample. If the sample is small, the contribution may be large. Obviously, all the factors that influence V_s influence κ_s, too.

One such factor capable of affecting the surface conduction is an external electric field applied normal to the surface of the semiconductor. Suppose F is the strength of the field applied to a plane-parallel capacitor one of whose plates is the semiconductor under investigation (the field effect method, or measurement); the other plate is known as the field electrode. Let δ be the electric charge induced in the semiconductor by field F per unit surface area. Obviously,

$$\delta = - \frac{\chi_0}{4\pi}\, F, \tag{4.49}$$

where χ_0 is the dielectric constant of the medium filling the capacitor. The charge δ is distributed between surface states and a space charge:

$$\delta = \sigma + \int_0^\infty \rho(x)\, dx. \tag{4.50}$$

This equation constitutes the law of conservation of electric charge. Under electronic equilibrium the first and second terms on the right-hand side of (4.50) are functions of V_s (see Section 4.1):

$$\sigma = \sigma(V_s), \quad \int_0^\infty \rho(x)\, dx = R(V_s). \tag{4.51}$$

Substituting (4.49) and (4.51) into (4.50), we arrive at an equation that relates F to V_s. This equation gives us the bending of the bands as a function of the field strength F (see Section 4.4.1).

Note that we can assume that the electronic equilibrium in the bulk of the semiconductor sets in instantly, while for equilibrium to establish itself between the bulk and the surface state level a certain time interval is required. When the external field is switched on, the entire induced charge gathers in the near-the-surface layer, while the charge on the surface remains the same as it was before the field was applied. The surface conductance κ_s rapidly changes in the process.

With the passage of time some of the carriers induced by the field are transferred from the near-the-surface layer to the surface, as a result of which both σ and κ_s change. In the process the value of κ_s approaches the initial value it had before the field was applied and in some cases even reaches it, provided the surface has a sufficient concentration of surface levels to absorb all of the induced carriers.

This relaxation process takes place, as experiments have shown, first rapidly and then slowly. This fact can often be explained by the presence of two types of levels with very different relaxation times κ_s: fast levels, for which $\tau_s \approx 10^{-8}$ sec, and slow levels, for which $\tau_s > 10^{-3}$. These two levels have very different free carrier capture cross sections or, in other words, the wave functions characterizing these levels have different spreads. Thus, these two levels correspond to two kinds of surface defects. In the case of germanium and silicon covered by an oxide film it is assumed that the fast levels lie on the inner side of the film and the slow levels on the outer side.

The origin of slow levels for a broad class of semiconductors was studied by Kiselev, Kozlov, and Zarif'yants [18].

As first established by Rzhanov with collaborators [19–24] and by Many and Gerlich [25], adsorption and desorption processes, in varying gaseous media and during low-temperature vacuum heating, drastically change the concentrations and capture cross sections of surface levels. In this way adsorption, as already noted in Section 4.1.3, is another factor capable of changing the surface potential V_s and, consequently, κ_s.

Suppose that V_{s0} is the value of the surface potential prior to adsorption, and V_s the value of the surface potential when there are adsorbed particles on the surface conductance. Obviously $\Delta V = V_s - V_{s0}$ and $\Delta\kappa_s = \kappa_s - \kappa_{s0}$ are the changes in the surface potential and conductance brought on by adsorption.

When the adsorbed particles are acceptors, the surface layer is enriched by holes and depleted of electrons and ΔV_s is positive (see Section 4.1.3); consequently, as shown by Fig. 4.3 (if we remain in a region of values of V_s that is far from the inversion region),

$\Delta\kappa_s < 0$ for an n-type semiconductor,
$\Delta\kappa_s > 0$ for a p-type semiconductor.

When donor particles are adsorbed, ΔV_s is negative and, hence (on the same assumption that we are far from the inversion region),

$\Delta\kappa_s > 0$ for an n-type semiconductor,
$\Delta\kappa_s < 0$ for a p-type semiconductor.

We see that from the sign of $\Delta\kappa_s$ (in the case of an extrinsic semiconductor) we can infer the acceptor or the donor nature of the adsorbed particles if we know the type of conduction (n- or p-type semiconductor). And vice versa, if we know the nature of the adsorbed particles (acceptors or donors), then the sign of $\Delta\kappa_s$ enables us to infer the nature of the semiconductor (n- or p-type semiconductor). The conclusions require the assumption that the surface layer is not inverted from n-type to p-type or vice versa.

Table 4.1

Adsorbate \ Adsorbent	Cu$_2$O	CuO	NiO	ZnO	MnO$_2$	PbS	PbSe	Ge
O$_2$	−[33,34]	−[7]	−[8,10]	−[28,35]	−[11]	−[36]	−[34]	−[9,37]
H$_2$			+[10]					−[38] +[37]
H$_2$O	+[34,35]				−[11]			−[39] +[9,40]
CO	+[10,42]	−[7] +[43]	+[8,10, 43]	+[35]	−[11, 12]			−[37] +[9]
CO$_2$		−[7]	−[8,10]		−[10] +[11]			+[9]
N$_2$O			−[10]	−[41]				
C$_2$H$_5$OH	+[34]			+[34]			+[34]	+[9]
C$_2$H$_6$CO	+[34]			+[34]			+[34]	+[9]
C$_6$H$_6$	+[34]			+[34]			+[34]	+[9]

The influence of adsorption on the electrical conduction of semiconductors was first theoretically studied by Sandomirskii and the present author [26–28]. Experimentally this effect was observed by many researchers for various adsorbates an adsorbents. For instance, according to the data of Myasnikov and Pshezhetskii [29], the adsorption of O$_2$ (acceptor) on ZnO (an n-type semiconductor) reduces electrical conductivity, while, according to the data of Lyashenko and Stepko [7], the adsorption of O$_2$ on Cu$_2$O increases the conductivity. The adsorption of ethyl alcohol C$_2$H$_5$OH (donor), according to the date of Lyashenko with collaborators [9, 30], increases the conductivity in n-type germanium but decreases it in p-germanium. Simultaneous studies of the variations of conductivity and the extent of adsorption of acceptor molecules on an n-PbS film were first done by Bazhanova and Zarif'yants [31].

Since both $\Delta\kappa_s$ and $\Delta\varphi_T$ are related to ΔV_s, there must be a correlation between $\Delta\kappa_s$ and $\Delta\varphi_T$ caused by adsorption. They have the same sign for p-type semiconductors and different signs for n-type semiconductors; in absolute value $\Delta\varphi_T$ increases with $\Delta\kappa_s$. Such behavior has indeed been observed in every case studied (e.g., see [4]).

We see that measuring $\Delta\kappa_s$ and measuring $\Delta\varphi_T$ constitute two independent ways of determining the sign of the charge on the surface in adsorption. This was done for a number of adsorbents and adsorbates. Some of the experimental data are summarized in Table 4.1, in which the + or − sign denotes positive or negative surface charging in the course of adsorption, determined by measuring the electrical conductivity or work function or the two quantities simultaneously. The figures in square brackets denote references to the appropriate papers.

Note that chemisorption does not always influence conductivity. For instance, this is the case during adsorption on a surface with strong defects, when the charge introduced by the chemisorbed particles is negligible compared with the intrinsic charge. This is also the case with the "weak" form of chemisorption or when the "strong" donor and "strong" acceptor forms of chemisorption are represented on the surface to approximately the same extent. This does not mean, however, that in these cases we are dealing with physisorption, as some authors insist (e.g., see [11]). At the same time, the fact that electrical conductivity is influenced by adsorption does not necessarily mean that we are dealing with chemisorption. Indeed, physisorbed molecules, by polarizing in the process and forming shallow traps for free carriers by their field, may charge the surface and, hence, change the conductivity. Such a change in electrical conductivity under the influence of physisorption was observed by Figurovskaya and Kiselev [32], who studied the adsorption of xenon and argon on TiO_2.

Note, in conclusion, that the present mechanism by which adsorption affects electrical conductivity, in which the bending of the bands changes, can be applied, strictly speaking, only to single crystals. However, the same mechanism can be employed for polycrystalline and powdery samples if we ignore the intergranular contact resistance. Some specific features of the electrical conductivity of such samples were studied by Petritz [44].

4.3. THE "QUASIISOLATED" SURFACE

4.3.1. The Notion of a "Quasiisolated" Surface

Here we will deal with a semiconductor whose density of surface states is great. This case is certain to occur when the sample is bound by a real rather than an ideal surface. The concentrations of holes and electrons localized at the surface may be considerable. But if the difference between these two concentrations is small (in absolute value) compared to their sum, the surface of the semiconductor possesses some special features (e.g., see [1]).

We denote, as in Sections 4.1.1 and 4.2.1, the distance between the Fermi level and the bottom of the conduction band in the bulk of the crystal as ϵ_v while ϵ_s is the same quantity but at the surface. Next we assume that as ϵ_v changes by $\Delta\epsilon_v$, with

$$\frac{|\Delta\epsilon_v|}{kT} \geqslant 1, \qquad (4.52)$$

ϵ_s changes by $\Delta\epsilon_s$ [according to (4.11)] in such a way that

$$\frac{|\Delta\epsilon_s|}{kT} \ll 1. \qquad (4.53)$$

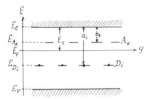

Fig. 4.6. Energy spectrum of a semiconductor in the plane of its surface.

In this case the position of the Fermi level at the crystal surface, i.e., ϵ_s, is weakly dependent (practically independent) on its position inside the crystal, ϵ_v. This means that when the Fermi level is displaced inside the crystal, its position at the surface remains practically unchanged (i.e., changes by an amount small compared with kT). Thus, the surface properties of a semiconductor are independent of its bulk properties. The bulk of the crystal has no effect on the surface in this case and the latter can be said to be "quasiisolated" in a given interval of values of ϵ. The condition is also often described by saying "the Fermi level is pinned at the surface."

It can be shown that a surface is "quasiisolated" inside the interval $\Delta\epsilon_v$ if for all values of ϵ_v in this interval the following condition is met:

$$|R(\epsilon_s, \epsilon_v)| \ll Q(\epsilon_s), \tag{4.54}$$

where $R(\epsilon_s, \epsilon_v)$ has the form (4.4) and where we have used the following notation:

$$Q(\epsilon_s) = -kT \frac{d}{d\epsilon_s} \sigma(\epsilon_s), \tag{4.55}$$

with ϵ_s being a function of ϵ_v, according to (4.11). Here σ, as usual, is the surface charge density.

Let us show that (4.53) does follow from (4.54). We will start by solving Eq. (4.10), i.e.,

$$\sigma(\epsilon_s) + R(\epsilon_s, \epsilon_v) = 0 \tag{4.56}$$

and introduce the following notation, where the meaning is clear from Fig. 4.6:

$$a_i = E_C - E_{D_i}, \quad b_k = E_C - E_{A_k}.$$

Figure 4.6 shows the energy spectrum of a semiconductor in the plane surface (for the sake of simplicity only one type of acceptor level and one type of donor level are shown; the y axis is directed along the surface and is assumed to coincide with the Fermi level).

Obviously we have

$$\sigma(\epsilon_s) = e[\sum_i p_s(\epsilon_s, a_i) - \sum_k n_s(\epsilon_s, b_k)], \tag{4.57}$$

where $p_s(\epsilon_s, a_i)$ and $n_s(\epsilon_s, b_k)$ are the surface concentrations of holes and electrons, respectively, localized at the surface donor and acceptor levels of the ith and kth type, and

$$\begin{cases} p_s(\epsilon_s, a_i) = \dfrac{X_i}{1 + A_i(\epsilon_s)}, \text{ where } A_i(\epsilon_s) = \exp \dfrac{v_i - \epsilon_s}{kT}, \\[3mm] n_s(\epsilon_s, b_k) = \dfrac{Y_k}{1 + B_k(\epsilon_s)}, \text{ where } B_k(\epsilon_s) = \exp \dfrac{\epsilon_s - b_k}{kT}, \end{cases} \tag{4.58}$$

where X_i and Y_k are the concentrations (per square centimeter) of the surface donor and acceptor levels of the ith and kth type, respectively.

On the basis of (4.55), (4.57), and (4.58) we have

$$Q(\epsilon_s) = \epsilon \left[\sum_i p_s(\epsilon_s, v_i) \frac{A_i(\epsilon_s)}{1 + A_i(\epsilon_s)} + \sum_k n_s(\epsilon_s, w_k) \frac{B_k(\epsilon_s)}{1 + B_k(\epsilon_s)} \right]. \tag{4.59}$$

If we introduce the notation

$$\Delta \epsilon_s = kT \frac{R(\epsilon_s, \epsilon_v)}{Q(\epsilon_s)}, \tag{4.60}$$

which means that

$$R(\epsilon_s, \epsilon_v) = \frac{\Delta \epsilon_s}{kT} Q(\epsilon_s), \tag{4.61}$$

substitute (4.59) into (4.61), and then substitute (4.61) and (4.57) into (4.56), we obtain

$$\sum_i (1 - \delta_i) p_s(\epsilon_s, a_i) - \sum_k (1 + \delta_k) n_s(\epsilon_s, b_k) = 0, \tag{4.62}$$

where we have used the following notation:

$$\delta_i = \frac{A_i}{1 + A_i} \frac{\Delta \epsilon_s}{kT}, \qquad \delta_k = \frac{B_k}{1 + B_k} \frac{\Delta \epsilon_s}{kT}. \tag{4.63}$$

Suppose that condition (4.54) is met. Then, according to (4.60),

$$\frac{|\Delta \epsilon_s|}{kT} \ll 1,$$

which implies [see (4.63)] that certainly

$$|\delta_i| \ll 1 \quad \text{and} \quad |\delta_k| \ll 1 \tag{4.64}$$

for all i and k. Moreover, according to (4.58) and (4.63), we have

$$(1 - \delta_i) p_s(\epsilon_s, a_i) = X_i \frac{1 - \delta_l}{1 + A_i} = \frac{X_l}{(1 + \delta_i)(1 + A_i)}$$

$$= \frac{X_i}{1 + \left(1 + \dfrac{\Delta \epsilon_s}{kT}\right)} = \frac{X_i}{1 + A_i \exp \dfrac{\Delta \epsilon_s}{kT}} = p_s(\epsilon_s - \Delta \epsilon_s, a_i) \qquad (4.65a)$$

and, similarly,

$$(1 + \delta_k) n_s(\epsilon_s, b_k) = n_s(\epsilon_s - \Delta \epsilon_s, b_k). \qquad (4.65b)$$

Substituting (4.65a) and (4.65b) into (4.62), we arrive at Eq. (4.56) in the following form:

$$\sigma(\epsilon_s - \Delta \epsilon_s) = 0,$$

where $\Delta \epsilon_s = \epsilon_s - \epsilon_s^0$, with ϵ_s^0 the root of the equation $\sigma(\epsilon_s^0) = 0$, i.e., ϵ_s^0 is the position of the Fermi level at the crystal surface at which the surface retains its electrical neutrality.

Thus, for all values of ϵ_v for which condition (4.54) is met, we have (4.64) and, hence, ϵ_v differs little (in comparison with kT) from ϵ_s^0; i.e., the surface for such values of ϵ_v is "quasiisolated."

Let us write condition (4.54) in a different form. We introduce the following notation:

$$S(\epsilon_s) = e[\sum_i p_s(\epsilon_s, a_i) + \sum_k n_s(\epsilon_s, b_k)], \qquad (4.66)$$

$$\gamma = \frac{\sigma(\epsilon_s)}{S(\epsilon_s)} = \frac{\sum_i p_s(\epsilon_s, a_i) - \sum_k n_s(\epsilon_s, b_k)}{\sum_i p_s(\epsilon_s, a_i) + \sum_k n_s(\epsilon_s, b_k)} \qquad (4.67)$$

and rewrite (4.54) as follows [see (4.56)]:

$$|\gamma| \ll \frac{Q(\epsilon_s)}{S(\epsilon_s)}. \qquad (4.68)$$

A comparison of (4.59) with (4.66) yields

$$\frac{Q(\epsilon_s)}{S(\epsilon_s)} \leq 1; \qquad (4.69)$$

in particular, when the distribution of electrons and holes on all the surface levels is a Boltzmann distribution, $A_i \gg 1$ and $B_k \gg 1$ in (4.59) and

$$\frac{Q(\epsilon_s)}{S(\epsilon_s)} = 1.$$

Thus, on the basis of (4.67), (4.68), and (4.69) we see that a surface is "quasi-isolated" when the difference between the surface concentrations of electrons and holes is small (in absolute value) compared with their sum.

4.3.2. Some Properties of "Quasiisolated" Surfaces

A quasiisolated surface has the following property. From (4.67) we obtain

$$(1 - \gamma) \sum_i p_s(\epsilon_s, a_i) - (1 + \gamma) \sum_k n_s(\epsilon_s, b_k) = 0,$$

from which, on the basis of (4.57) (since always $|\gamma| \ll 1$ on a quasiisolated surface),

$$\sigma(\epsilon_s) = 0. \tag{4.70}$$

This means that in the case of a quasiisolated surface the quantities ϵ_v and ϵ_s are determined not from solving the two coupled equations, (4.56) and

$$\rho(\epsilon_v) = 0 \tag{4.71}$$

(see Section 4.1.1), but from solving two independent equations, (4.70) and (4.71), of which the first depends only on parameters that characterize the surface and the second only on parameters that characterize the bulk of the crystal. Consequently, now there is no interrelationship between ϵ_s and ϵ_v.

We also see that a surface that meets condition (4.54) is not only quasiisolated but quasineutral, so to say, since Eq. (4.70) is approximately fulfilled for it. However, the surface is not really neutral here. Equation (4.70) only indicates [as is evident from (4.67) at $|\gamma| \ll 1$] that the difference between the positive and negative charges concentrated on the surface is very small in absolute value compared with their sum. In itself, however, the absolute value of this difference may be considerable. In that case quasiisolation is only possible for large values of S, which can occur for a high density of surface states.

Note that there are three distinct types of surface states that can be responsible for the surface being quasiisolated:

(a) States that belong to surface energy bands (see Section 1.5.3). Such bands, if they exist (the surface conduction band and the surface valence band where an electron and, respectively, a hole may move freely about the surface but cannot penetrate the crystal), produce a high density of states. Both real and ideal surfaces may have such states.

(b) States due to various intrinsic structural defects, which are present on every real surface and act as imperfections in the strictly periodic structure of the surface (see Section 1.1.1). These may be vacant sites at the surface layer of the lattice, atoms or ions ejected to the surface, and foreign atom inclusions at the surface of the lattice (surface impurities).

(c) Finally, the atoms or molecules chemisorbed on the surface, which act as structural defects and contribute to the overall density of surface states.

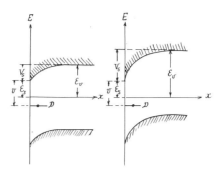

Fig. 4.7. Energy spectrum of a quasiisolated surface.

Thus we see that both real and ideal semiconductor surfaces under certain conditions have the property of quasiisolation. Estimates show that a surface is sure to be quasiisolated at values of S greater than 10^{12} cm^{-2}. Apparently, in the case of real semiconductors we are often concerned with such a surface. Theoretically the existence of quasiisolated surfaces was demonstrated by Bardeen [45] (in a qualitative manner), while experimentally their existence was proved for a number of semiconductors in experiments that show that the work function is independent of the position of the Fermi level in the bulk of a semiconductor.

In the case of a quasiisolated surface the influence of the bulk of the semiconductor (e.g., impurities introduced into the interior of the crystal) on its chemisorptive and, as we will subsequently see in Chapter 5, catalytic properties disappears, and only the dependence of these properties on the structure of the surface remains. This dependence is revealed in Eq. (4.70), according to which the position of the Fermi level at the surface, ϵ_s, proves to be dependent on the concentration and nature of the structural defects on the surface.

In the case of a quasiisolated surface ϵ_s does not depend on ϵ_v. For this reason when ϵ_v changes, the bending of the bands, V_s, varies by the same amount, since ϵ_s remains constant in the process. This fact is illustrated by Fig. 4.7, whose right part depicts the same semiconductor as the left except for a different value of ϵ_v but the same value of ϵ_s (the horizontal axis in Fig. 4.7 coincides with the Fermi level).

Although the bulk of a semiconductor does not act on the quasiisolated surface, the surface does act on the bulk. Indeed, when the state of the surface changes, which is characterized by a shift in the ϵ_s level (which is the case with chemisorption, for instance), the bending of the band V_s changes and so do all the bulk properties associated with this bending (e.g., the electrical conductivity of the sample; see Section 4.2.2).

We see that the influence of chemisorption on electrical conductivity, which is characteristic of semiconductors, is retained, strictly speaking, in the case of a quasiisolated surface. Here, however, the effect manifests itself only for high surface coverages, when the chemisorbed particles contribute considerably to the overall concentrations of electrons and holes localized at the surface.

Indeed, if the surface coverage is smaller than a certain critical value, the position of the Fermi level on the surface, ϵ_s, remains insensitive to chemisorption, which means that chemisorption in this case has no effect on the electrical conductivity and work function of the semiconductor. As shown in [1], this situation takes place if

$$eN \ll Q(\epsilon_s), \tag{4.72}$$

where N is the surface concentration of the chemisorbed particles, and Q the respective quantity for a clean surface (i.e., for a surface with $N = 0$). In the case of a quasiisolated surface, which obeys condition (4.54), Q is large and condition (4.72) is met even for large values of N.

4.3.3. The Continuous and Quasicontinuous Spectra of Surface States

We usually encounter "quasiisolated" surfaces when we deal with real surfaces, whose properties and special features will be discussed in Chapter 6.

A concept most often used in dealing with real surfaces of semiconductors is the continuous energy spectrum of the surface states. This term, it must be noted, is often used in two different senses, and one must always be careful to note which of the two meanings the author is using.

Strictly speaking, the surface energy spectrum is said to be continuous when to each (arbitrary) value of the energy there corresponds a wave function such that its squared modulus is periodic (with a period equal to the lattice period) in the plane of the surface and falls off more or less rapidly in both directions from the surface. A carrier described by such a wave function can be said to spread uniformly over the surface or, in other words, wanders freely over the surface. An energy spectrum of this type is characteristic of the surface energy bands of Goodwin [46], which in the three-dimensional case have the same origin as the Tamm surface levels [47] in the one-dimensional case.

We continue to call the energy spectrum continuous even when the wave function is restricted by the Born–von Karman condition (the periodicity condition) or by boundary conditions, which, as we know, transform the continuous spectrum into a system of an immense number of closely packed discrete levels.

Often, however, the concept of a continuous spectrum is used in another sense. A spectrum is said to be continuous when to each (arbitrary) value of the energy there correspond definite spatial regions in which the wave function is nonzero, while outside the regions it is practically zero. A carrier described by such a wave function is said to be localized at certain regions on the surface. In contrast to the previous definition of a continuous spectrum, we will call such a spectrum quasicontinuous. It would be more correct to speak not of a continuous surface energy spectrum but of a continuous set of surface local levels. Often such a set is discrete rather than continuous, i.e., consists of a set of discrete local levels closely spaced on the energy scale.

Such a quasicontinuous spectrum that fills the forbidden band between the energy bands was observed by Rzhanov and other authors. According to Rzhanov *et al.* [48], a quasicontinuous spectrum has different densities of states in different parts. Near the middle of the forbidden band the density is minimal. The density rapidly increases as we get closer to the edges of the energy band, remaining, however, negligible compared with the density of states inside the energy bands. This fact and also the fact that the distribution of the density of states over the spectrum depends essentially on the prehistory of the surface unambiguously indicate that these states are the result of the surface having defects.

4.4. ADSORPTIVE PROPERTIES OF A CHARGED SEMICONDUCTOR

4.4.1. The Adsorptivity of a Charged Semiconductor

Up till now we have considered only cases where the semiconductor together with particles adsorbed on its surface form an electrically neutral system as a whole. Here we will study the case where this system is electrically charged. Such an investigation makes it possible to study a number of effects, namely:

(1) How the adsorptivity of a semiconductor changes in an electric field applied to a capacitor one of whose plates is the sample being studied (the field effect measurement). The field induces a charge in the semiconductor, as a result of which the adsorptive properties of the semiconductor change (see Section 4.4.2).

(2) How the chemisorption of ions influences the properties of the semiconductor. Here the charge is introduced by the adsorbed particles proper, which leads to a number of features that are absent when neutral particles of the same nature are adsorbed (see Section 4.4.3).

(3) How external radiation or radioactive dopants influence the chemisorptive properties of the semiconductor when in certain conditions the semiconductor can hold onto a stationary charge (see Section 6.5).

But first let us see how charging influences the work function and adsorptivity of a sample [49]. Take a sample in the form of a plate whose thickness is $2L$ and the surface area of the lateral face is S; we neglect the surface area of the end planes and assume that the x axis is directed normal to the surface of the plate. Suppose that an electric charge Q is imparted to the sample as a whole, so that the charge per unit surface area is $\delta = Q/2S$. In the charging process the near-the-surface bending of the bands changes by $\Delta V_s = V_s - V_{s0}$, while the adsorptivity of the surface changes by $\Delta N = N - N_0$ and the surface charge by $\Delta\sigma = \sigma - \sigma_0$. The label 0 stands for the absence of an electric charge, i.e., $\sigma = 0$. The problem will be solved if we find ΔV_s, ΔN, and $\Delta\sigma$ as functions of σ.

As we saw in Section 4.1.2, V_{s0} is determined from the condition of electrical neutrality of the system, which for a semiinfinite crystal has the form (4.1):

$$\sigma + \int_0^\infty \rho(x)dx = 0.$$

Similarly, V_s can in the case at hand be found from the law of charge conservation:

$$\sigma + \int_0^L \rho(x)dx = \delta, \qquad (4.73)$$

where as usual $\rho(x)$ is the density of charge in the space charge region. The symmetry of the problem yields

$$\left(\frac{dV}{dx}\right)_{x=L} = 0$$

[cf., (4.8)]. Denoting the value of the potential at the center of the plate (at $x = L$) by V_c and assuming that

$$\frac{|V_c|}{kT} \ll 1, \qquad (4.74)$$

we obtain for V_c (just as we did in Section 4.1.2) a formula of the type (4.27), where, however, we must substitute $\sigma - \delta$ for σ. [Note that the condition $V_c \neq 0$, which strictly speaking is met in our case, means that the crystal's center is charged rather than being electrically neutral and the Fermi level at the center of the crystal is shifted by a distance V_c compared with its position in an infinitely thick neutral crystal; condition (4.74) means that the charge density at the crystal's center is assumed to be low.]

We will restrict ourselves to the Boltzmann distribution of the electrons and holes inside the crystal (in the general case of a Fermi distribution the results are qualitatively the same). Instead of (4.29) we have

$$2\sinh\frac{V_s}{kT} = -\frac{\sigma-\delta}{\sigma*}, \qquad (4.75)$$

where $\sigma*$ is defined in (4.28). Assuming that the intrinsic charge of the surface is much greater than the adsorption charge and that, in addition, all the intrinsic defects are almost completely ionized, we can ignore the dependence of σ on N and on δ. In this case we can put

$$\sigma = \sigma_0. \qquad (4.76)$$

Combining (4.75) with (4.76), we see that if the semiconductor is charged positively ($\delta > 0$) and all other conditions remain unchanged, the bands bend upward (V_s increases), i.e., the work function increases, while if the semiconductor is charged negatively ($\delta < 0$), the bands are bent downward (V_s decreases), i.e., the work function decreases, as we expected. The variation in the band bending leads to a change in the conductivity of the sample; i.e., for δ positive the n-type conduction increases, while the p-type conduction drops, and for δ negative the p-

type conduction increases, while the *n*-type drops. This effect was studied in Section 4.2.3.

Let us now examine the influence of charging on the adsorptivity N of a semiconductor. Here N is the number of molecules of a given type retained by a surface of unit area with given external conditions (pressure P and temperature T). If the surface coverage is not too great (the Henry region), then, according to (3.10), (3.11), (3.6a), and (3.6b),

$$N = \frac{\alpha P}{\eta^0} = \alpha P \left[1 + \exp\left(\mp \frac{\epsilon_v + V_s - v}{kT} \right) \right]. \tag{4.77}$$

Here and in what follows the upper sign corresponds to the case of acceptor molecules and the lower sign to donor molecules. The meaning of the various notations is clear from Fig. 4.7, and the surface local level depicted in the figure may be either a donor level or an acceptor level. On the basis of (4.77) we obtain the following formula for the fractional change of the adsorptivity:

$$\frac{\Delta N}{N} = \eta_0^{\mp} \left[\exp\left(\mp \frac{\Delta V_s}{kT} \right) - 1 \right], \tag{4.78}$$

where $\eta_0^{\mp}$ is the fraction of the "strongly" bound particles on the neutral semiconductor. According to (3.6a) and (3.6b) we have

$$\eta_0^{\mp} = \frac{N_0^{\mp}}{N_0} = \frac{1}{1 + \exp\left(\pm \dfrac{\epsilon_v - v + V_{s0}}{kT} \right)}. \tag{4.79}$$

From (4.75) we obtain

$$\exp\frac{V_s}{2kT} = \frac{\delta - \sigma}{2\sigma^*} + \sqrt{\left(\frac{\delta - \sigma}{2\sigma^*} \right)^2 + 1}. \tag{4.80}$$

Putting $\sigma = \sigma_0$ and substituting (4.80) into (4.78), we find $\Delta N/N_0$ as a function of δ. We see that positive charging ($\delta > 0$) increases the adsorptivity of the sample surface with respect to donor particles ($\Delta N/N_0 > 0$) and decreases it with respect to acceptor particles ($\Delta N/N_0 < 0$). Negative charging has the opposite effect.

4.4.2. The Electroadsorptive Effect

Let us assume that a semiconductor in the form of a plate is placed in a uniform electric field normal to the plate surface. The effect of the field on the adsorptivity of the semiconductor is called the electroadsorptive effect [50, 51]. Suppose that δ is the charge induced in the semiconductor by the field on a unit surface area. We assume that the induced charge is distributed symmetrically

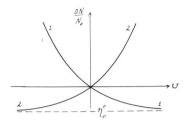

Fig. 4.8. The eletroadsorptive effect.

with respect to the center of the plate. This situation is realized if we place two electrodes with the same potential (called field electrodes) symmetrically with respect to the sample electrode. Note, however, that all our reasoning remains valid for the case of one field electrode (the field effect measurement). According to (4.49),

$$\delta = CU = -\frac{\chi_0}{4\pi} F \tag{4.81}$$

where U is the potential difference between the sample and field electrodes, C the capacitance of the system per unit area of the sample surface, F the strength of the applied field, and χ_0 the dielectric constant of the medium between the sample and the field electrode. Substituting (4.81) and (4.76) into (4.80) and then (4.80) into (4.78), we arrive at the dependence of $\Delta N/N_0$ on F or on U, schematically depicted in Fig. 4.8 (curve 1 corresponds to acceptor particles and curve 2 to donor particles).

From Fig. 4.8 we can arrive at the following conclusions:

(a) A positive potential on the semiconductor decreases the adsorption of acceptors and increases the adsorption of donors. A negative potential acts in the opposite direction.

(b) The effect is unsymmetric with respect to the sign of the field; i.e., as U increases in absolute value, the positive effect increases without limit, while the negative effect asymptotically approaches $\Delta N/N_0 = -\eta_0^{\mp}$, where the value of $\eta_0^{\mp}$ is fixed by the prehistory of the sample [see (4.79)]. If in the absence of a field the neutral form of chemisorption is predominant on the surface, i.e., $\eta_0^{\mp} \ll 1$, the negative effect is practically nil.

Thus, an electric field changes not only the occupancy of the surface states but, in certain conditions, changes the adsorptivity of the surface and in this way may considerably change the density of the surface states. This fact may prove to be important in interpreting field effect measurements.

The variation of the adsorptivity of the surface in an external field is due to the variation in the fraction of the charged form of chemisorption on that surface. The fraction of the neutral form remains constant in the process, as shown by (4.77), and is fixed for a definite pressure and temperature. Since the charged form, as we know (see Section 3.4.3), can in some cases be taken as practically irreversible, the increase in the chemisorptivity caused by the field must be irreversible in such cases, too, i.e., must remain constant after the field has been switched off (other conditions remaining the same). This means that the chemisorption that occurs when the field is switched on must lead to irreversible changes in the work function and conductivity of the sample, which opens up the possibility of detection and investigation of the electroadsorptive effect on single crystals, on which direct adsorption measurements are difficult.

The electroadsorptive effect was observed by Rzhanov with collaborators [52–54] in experiments in measuring the concentration of surface states on germanium in a transverse field and was called the "buildup effect." Later Lyashenko, Serba, and Stepko [55] by direct adsorption methods (pressure measurements) established the presence of desorption of an acceptor gas from a germanium plate under a positive potential. Mikheeva and Keier [56], in complete agreement with the theory, observed irreversible chemisorption of donor molecules of methanol on germanium when a positive potential was imparted to the sample and a nil effect for a negative potential or when the field was switched off. We note also the work of Ivankiv [57], who studied the electroadsorptive effect for O_2 and CO_2 on HgS.

Hoenig and Lane [58] discovered the electroadsorptive effect for O_2 on ZnO. It manifested itself in the change of the electrical conductivity of the sample and directly in the change in pressure in the adsorption volume. The researchers studied the effect of the field on the kinetics of chemisorption. A positive potential applied to the semiconductor hindered the chemisorption of oxygen, while a negative potential accelerated the process, as was to be expected.

Two works of Rumanian scientists are also of interest. Constantinescu, Segal, Vass, and Teodorescu [59, 60] studied the electroadsorptive effect on thin films of zinc oxide in an atmosphere of oxygen. The researchers showed that the absolute value of the electroadsorptive effect, found from the variation in the resistance of the sample, depends on oxygen pressure (for a fixed value of the applied electric field); i.e., the fractional change in resistance increases with oxygen pressure. After the field is switched off, the resistance of the film tends to its initial value but does not attain it. The higher the pressure of oxygen, the higher the final resistance of the oxygen film after the field has been switched off.

Note in conclusion that field effect measurements can be used not only for detecting and measuring the electroadsorptive properties of semiconductors, but for other purposes as well:

(1) By fixing the change in the adsorptivity (its increase or decrease) when the polarity of the applied voltage is reversed, we can infer whether the adsorbed gas consists of acceptors or donors; i.e., when we change the potential on the adsorbent from positive to negative, the adsorption of an acceptor gas increases and that of a donor gas decreases.

(2) As noted in Section 4.2.1, field effect measurements of the surface potential V_s (finding the variation of the conductivity with the field; see Section 4.2.2) conducted simultaneously with cpd measurements of the work function enable the researcher to separate in the overall change in the work function due to chemisorption the part caused by surface charging from the part caused by the dipole moment of the chemisorbed molecules.

(3) Field effect measurements constitute a sensitive method for finding the fraction of the charged form of chemisorption on the surface and may be used to measure adsorption on single crystals. Let us denote the strength of the external field by F^*, and by V_s^* the corresponding value of the surface potential at which the sample conductivity in vacuum (i.e., in the absence on the surface of states of adsorption origin) has the value $\kappa^* = \kappa(V_s^*)$. Let $F = F^* + \Delta F$ be the field strength at which in the gaseous atmosphere (with a given pressure) the conductivity of the sample and hence the surface potential return to their vacuum values $\kappa = \kappa^*$ and $V_s = V_s^*$. According to (4.81) and (4.73),

$$-\frac{1}{4\pi} F^* = \sigma_B(V_s^*) + R(V_s^*),\tag{4.82a}$$

$$-\frac{\chi_0}{4\pi}(F^* + \Delta F) = \sigma_A(V_s^*) + \sigma_B(V_s^*) + R(V_s^*),\tag{4.82b}$$

where $\sigma_A(V_s^*)$ and $\sigma_B(V_s^*)$ are the surface charges of adsorption and intrinsic origins at $V_s = V_s^*$ and where we introduced the notation [cf. the notation (4.3)]

$$R(V_s) = \int_0^L \rho(x)\,dx.$$

Comparing (4.82a) and (4.82b), taking into account that $\sigma_A = \pm eN^{\mp}$, and putting $\chi_0 = 1$, we obtain

$$N^{\mp} = \mp\frac{1}{4\pi e}\Delta F.$$

Thus, knowing the "compensating" field ΔF and the dependence of ΔF on F^*, we can unambiguously determine both the surface coverage by a charged form of chemisorption, $N^{\mp}$, and the variation in the surface coverage caused by the external field, $N^{\mp} = N^{\mp}(F^*)$ (the electroadsorptive effect).

4.4.3. Adsorption of Ions on a Semiconductor

The adsorption of ions on semiconductors has a number of specific features in comparison with the adsorption of electrically neutral particles of the same nature. Take a semiconductor in the form of an infinite plane-parallel plate on both sides of which particles of one type are chemisorbed. Let N be the number of particles per unit surface area. We wish to find the V_s vs. N dependence for the case where the particles are neutral and the case where the particles are singly charged ions. For the sake of definiteness we will consider only weakly ionized impurities in the crystal (the Boltzmann statistics for the electrons and holes on the impurities). Similar results are obtained for almost completely ionized impurities (the Fermi statistics). Finally, we assume that

$$\left(\frac{\delta - \sigma}{\sigma^*}\right)^2 \gg 1. \tag{4.83}$$

Under (4.83) Eq. (4.80) yields

$$\exp\left(\frac{V_s}{kT}\right) = \begin{cases} \left(\dfrac{\delta - \sigma}{\sigma^*}\right)^2 & \text{for} \quad \delta > \sigma, \\[3mm] \left(\dfrac{\sigma^*}{\delta - \sigma}\right)^2 & \text{for} \quad \delta < \sigma, \end{cases} \tag{4.84}$$

or, to put it differently,

$$V_s = \pm kT \ln\left(\frac{\delta - \sigma}{\sigma^*}\right)^2, \tag{4.85}$$

where the upper sign corresponds to $\delta > \sigma$ and the lower to $\delta < \sigma$. On the basis of (4.84) we can easily see that condition (4.83) implies that

$$\frac{|V_s|}{kT} \gg 1.$$

Let us first assume that we are dealing with neutral particles. Then

$$\sigma = \sigma_B \mp eN^{\mp} \quad \text{and} \quad \delta = 0, \tag{4.86}$$

where, as usual, σ_B is the density of the intrinsic charge on the surface and where the upper signs correspond to acceptor particles and the lower to donor particles. We will restrict ourselves to a high intrinsic disorder (or a low surface coverage), i.e., assume that

$$eN \ll |\sigma_B|, \tag{4.87}$$

and, hence, all the more

$$eN^+ \ll |\sigma_B|.$$

Then, if we substitute (4.86) into (4.85), we find that

$$V_s(N) - V_s(0) = \pm \frac{2kTe}{|\sigma_B|} N^{\mp}. \tag{4.88}$$

This formula coincides with (4.36a), which was obtained in Section 4.1.3, if we take all the adsorbed particles to be ionized, i.e., if $N^{\mp} = N$.

Now let us assume that the adsorbed particles are singly charged ions. Then

$$\sigma = \sigma_B \mp eN^{\mp} \quad \text{and} \quad \delta = \mp eN$$

and, hence,

$$\sigma - \delta = \sigma_B \pm e(N - N^{\mp}) = \sigma_B \pm eN^0, \tag{4.89}$$

where $N^{\mp}$ is the surface concentration of the ions that retain their charge, and N^0 the surface concentration of the ions neutralized in the adsorption process. The interplay between N^0 and $N^{\mp}$ is determined by the position of the Fermi level at the crystal surface. In (4.89) (and in what follows) the upper sign corresponds to negative ions and the lower sign to positive ions. Substituting (4.89) into (4.85) and allowing for (4.87), instead of (4.88) we obtain

$$V_s(N) - V_s(0) = \mp \frac{2kTe}{|\sigma_B|} N^0. \tag{4.90}$$

Comparing (4.90) with (4.80), we see that the adsorption of negative ions [the upper sign in (4.90)] bends the bands downward at the surface, i.e., negative ions act as donors (decrease the work function, increases the n-type conduction, and decreases the p-type conduction), while the adsorption of positive ions [the lower sign in (4.90)] bends the bands upward at the surface, i.e., positive ions act as acceptors (increase the work function, decrease the n-type conduction, and increase the p-type conduction). The charge of the surface is determined here by the particles chemisorbed on the surface that are in the neutral rather than charged state.

4.5. THE INFLUENCE OF THE SURFACE ON THE IMPURITY DISTRIBUTION INSIDE A SEMICONDUCTOR

4.5.1. Statement of the Problem

It is a known fact that many of the properties of a semiconductor depend on the nature and amount of impurity in it. We recall that by an impurity (see Section 1.2.1) we mean not only a chemically foreign atom introduced into the lattice, but also any microscopic lattice defect, i.e., local imperfection in the strict periodicity of the lattice.

For a given impurity (its nature and concentration) the properties of the semiconductor will depend on the distribution of this impurity. For instance, a concentration gradient of the impurity may lead to rectifying properties manifesting themselves, where the conductivity of the semiconductor in the opposite direction is different. A nonuniform distribution of acceptor and donor impurities may lead to the formation of p–n junctions, i.e., formation of regions in the semiconductor with different types (n- and p-type) of conduction. A semiconductor inside of which the impurity is distributed nonuniformly is known as a disordered semiconductor. Some specific features of adsorption on such a disordered semiconductor will be studied in Section 6.3.3. Here we will only discuss the case where such a disorder is due solely to the presence of a surface.

Let us assume that the semiconductor contains an impurity of only one type. In the case of an infinite semiconductor the equilibrium distribution of the impurity will be a uniform distribution (concentration is constant throughout the volume). The situation is different when the semiconductor is finite. The presence of surface states (whatever their nature) leads to the formation of a surface charge. The electric field of this charge propagates to a certain depth into the crystal. In the surface layer of the semiconductor, where this field is not entirely compensated for by the space charge originating in the bulk, the uniform distribution of the impurity is violated.

Indeed, ionized impurity particles whose charge is of a sign opposite to that of the surface charge will be pulled to the surface, while ionized impurity particles whose charge is of a like sign with that of the surface charge will leave the surface for the bulk. The resulting concentration gradient will result in a diffusion current opposing the concentration gradient. This diffusion current is opposite in direction to the ion current generated by the electric field. An equilibrium distribution of the impurity will be achieved when the two currents become equal in magnitude. The impurity concentration in the surface layer is elevated or lowered depending on whether the charge localized at the semiconductor surface and the charges of the ionized impurity particles are of like or unlike sign.

We wish to determine the equilibrium distribution of an impurity in the surface layer of a semiconductor for a given surface charge density σ or, in other words, for a given surface potential V_s. This problem was studied by Kuznetsov and Sandomirskii [61].

We take a semiinfinite semiconductor. The half-space $x \geq 0$ is occupied by the semiconductor and the half-space $x < 0$ by the gaseous phase. For the sake of definiteness we will assume that the semiconductor contains only a donor impurity and is of a single type (an n-type semiconductor). We denote the impurity concentration by X, as usual, and the concentrations of the neutral and ionized impurity particles by X^0 and X^+, respectively. Obviously,

$$X = X(x), \quad X^0 = X^0(x), \quad X^+ = X^+(x), \tag{4.91}$$

with

$$X(x) = X^0(x) + X^+(x).$$

We will introduce the following notations:

$$X_v = X(\infty), \quad X_v^0 = X^0(\infty), \quad X_v^+ = X^+(\infty).$$

The problem consists in finding the functions in (4.91), where σ (or V_s) enters as a parameter. We will consider this problem in Section 4.5.2.

Note that adsorption, as we have seen, changes σ and V_s (see Sections 3.2.2 and 4.1.3) and, hence, may lead to a redistribution of the impurity in the semiconductor. On the other hand, as we will see below, redistribution of the impurity changes the adsorptivity of the surface. Therefore, we are dealing here with a self-consistent problem, which is one more example of the interaction of the surface with the bulk of a semiconductor. Section 4.5.3 is devoted to the problem of the effect of the impurity distribution in a semiconductor on its adsorptivity.

Note further that the redistribution of the impurity in a semiconductor caused by the surface charge (for one, by adsorption) may be accompanied by ejection of the impurity to the surface, which leads to a change in the adsorptive properties of the latter. This effect will be studied in Section 4.5.4.

Of course, all the effects we have mentioned which are caused by the redistribution of the impurity because of a change in the surface charge manifest themselves in a noticeable manner only if this redistribution process proceeds very rapidly—in other words, if the time τ in which equilibrium in the impurity distribution is established is shorter than the duration of the experiment. Obviously, the greater the diffusion coefficient D for the impurity and, hence, the higher the temperature T, the shorter the value of τ. Numerical estimates of τ done for some cases [61] yield the following: for germanium doped with copper $\tau = 10^{-4}$ sec, for germanium doped with lithium $\tau = 0.3$ sec, and for zinc oxide doped with zinc $\tau = 1.3$ sec. (Here we have taken $l = 10^{-5}$ cm, $T = 200$–$300°C$, and D from [62].) We see that in some cases the redistribution of the impurity accompanying the bending of the bands can and must be taken into consideration. Note that the equilibrium (but not uniform) distribution of the impurity is inherent in the initial samples (prior to any measurements related to changes in σ), which as usual have undergone thermal treatment and possess an intrinsic charge on their surface.

4.5.2. Impurity Distribution in the Surface Layer of Semiconductors

When there is electronic equilibrium in the sample, the functions in (4.91), according to the Fermi statistics, are as follows [see (4.17)]:

$$X^0(x) = \frac{X(x)}{1 + \exp\left[-\dfrac{E_F - E_D(x)}{kT}\right]},$$

$$X^+(x) = \frac{X(x)}{1 + \exp\left[\dfrac{E_F - E_D(x)}{kT}\right]},$$

(4.92)

and, hence,

$$\frac{X^+(x)}{X^0(x)} = \exp\left[-\frac{E_F - E_D(x)}{kT}\right], \quad \frac{X_v^+}{X_v^0} = \exp\left[-\frac{E_F - E_D(\infty)}{kT}\right].$$

(4.93)

Here $E_D(x)$ is the local level of the donor impurity in the plane x, and E_F the Fermi level.

Let us denote the fluxes of neutral and ionized impurity particles by j^0 and j^+, respectively. We have

$$j^0 = -D \frac{dX^0}{dx},$$ (4.94a)

$$j^+ = -D^+ \left(\frac{dX^+}{dx} - X^+ \frac{eF}{kT} \right),$$ (4.94b)

where D^0 and D^+ are the diffusion coefficients for the neutral and ionized impurity particles, $F = F(x)$ the electric field generated by the presence of the surface charge σ, and e the absolute value of the electron charge. In an equilibrium distribution of the impurity we have

$$j^0 = 0,$$ (4.95a)

$$j^+ = 0.$$ (4.95b)

Combining (4.95a) with (4.94a) yields

$$X^0(x) = X^0_v = \text{const}$$

and, hence, according to (4.93),

$$X^+(x) = X^+_v \exp \frac{V(x)}{kT},$$ (4.96a)

$$X(x) = X^0_v + X^+_v \exp \frac{V(x)}{kT},$$ (4.96b)

where (see Fig. 4.1)

$$V(x) = E_c(x) - E_c(\infty).$$

Equation (4.95a) yields (4.95b), which we can easily verify by substituting (4.96) into (4.94b) and bearing in mind that $dV/dx = eF$.

To write the functions (4.96a) and (4.96b) explicitly, we must still find the function $V = V(x)$ in (4.96a) and (4.96b). This can be done by integrating Poisson's equation

$$\frac{d^2 V}{dx^2} = \frac{4\pi e}{\chi} \rho(x),$$ (4.97)

where χ is the dielectric constant of the crystal, and $\rho(x)$ the space charge density. In our case Eq. (4.97) can be evaluated completely. Restricting our discussion to reasonably high temperatures, at which we can ignore the hole component of the conduction, we have [see (4.15)]

$$\rho(x) = e[X^+(x) - n(x)],$$ (4.98)

where [see (4.16)]

$$n(x) = n_v \exp\left[-\frac{V(x)}{kT}\right]$$ (4.99)

is the free electron concentration (the electron gas is nondegenerate), and $X^+(x)$ is given by (4.96b). Substituting (4.96b) and (4.99) into (4.98) and then (4.98) into (4.97), integrating Eq. (4.97) from x to ∞, and bearing in mind that

$$X_v^+ = n_v \quad \text{and} \quad \left(\frac{dV}{dx}\right)_{x = \infty} = 0,$$ (4.100)

we get [see (4.22)]

$$\frac{dV}{dx} = -\frac{kT}{l} 2 \sinh \frac{V(x)}{2kT},$$ (4.101)

where l (the screening length) has the form (4.24). Integrating once more, we get [see (4.23)]

$$\exp\left[\frac{V(x)}{2kT}\right] = \frac{\exp\dfrac{x}{l} + \tanh\dfrac{V_s}{4kT}}{\exp\dfrac{x}{l} - \tanh\dfrac{V_s}{4kT}},$$

where, introducing the notation

$$\tanh \frac{V_s}{4kT} = \pm \exp(-\gamma),$$ (4.102)

where the upper sign is taken for $V_s \geq 0$ and the lower for $V_s \leq 0$ and where $\gamma \geq 0$, we obtain

$$\exp\left[\frac{V(x)}{kT}\right] = 1 - \frac{2}{1 \mp \cosh\left(\dfrac{x}{l} + \gamma\right)} = \begin{cases} 1 + \dfrac{1}{\sinh^2\dfrac{1}{2}\left(\dfrac{x}{l} + \gamma\right)} & \text{for} \quad V_s \geq 0, \\[4mm] 1 - \dfrac{1}{\cosh^2\dfrac{1}{2}\left(\dfrac{x}{l} + \gamma\right)} & \text{for} \quad V_s \leq 0. \end{cases}$$ (4.103)

Now let us return to Eq. (4.96b). We will assume [see (4.100)] that

$$X_v^0 = X_v - X_v^+ \quad \text{and} \quad X_v^+ = n^*,$$

where, as before, n^* is the majority carrier concentration in the bulk of the crystal (in our case $n^* = n_v$). We rewrite Eq. (4.96b) as follows:

$$X(x) = X_v - n^* \left\{ 1 - \exp\left[\frac{V(x)}{kT} \right] \right\},$$

or, substituting (4.103),

$$X(x) = \begin{cases} X_v + \dfrac{n^*}{\sinh^2 \dfrac{1}{2}\left(\dfrac{x}{l} + \gamma\right)} & \text{for} \quad V_s \geqslant 0, \\[30pt] X_v - \dfrac{n^*}{\sinh^2 \dfrac{1}{2}\left(\dfrac{x}{l} + \gamma\right)} & \text{for} \quad V_s \leqslant 0. \end{cases} \tag{4.104}$$

Note that the surface potential V_s and the surface charge σ are coupled by the following relationship (see Section 4.1.1):

$$\left(\frac{dV}{dx} \right)_{x=0} = \frac{4\pi e}{\chi} \cdot \sigma,$$

which may be taken as one of the boundary conditions for Poisson's equation or as the condition of electroneutrality of the system [see (4.1) and (4.9)]. Indeed, if we substitute (4.101), we obtain

$$\left(\frac{dV}{dx} \right)_{x=0} = - \frac{kT}{l} \, 2 \sinh \frac{V_s}{2kT} = \frac{4\pi e}{\chi} \, \sigma,$$

where we arrive at (4.29):

$$2 \sinh \frac{V_s}{2kT} = - \frac{\sigma}{\sigma^*} , \tag{4.105}$$

where, according to (4.24), σ^* has the form (4.28).

On the basis of (4.104) we can say that at $\sigma < 0$ [or $V_s > 0$, according to (4.105)] the surface layer of the semiconductor proves to be enriched by the impurity [$X(x) \geq X_v$], while at $\sigma > 0$ (or $V_s < 0$) the surface layer proves to be depleted [$X(x) \leq X_v$]. At $\sigma = 0$ (or $V_s = 0$) we have $\gamma = \infty$ [according to (4.104)] and formula (4.104) yields

$$X(x) = X_v = \text{const},$$

i.e., we have a uniform distribution of the impurity, as was to be expected.

Note that when the semiconductor has an acceptor impurity instead of a donor impurity (a p-type semiconductor), we can easily arrive at a formula that is like (4.104) with $n^* = p_v$, where p_v is the free hole concentration at $x = \infty$.

4.5.3. Effect of Impurity on the Adsorptivity of Semiconductors

We already know (see Section 3.2.1) that the adsorptivity of a semiconductor depends, all other things being equal, on the position of the Fermi level at the surface. For an acceptor gas the adsorptivity is lower the lower the position of the Fermi level, i.e., the greater ϵ_s is (see Fig. 4.1). For a donor gas the opposite is true. We can find ϵ_s by solving Eq. (4.10),

$$\sigma(\epsilon_s) + R(\epsilon_s, \epsilon_v) = 0, \qquad (4.106)$$

where, according to (4.3) and (4.4)

$$R(\epsilon_s, \epsilon_v) = \int_0^\infty \rho(x)\,dx = \pm \sqrt{\frac{\chi}{2\pi e} \int_{\epsilon_v}^{\epsilon_s} \rho(\epsilon)\,d\epsilon}. \qquad (4.107)$$

Here σ is the surface charge density, and ρ the space charge density. Equation (4.106) constitutes the condition for electrical neutrality of the system. The upper sign in (4.107) corresponds to $\sigma < 0$, the lower sign to $\sigma > 0$.

As we have already seen (see Section 4.1.1), from (4.106) we can arrive at (4.13):

$$\frac{d\epsilon_s}{d\epsilon_v} \geqslant 0.$$

This, firstly, provides some hints about how the impurity acts on adsorptivity. Indeed, a donor impurity introduced into a semiconductor crystal always reduces ϵ_v, while an acceptor impurity always increases ϵ_v. For this reason a donor impurity diminishes the adsorptivity of a surface with respect to a donor gas and increases the adsorptivity with respect to an acceptor gas. An acceptor impurity acts in the opposite manner.

This result agrees with experimental data. As an example we cite the work of Enikeev, Margolis, and Roginskii [13]. The researchers studied the adsorption of O_2 (acceptor) on ZnO. Introduction of the dopant Li_2O (donor) increased the adsorptivity of the semiconductor surface with respect to O_2.

Note that for a given impurity (i.e., type and concentration) the second term on the left-hand side of Eq. (4.106) as a function of ϵ_s varies with the distribution of the impurity inside the semiconductor. Thus, the impurity distribution influences ϵ_s calculated via (4.106) and, in this way, influences the adsorptivity.

We will study two cases here, the case of a uniform distribution of the impurity and the case where this uniform distribution is violated by the electric field created by the surface charge (Section 4.5.2). The problem is to determine how the redistribution of the impurity caused by the surface charge acts on the adsorptivity of the surface. We take the case where the entire surface charge is of an intrinsic origin, i.e., remains practically unaltered in adsorption, and the case where this charge is produced in adsorption, i.e., changes in the adsorption process. Just as in Section 4.5.2 we assume, for the sake of definiteness, that the

semiconductor is the host of a donor impurity of a single type (an n-type scmicon-
ductor).

We start with the uniform distribution of the impurity. Substituting (4.99) and
(4.92), where we take $X(x) = X_v = $ const, into (4.98) and then (4.98) into (4.107),
we obtain [see (4.19)]

$$R(\epsilon_s, \epsilon_v) = \pm \sqrt{\frac{\chi k T}{2\pi}} \left\{ X_v \ln \frac{1 + \exp\dfrac{\epsilon_s - v}{kT}}{1 + \exp\dfrac{\epsilon_v - v}{kT}} - n_v \left[1 - \exp\frac{\epsilon_v - \epsilon_s}{kT} \right] \right\} \equiv R_1(\epsilon_s, \epsilon_v),$$

(4.108a)

where (in other notations; see Fig. 4.1)

$$\epsilon_s = E_C^s - E_F, \qquad \epsilon_v = E_C^v - E_F, \qquad v = E_C - E_D.$$

Now let us assume that the surface charge breaks the equilibrium, i.e., the distribution becomes nonequilibrium. Then instead of (4.108a) we have, according
to (4.96b) and (4.99),

$$R(\epsilon_s, \epsilon_v) = \sqrt{\frac{2\chi k T}{\pi}} \sinh \frac{\epsilon_s - \epsilon_v}{2kT} \equiv R_2(\epsilon_s, \epsilon_v).$$

(4.108b)

Substituting (4.108a) or (4.108b) into Eq. (4.106), we obtain

$$\sigma(\epsilon_s) + R_1(\epsilon_s, \epsilon_v) = 0$$

(4.109a)

or

$$\sigma(\epsilon_s) + R_2(\epsilon_s, \epsilon_v) = 0.$$

(4.109b)

respectively. Let us denote the roots of Eq. (4.109a) and Eq. (4.109b) by ϵ_{s1} and
ϵ_{s2}, respectively. Obviously, ϵ_{s1} and ϵ_{s2} represent the positions of the Fermi level
at the surface of the crystal, respectively, for a uniform distribution of the impurity
and for the distribution that takes into account the redistribution caused by the
surface charge. The system attains equilibrium when ϵ_s becomes ϵ_{s2} (and not ϵ_{s1}).

Starting from (4.108a) and (4.108b), we can show (see [61]) that

$$\epsilon_{s1} < \epsilon_{s2} < \epsilon_v, \quad \text{if} \quad \sigma > 0,$$

$$\epsilon_{s1} = \epsilon_{s2} = \epsilon_v, \quad \text{if} \quad \sigma = 0,$$

$$\epsilon_{s1} > \epsilon_{s2} > \epsilon_v, \quad \text{if} \quad \sigma < 0.$$

(4.110)

Note that the same result can be obtained for a semiconductor with an acceptor
impurity instead of a donor impurity (a p-type semiconductor).

Thus, redistribution of the impurity leads, as we have seen, to a mild straightening of the energy bands both when they are bent upward ($\sigma < 0$) and when they are bent downward ($\sigma > 0$). The reason for this is obvious. Indeed, whereas under a uniform impurity distribution the entire space charge in the surface layer is due solely to the free electrons and holes, when the charge is redistributed, the surface layer proves to be depleted or enriched by the impurity, and the impurity proper contributes to the space charge of the surface layer.

Obviously, allowing for redistribution of the impurity changes the adsorptivity of the surface. Let us assume, as usual, that the surface charge σ consists of two parts, the adsorption charge σ_A and the intrinsic charge σ_B, i.e., $\sigma = \sigma_A + \sigma_B$. First we assume that $|\sigma_A| \ll |\sigma_B|$. The deviation from the uniform distribution caused by σ_B leads to greater adsorptivity with respect to an acceptor gas and to lesser adsorptivity with respect to a donor gas if $\sigma_B < 0$. But if $\sigma_B > 0$, the adsorptivity with respect to an acceptor gas proves to be less, and that with respect to a donor gas more.

Now let us assume that $|\sigma_B| \ll |\sigma_A|$. In this case redistribution of the impurity takes place in the adsorption process itself. From (4.110) it follows that the redistribution concomitant to the adsorption of an acceptor gas ($\sigma < 0$) moves the Fermi level upward, while redistribution concomitant to the adsorption of a donor gas ($\sigma > 0$) moves it downward. Thus, in both cases (irrespective of the nature of the gas), the redistribution of the impurity that takes place in adsorption increases the adsorptivity of the surface. Obviously, the effect is stronger the greater the "strong" fraction of adsorption. Estimates show that in some cases there is a severalfold increase in adsorptivity. We see that redistribution of an impurity in a semiconductor accompanying adsorption may lead to a significant increase in adsorption, which in turn slows down the advance toward adsorption equilibrium.

Straightening of the energy bands due to impurity redistribution must affect the dependence of the surface conductivity κ_s of the sample on σ, making this dependence weaker. Indeed, as we know (see Section 4.2.2), the surface conductivity κ_s changes with σ, all other conditions being the same. This is the essence of the mechanism by which the adsorption acts on electrical conduction (see Section 4.2.3). We will assume that as σ varies from $\sigma = \sigma_0$ to $\sigma = \sigma_0 + \Delta\sigma$, the surface conductivity of the sample, κ_s, varies from $\kappa_s = \kappa_{s0}$ to $\kappa_s = \kappa_{s0} + \Delta\kappa_s$. For a given value of $\Delta\sigma$ the value of $\Delta\kappa_s$ will be smaller (in absolute value) when the variation in the impurity distribution follows the variation in σ rather than when the impurity distribution remains unaltered. Thus, allowing for redistribution of the impurity in the adsorption process that accompanies the redistribution leads to a weakening of the effect of adsorption on the adsorbent electrical conductivity.

A final remark is in order here. Under an equilibrium impurity distribution in a semiconductor, which establishes itself for a given σ or V_s, Poisson's equation (4.97), as we saw in Section 4.5.2, may be integrated completely (in the case of the Fermi statistics on local levels), i.e., we can find the behavior of the potential V with x in the surface layer [see (4.23) or (4.103)]. But if the distribution is uniform, then (see Section 4.1.2) this integration may be completed only for the Boltzmann statistics on local levels or in the case of an intrinsic semiconductor.

4.5.4. Irreversible Adsorption

The redistribution of an impurity in a semiconductor occurring during adsorption may in certain conditions be accompanied by ejection of the impurity to the surface and subsequent reaction with the chemisorbate. This effect has been repeatedly cited by Garner [63], Barry and Stone [64], and other authors. Thus, an impurity in a semiconductor may affect the chemisorption process not only by shifting the electronic equilibrium but in a more direct way, i.e., by reacting chemically with the chemisorbed particles.

Let us take the example of chemisorption of an acceptor gas R on a crystal $M^{++}R^{--}$ that contains the metal M stoichiometrically in excess. Let us assume that the donor impurity M, which we take to be completely ionized (M^+), when ejected to the crystal surface reacts with the charged chemisorbed atoms R^- according to the scheme

$$M^{++} + R^- \rightarrow M^{++}R^{--}. \tag{4.111}$$

An example of such a reaction is the adsorption of oxygen on zinc oxide.

The absorption of the gas via reaction (4.111) is not a chemisorption process in the proper sense of the word. It represents completion of the lattice, a process that must be considered irreversible. Here we will examine how this process acts on the chemisorption kinetics and on the partial irreversibility of chemisorption [65, 66].

Just as we did before, we will denote the surface concentrations of the neutral and charged forms of chemisorption by N^0 and N^-, the number of gas particles absorbed via the reaction (4.111) per unit surface area by N_L, and the total number of absorbed gas particles by N, so that

$$N = N^0 + N^- + N_L.$$

In the absence of reaction (4.111) the adsorption kinetics is described by the system of equations (3.19). With (4.111) the system of equations is

$$\frac{dN^0}{dt} = \frac{N_\infty^0 - N^0}{\tau} - \left(\frac{N^0}{\tau^0} - \frac{N^-}{\tau^-}\right), \tag{4.112a}$$

$$\frac{dN^-}{dt} = \left(\frac{N^0}{\tau^0} - \frac{N^-}{\tau^-}\right) - \frac{N^-}{\tau_L}, \tag{4.112b}$$

$$\frac{dN_L}{dt} = \frac{N^-}{\tau_L}, \tag{4.112c}$$

with the initial condition

$$N^0 = N^- = N_L = 0 \qquad \text{at} \quad t = 0.$$

Here $1/\tau_L$ is the probability of the particle R^- being absorbed by the impurity per unit time. The meaning of the other notations is the same as in Section 3.3.

The gas absorption rate accompanying reaction (4.111) may be limited by any of the following factors: the input of component R^- (chemisorption), the input of component M^+, and finally the chemical transformation (4.111). We will assume that it is the last factor that limits the absorption rate. It can be demonstrated that this happens when

$$\tau_2 \ll \tau_L, \tag{4.113}$$

where τ_2 has the form (3.34):

$$\tau_2 = \frac{1/\tau + 1/\tau^0 + 1/\tau^-}{(1/\tau)(1/\tau^-)}. \tag{4.114}$$

If condition (4.113) is met, then $\tau^- \ll \tau_L$ [according to (4.114)], which justifies our neglecting the third term on the right-hand side of Eq. (4.112b). Equations (4.112a) and (4.112b) transform into Eqs. (3.19), which enable finding $N^0 = N^0(t)$ and $N^- = N^-(t)$ (which is what we did in Section 3.3).

It can be demonstrated that at $0 \leq t \leq \tau_2$ we have

$$N_L(t) \ll N^0(t), \quad N_L(t) \ll N^-(t),$$

i.e., allowing for reaction (4.111) has practically no effect on the chemisorption kinetics, namely, chemisorption proceeds in the same manner as without reaction (4.111) (see Section 3.3). At $\tau_2 \ll t$ the concentrations of the neutral and charged forms of chemisorption attain their steady-state values, which for all practical purposes coincide with the equilibrium values in the absence of reaction (4.111):

$$N^0 = N^0_\infty, \quad N^- = N^-_\infty. \tag{4.115}$$

Chemisorption, then, does not influence the charge of the surface any more, and the absorption of the gas is due solely to reaction (4.111). If, in addition, we assume that the impurity content is high and remains practically constant, i.e., the impurity is not depleted in reaction (4.111), then, as we can show,

$$\tau_L = \text{const.} \tag{4.116}$$

On the basis of (4.115) and (4.116) we can rewrite Eq. (4.112c) as follows:

$$N_L(t) = \frac{N^-_\infty}{\tau_L} t.$$

In desorption all neutral and charged chemisorbed particles finally leave the surface, while the particles that have entered into reaction (4.111) are left in the crystal.

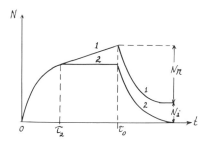

Fig. 4.9. Kinetics of adsorption and desorption.

Figure 4.9 shows the kinetics of adsorption ($t < t_0$) and desorption ($t_0 < t$) with (curve 1) and without (curve 2) surface reactions of the type (4.111). If desorption proceeds with electronic equilibrium, the irreversible fraction of adsorption, N_i, is the amount of gas that has reacted with the impurity (N_r in Fig. 4.9 stands for the surface concentration of reversibly chemisorbed particles). This is the so-called true irreversibility mentioned in Section 3.4.3. If the electronic equilibrium is violated, then on the true irreversibility there may be superimposed the so-called apparent irreversibility, which is caused by the hindered discharge of the charged form of chemisorption (see Section 3.3.3).

The chemisorption kinetics affects the kinetics of electrical conduction and the work function. If the behavior of adsorption and desorption is studied against the variations of electrical conductivity κ and the work function φ_T by comparing the N vs. t, κ vs. t, and φ_T vs. t curves, true irreversibility can be separated from apparent irreversibility. Indeed, in the case of a truly irreversible form of chemisorption both the electrical conductivity and the work functions must return to their initial values (curves 1 in Fig. 4.10). Here both neutral and charged particles are removed from the surface during desorption, while the absorption of the charged particles by impurity is equivalent, from the standpoint of their influence on the electrical conductivity and the work function, to their leaving the surface. But in the case of apparent irreversibility in contrast to true irreversibility, there must be complete irreversibility in electrical conductivity and only partial reversibility in the work function, as shown in Fig. 4.10 (curves 2). Indeed, according to (4.40),

$$\Delta\varphi_T = \Delta V_s + \Delta V_D, \tag{4.117}$$

i.e., both the neutral chemisorbed particles, which leave the surface in desorption and are reflected by the term ΔV_D in (4.117), and the charged chemisorbed particles, which remain on the surface in desorption and are reflected by the term ΔV_s in (4.117), contribute to the change in the work function caused by adsorption. On the other hand, the change in electrical conductivity is caused by the charged (and only the charged) form of chemisorption and, therefore, must remain totally irreversible.

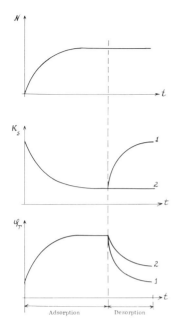

Fig. 4.10. Separation of true irreversibility from apparent irreversibility.

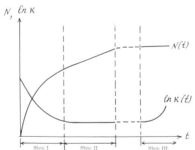

Fig. 4.11. Electrical conductivity and adsorption of oxygen on zinc oxide as functions of time.

Apparently, Glemza and Kokes [67] dealt with true irreversibility when they measured the time dependence of the electrical conductivity and the kinetics of adsorption, $N = N(t)$, of oxygen on zinc oxide. Figure 4.11 presents a sketch of the picture they observed when oxygen was admitted (at 353°C). At the first stage (step I in Fig. 4.11) there is chemisorption in the proper sense of the word, and the concentrations of the neutral and charged forms reach their steady-state values, which corresponds to the electrical conductivity reaching its steady-state value. At the second stage (step II in Fig. 4.11) gas absorption continues only by extracting the impurity from the semiconductor's bulk. When the admitted oxygen is spent, there is no more absorption of oxygen, obviously, but the im-

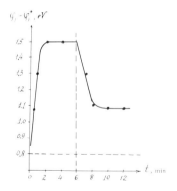

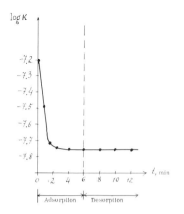

Fig. 4.12. Electrical conductivity and work function in the adsorption and desorption of oxygen on titanium dioxide at room temperature.

purity continues to absorb charged chemisorbed particles. This is accompanied by a decrease (in absolute value) in the surface charge, which sends the electrical conductivity to its initial value (step III in Fig. 4.11).

Figurovskaya, Kiselev, and Wolkenstein [70] dealt with apparent irreversibility when they studied the time dependence of the work function φ_T and of the electrical conductivity κ in the adsorption and desorption of oxygen on titanium dioxide at room temperature. They observed partial reversibility in the work function and complete irreversibility in electrical conductivity. The experimental curves are presented in Fig. 4.12 taken from [68] (cf. curves 1 in Fig. 4.10; in Fig. 4.12 φ_T* stands for the work function of the reference gold electrode).

Apparent irreversibility was dealt with in the work of Derlyukova [69]. In complete agreement with the theory she showed (see Section 3.4.3) that the irreversible form of chemisorption may be identified with the charged form, while the reversible form may be identified with the neutral form. Derlyukova studied the chemisorption (and its effect on electrical conductivity) of oxygen (acceptor) and various hydrocarbons (donors) on SnO_2, Cr_2O_3, V_2O_5, and other oxide catalysts.

4.6. THE ADSORPTIVITY OF SEMICONDUCTOR FILMS ON METALS

4.6.1. The Variation of the Potential in the Film

It is a well-known fact that many metals are often covered with a film of a binary compound, as if they were dressed in a semiconductor jacket. Consequently, processes that we assume to take place on the surface of the metal actually take place at the surface of a semiconductor. This fact is often noted (e.g., [50]). If the film thickness is small compared with the screening length, then the chemisorptive properties of the film (as well as its catalytic properties; see Chapter 5) must depend on its thickness and the properties of the substrate metal. Let us study the laws that govern this dependence [51].

We will take a metal covered with a plane-parallel film of a homogeneous semiconductor that in general has both donor and acceptor impurities. We will restrict our discussion to the idealized case where the impurity is distributed uniformly over the volume of the film. Suppose that the half-space $x < 0$ is occupied by the metal, the half-space $x > L$ by the gaseous phase, and the region $0 \leq x \leq L$ by the semiconductor film. On the free surface of the film there is gas chemisorption. We will assume that there are no surface levels at the metal–semiconductor interface, while on the outer surface of the semiconductor there are surface levels of, generally speaking, both adsorption and intrinsic nature, which result in a surface charge σ_L appearing on the outer surface of the semiconductor film. As a result of contact with the semiconductor film the metal acquires a charge, which we denote by σ_0.

The position of the Fermi level inside the film will be described by the distance of the Fermi level, ϵ, from the bottom of the conduction band. Obviously ϵ depends on x (where x may vary between 0 and L). We introduce the following notation (see Fig. 4.13, where the horizontal axis is assumed to coincide with the Fermi level):

$$\epsilon_0 = \epsilon(0), \quad \epsilon_L = \epsilon(L).$$

These quantities are (to within constant terms) respectively the work function of an electron leaving the metal and the work function of an electron leaving the outer surface of the film. In addition we will write ϵ_v for ϵ_L when $L \gg l$ and $\sigma_L = 0$ (the case where the metal is covered with a thick layer of semiconductor, which has no surface levels; see Fig. 4.14). Obviously [see (4.6)],

$$V(x) = \epsilon(x) - \epsilon_v \tag{4.118}$$

is the potential energy of the electron inside the film in the field created by the surface charges σ_L and σ_0. The quantity

$$\frac{V_0}{e} = \frac{\epsilon_0 - \epsilon_v}{e},$$

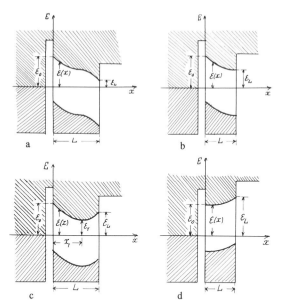

Fig. 4.13. Position of the Fermi level vs. the distance from the interface between the gaseous phase and the metal (L is the thickness of the film): a) positively charged surface, negatively charged metal, and not thin films; b) positively charged surface, negatively charged metal, and thin film; c) negatively charged surface, negatively charged metal, and not thin film; d) negatively charged surface, positively charged metal, and thin film.

where $V_0 = V(0)$, is the contact potential difference between the metal and the semiconductor.

Let us study the behavior of the potential with x, i.e., the function (4.118). We can gain some insight by studying Poisson's equation [see (4.5)]

$$\frac{d^2 V}{dx^2} = \frac{4\pi e}{\chi} \rho(V). \tag{4.119}$$

As the boundary conditions we take the following:

$$\left(\frac{dV}{dx}\right)_{x=L} = -\frac{4\pi e}{\chi} \sigma_L, \qquad V(0) = V_0. \tag{4.120}$$

Equations (4.119) and (4.120) provide a complete system of equations for finding the function $V = V(x)$. We will study this function only qualitatively.

Note that the condition of the electrical neutrality of the system as a whole (i.e., the metal together with the film), which has the form

$$\int_0^L \rho(x)\,dx + \sigma_0 + \sigma_L = 0, \tag{4.121}$$

yields

$$\left(\frac{dV}{dx}\right)_{x=0} = \frac{4\pi e}{\chi}\,\sigma_0. \tag{4.122}$$

Indeed, integrating Poisson's equation (4.119) from 0 to L, we obtain

$$\int_0^L \rho(x)\,dx = \left(\frac{dV}{dx}\right)_{x=L} - \left(\frac{dV}{dx}\right)_{x=0} \tag{4.123}$$

This, together with (4.120) and (4.121), yields (4.122). Note further that

$$\rho(0) \geqslant 0, \text{ and consequently, } \left(\frac{d^2V}{dx^2}\right)_{x=0} \geqslant 0, \quad \text{if} \quad V_0 \geqslant 0,$$

$$\rho(0) \leqslant 0, \text{ and consequently, } \left(\frac{d^2V}{dx^2}\right)_{x=0} \leqslant 0, \quad \text{if} \quad V_0 \leqslant 0. \tag{4.124}$$

Note, finally, that the higher V is, i.e., the lower the position of the Fermi level, the greater the algebraic value of ρ, i.e.,

$$\frac{d\rho}{dV} > 0. \tag{4.125}$$

It must be noted, however, that (4.125) is valid only if the impurity is uniformly distributed in the semiconductor, and below we will assume such uniform distribution.

Let us show that there cannot be more than one extremum on the V vs. x curve in the interval from 0 to L. Suppose that there are two extrema, a maximum at $x = x_1$ and a minimum at $x = x_2$, i.e.,

$$V(x_1) = V_{max}, \qquad V(x_2) = V_{min},$$

where, obviously,

$$V_{max} > V_{min}.$$

Then, according to (4.119), we have

$$\rho(V_{max}) < 0, \qquad \rho(V_{min}) > 0,$$

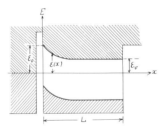

Fig. 4.14. Position of the Fermi level vs. the distance from the interface between the gaseous phase and the metal (the thickness of the film is large and the semiconductor has no surface levels).

and, consequently,

$$\rho(V_{\max}) < \rho(V_{\min}),$$

which contradicts (4.125).

Now let us show that the only possible extremum on the V vs. x curve when V_0 is nonnegative is a minimum. Suppose that this is not so, i.e., at a point $x = x_0$ the curve passes through a maximum:

$$V(x_0) = V_{\max}.$$

Then, obviously, according to (4.119),

$$V_{\max} > V_0, \quad \rho(V_{\max}) < 0$$

and, hence, according to (4.125),

$$\rho(V_0) < \rho(V_{\max}) < 0,$$

which contradicts (4.124). Similarly, we can show that when $V_0 \leq 0$, the only possible extremum on the V vs. x curve is a maximum.

In Sections 4.6.1, 4.6.2, and 4.6.3 we will assume, for the sake of definiteness, that V_0 is nonnegative. The curves (4.118) for this case are depicted in Fig. 4.13 (the heavy lines). As we see, the four types of curves in this figure exhaust all the possible cases. On the basis of (4.120) and (4.122) we can conclude that Figs. 4.13a and 4.13b refer to a positively charged surface ($\sigma_L > 0$) and Figs. 4.13c and 4.13d to a negatively charged surface ($\sigma_L < 0$); in the case depicted in Figs. 4.13a, 4.13b, and 4.13c the metal is charged negatively in relation to the film ($\sigma_0 < 0$; electrons are moved from film to metal), while in the case depicted in Fig. 4.13d the metal is charged positively in relation to the film ($\sigma_0 > 0$; electrons are moved from metal to film). Subsequently we will show that Figs. 4.13a and 4.13c correspond to a not very thin film and Figs. 4.13b and 4.13d to a thin film.* For comparison we show Fig. 4.14, where the curve (4.118) is depicted for the case of a thick film ($L \gg l$), which has no surface states.

4.6.2. The Adsorptivity of the Film for a Positively Charged Surface

Let us study the dependence of the film's adsorptivity on thickness L and work function ϵ_0 of the substrate metal. Since adsorptivity is determined by the position of the Fermi level at the surface of the crystal, the problem is reduced to finding the position of the Fermi level at the outer surface of the film, ϵ_L, as a function of L and ϵ_0. In other words, we must study, according to (4.118), the function $V_L = V_L(L, V_0)$.

We start with the case where the function $V = V(x)$ is monotonic (Figs. 4.13a, 4.13b, and 4.13d). Integration of Poisson's equation over the entire length of the film yields, as can easily be shown if we use (4.120) and (4.122), the following:

$$\left[2 \int_{V_L}^{V_0} \rho(V)dV + \left(\frac{4\pi e}{\chi} \sigma_L \right)^2 \right]^{1/2} = \mp \frac{4\pi e}{\chi} \sigma_0, \tag{4.126a}$$

$$\int_{V_L}^{V_0} \left[2 \int_{V_L}^{V} \rho(V)dV + \left(\frac{4\pi e}{\chi} \sigma_L \right)^2 \right]^{-1/2} dV = \pm L, \tag{4.126b}$$

where the upper sign corresponds to $\sigma_L > 0$ (Figs. 4.13a and 4.13b) and the lower sign to $\sigma_L < 0$ (Fig. 4.13d).

Now let us turn to the case where the V vs. x curve has a minimum at point $x = x_1$, with $0 \leq x_1 \leq L$ (See Fig. 4.13c). In this case Poisson's equation (4.119) yields

$$[2 \int_{V_1}^{V_0} \rho(V)dV]^{1/2} = - \frac{4\pi e}{\chi} \sigma_0,$$

$$[2 \int_{V_1}^{V_L} \rho(V)dV]^{1/2} = - \frac{4\pi e}{\chi} \sigma_L,$$

$$\int_{V_L}^{V_0} [2 \int_{V_1}^{V} \rho(V)dV]^{-1/2} dV = x_1, \tag{4.127}$$

$$\int_{V_1}^{V_L} [2 \int_{V_1}^{V} \rho(V)dV]^{-1/2} dV = L - x_1,$$

---

*The behavior of the potential depicted in Fig. 4.13a for $V_0 - V_L \ll kT$, where $V_L = V(L)$, was calculated by Butler [71].

where $V_1 = V(x_1)$. Differentiating the left-hand and right-hand sides of Eqs. (4.126a), (4.126b), and (4.127) with respect to L and V_0 (differentiation with respect to a parameter) and investigating the resulting derivatives, we arrive at the following results [70]:

for Fig. 4.13a, b

$$\left(\frac{dV_L}{dL}\right)_{V_0} < 0, \qquad \left(\frac{dV_L}{dV_0}\right)_L > 0,$$

for Fig. 4.13c

$$\left(\frac{dV_L}{dL}\right)_{V_0} < 0, \qquad \left(\frac{dV_L}{dV_0}\right)_L > 0,$$

$$\left(\frac{dV_1}{dL}\right)_{V_0} < 0, \qquad \left(\frac{dV_1}{dV_0}\right)_L > 0, \qquad (4.128)$$

for Fig. 4.13d

$$\left(\frac{dV_L}{dL}\right)_{V_0} > 0, \qquad \left(\frac{dV_L}{dV_0}\right)_L > 0.$$

Moreover, taking into account the fact that

$$\frac{d\sigma_L}{dL} = \frac{d\sigma_L}{dV_L}\frac{dV_L}{dL}$$

and that always

$$\frac{d\sigma_L}{dV_L} > 0, \qquad (4.129)$$

we arrive at the following results [70]:

for Fig. 4.13a

$$\sigma_L > 0, \qquad \left(\frac{d\sigma_L}{dL}\right)_{V_0} < 0,$$

$$\sigma_0 < 0, \qquad \left(\frac{d\sigma_0}{dL}\right)_{V_0} > 0, \qquad (4.130a)$$

for Fig. 4.13c

$$\sigma_L < 0, \qquad \left(\frac{d\sigma_L}{dL}\right)_{V_0} < 0,$$

$$\sigma_0 < 0, \qquad \left(\frac{d\sigma_0}{dL}\right)_{V_0} < 0, \qquad (4.130b)$$

for Fig. 4.13d

$$\sigma_{L'} < 0, \qquad \left(\frac{d\sigma_L}{dL}\right)_{V_0} > 0,$$

$$\sigma_0 > 0, \qquad \left(\frac{d\sigma_0}{dL}\right)_{V_0} < 0.$$

(4.130c)

On the basis of (4.128) we can estimate the behavior of V_L with L. Two cases are present here: $\sigma_L > 0$ (Figs. 4.13a and 4.13b) and $\sigma_L < 0$ (Figs. 4.13c and 4.13d). Let us study the case where $\sigma_L > 0$ (see Section 4.6.3 for the case where $\sigma_L < 0$).

If the film is not too thin (see Fig. 4.13a), then

$$V_L^0 < V_L < 0, \qquad \rho(V_L) < 0,$$

where V_L^0 is the root of the equation

$$\sigma_L(V_L^0) = 0.$$

(4.131)

Obviously, the value of V_L^0 is fixed by the system of surface levels on the outer surface of the film, i.e., it depends on the nature of the surface and the partial pressures of the gases in contact with the surface.

As the film gets thinner, V_L increases, as can be seen from (4.128), and approaches $V_L = 0$, i.e., the Fermi level on the film surface moves downward (the work function increases). In the process $\rho(V_L)$ grows [see (4.125)]. According to (4.130), σ_L grows, too (remaining positive), while σ_0 drops (remaining negative); i.e., electrons are "pumped" from the outer surface of the film into the bulk of the film and from there into the metal.

We will denote the thickness of the film at which V_L and $\rho(V_L)$ vanish, i.e., $V_L = 0$ and $\rho(0) = 0$, by l_1. The point of inflection on the ϵ vs. x curve (Fig. 4.13a), for this thickness, reaches the outer boundary of the film, and we transform from Fig. 4.13a to Fig. 4.13b. Obviously, the critical thickness l_1 at which this happens can be found from Eq. (4.126b) in which we put $V_L = 0$ and $L = l_1$.

As the thickness of the film further increases ($L < l_1$; see Fig. 4.13b), we have

$$V_L^0 < 0 < V_L, \qquad \rho(V_L) > 0.$$

In the process V_L continues to grow, as we can see from (4.128), and the Fermi level continues to move downward (the work function continues to increase). In the limiting case where $L = 0$ we have $V_L = V_0$ and, according to (4.126a), $\sigma_0 = -\sigma_L$ (but here $\sigma_0 = -\sigma_L \neq 0$ since there remain surface states on the surface of the metal injected by the film).

The behavior of V_L with the film thickness L at $\sigma_L > 0$ is sketched in Fig. 4.15a. Different curves correspond to different values of V_0. The critical thickness l_1 at which we must go over from Fig. 4.13a to Fig. 4.13b is determined from the intersection point of the V_L vs. L curve with the straight line $V_L = 0$.

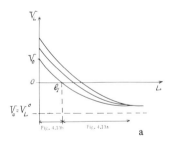

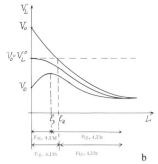

Fig. 4.15. Potential energy of the electron as a function of film thickness: a) for positive σ_L; b) for negative σ_L.

Now let us turn to the adsorptivity of the film. Let us assume that the outer surface of the film has a high density of surface state of nonadsorptive nature. In this case the adsorption of a gas will lead to a change in the absolute value of the surface charge but not in its sign. In other words, the sign of σ_L will be determined not by the nature of the adsorbed gas but by the prehistory of the adsorbing surface.

We recall that as the Fermi level becomes lower, the adsorptivity of a semiconductor surface with respect to a donor gas grows and that with respect to an acceptor gas drops (see Section 3.2.1). This enables us to conclude that the adsorptivity of the surface with respect to a donor gas steadily increases and that with respect to an acceptor gas steadily decreases as the film becomes thinner (for $\sigma_L > 0$).

Let us take the opposite limiting case, i.e., when the outer surface of the film has no nonadsorptive surface states. In this case the sign of the surface charge is determined solely by the nature of the adsorbed gas. Suppose that adsorption of an acceptor gas provides the positive charge on the surface. Obviously, the adsorptivity of the film with respect to this gas will monotonically increase as the film gets thinner.

4.6.3. The Adsorptivity of the Film for a Negatively Charged Surface

We have studied the dependence of V_L on L when the outer surface of the film is positively charged ($\sigma_L > 0$). Now we turn to the case where this charge is negative ($\sigma_L < 0$). In this case we have

$$0 < V_L < V_L^0, \qquad \rho(V_L) > 0,$$

where V_L^0 is found by solving Eq. (4.131). If the film is not too thin (Fig. 4.13c), then, as the film get thinner, V_L increases, just as when σ_L was positive, which is seen from (4.128), and, hence, the Fermi level moves downward (the work function increases). The behavior of the film in this case differs in the two cases defined below, which depend on the nature of the metal, the prehistory of the surface, and the pressure and composition of the gaseous phase, namely,

(a) when $V_L^0 < 0$,
(b) when $V_0 < V_L^0$.

We will consider each case separately. (Note that if the surface has levels only of the acceptor type, then $V_L^0 = \infty$ and, consequently, case (a) is not realized.)

(a) If $0 < V_L < V_L^0 < V_0$, then, as the film gets thinner, V_L increases and tends to the value V_L^0. It follows from (4.130), (4.127), and (4.131) that here $\sigma_0 < \sigma_L < 0$, with σ_0 and σ_L increasing (as the film gets thinner) and σ_L tending to zero. From (4.127) and (4.130) it also follows that at the minimum point x_1 on the ϵ vs. x curve in Fig. 4.13c we have $L/2 < x_1 < L$, and as V_L approaches V_L^0, the minimum in Fig. 4.13c shifts upward and to the right, approaching the surface of the film.

The thickness of the film at which V_L becomes V_L^0 and σ_L vanishes we denote by l_2. At this thickness the minimum point in Fig. 4.13c reaches the outer boundary of the film ($x_1 = L$ and $V_1 = V_L$). If the film is made still thinner ($L < l_2$), σ_L changes sign, and from Fig. 4.13c we have to go back to 4.13b. Obviously, the critical length l_2 can be found from Eqs. (4.127), where we put $V_1 = V_L$ and $L = l_2$.

(b) If $0 < V_L < V_0 < V_L^0$, then, as the film gets thinner, V_L increases and approaches the value V_0. Here, just as in the previous case, $\sigma_0 < \sigma_L < 0$, with σ_0 and σ_L increasing and approaching each other. For the minimum point on the ϵ vs. x curve we have $L/2 < x_1 < L$, and as V_L approaches V_0, the minimum point in Fig. 4.13c moves upward and to the left, approaching the center of the film, as it should according to (4.128) and (4.127). When V_L becomes equal to V_0 and σ_0 to σ_L, we have $x_1 = L/2$ [see (4.127)]; i.e., the minimum point proves to be at the center of the film.

As L decreases still further, we have $0 < V_0 < V_L < V_L^0$, and V_L continues to grow. Here, as follows from (4.130) and (4.127), $\sigma_L < \sigma_0 < 0$, with σ_0 and σ_L continuing to increase. For this minimum point in Fig. 4.13c we have $0 < x_1 < L/2$, and this point continues to move upward and to the left, which follows directly from (4.128) and (4.127).

The thickness of the film at which the minimum point in Fig. 4.13c reaches the film's inner boundary, i.e., where x_1 vanishes and $V_1 = V_0$, we denote by l_3. Here $\sigma_0 = 0$ [as follows from (4.127)]; i.e., the metal proves to be electrically neutral (the positive charge of the film's bulk is counterbalanced by the negative surface charge on the outer surface of the film). At $L = l_3$ we go from Fig. 4.13c to Fig. 4.13d, with V_L attaining its maximum value V_L^*, where $V_0 < V_L^* < V_L^0$. Obviously, l_3 and V_L^* can be found from Eqs. (4.127) if we put $V_1 = 0$ and $L = l_3$.

If the film becomes still thinner ($L < l_3$; see Fig. 4.13d), we have $0 < V_0 < V_1 < V_L^0$, with V_L beginning to drop, as is seen from (4.128), and the Fermi level, consequently, beginning to move upward (the work function drops). Here, as follows from (4.130), $\sigma_L < 0 < \sigma_0$, with σ_L decreasing and σ_0 increasing. Thus, the metal is charged positively with respect to the film; i.e., the metal gives off electrons to the film rather than acquires them, and this effect is the greater the thinner the film. In the limiting case of $L = 0$ we have $V_L = V_0$ and $\sigma_0 = -\sigma_L$ (with $\sigma_0 = -\sigma_L \neq 0$).

We see that as the thickness of the film steadily decreases, V_L passes through a maximum whose physical meaning is as follows: a film that is not too thin ($L > l_3$) gives off electrons from its bulk to its outer surface (causing the latter to be charged negatively) and to the metal, while if a film is thin ($L < l_3$), its outer surface draws electrons from both the film's bulk and the metal.

Figure 4.15b gives a rough sketch of the function $V_L = V_L(L)$ for various values of V_0 (when V_L^0 is smaller than V_0 and when it is larger). The critical thickness l_2 at which we must go over from Fig. 4.13c to Fig. 4.13b is determined by the intersection point of the V_L vs. L curve with the straight line $V_L = V_L^0$. The critical length l_3 at which we must go over from Fig. 4.13c to Fig. 4.13d is determined by the position of the maximum on the V_L vs. L curve.

We will now discuss the adsorptivity of the film. Let us first assume that the charge on the outer surface of the film is intrinsic. If, in addition, $V_L^0 < V_0$, then the adsorptivity of the film with respect to a donor gas will steadily grow as the film gets thinner, while the adsorptivity of the film with respect to an acceptor gas will steadily drop, just as in the case where $\sigma_L > 0$ (see Section 4.6.2.). But if $V_0 < V_L^0$, then, as the film gets thinner, its adsorptivity with respect to a donor gas will pass to a minimum.

Now let us assume that adsorption of an acceptor gas causes the surface of the film to be charged negatively. In this case the adsorptivity of the film with respect to this gas passes through a minimum, as the film gets thinner, and then, as the film gets still thinner, increases.

We have thus examined the dependence of the adsorptivity of a film on its thickness L for both $\sigma_L > 0$ and $\sigma_L < 0$. As to the dependence of the adsorptivity on V_0, in both of these cases it has the same shape, as shown by (4.128): for a given film (its nature and thickness L, where $L < l$) its adsorptivity with respect to a donor gas will be greater (and with respect to an acceptor gas, smaller) the greater the value of V_0, i.e., the higher the value of the work function of the metal that lies under the film, irrespective of the sign of the surface σ_L on the outer surface of the film.

In conclusion we note that not only the chemisorptive but the catalytic properties of the film as well prove to depend on the thickness of the film and the nature of the substrate metal. We will return to this problem in Section 5.5.3.

4.7. GROWTH OF A SEMICONDUCTOR FILM ON A METAL*

4.7.1. Statement of the Problem

Let us consider a metal that is in a gaseous atmosphere that contains oxygen. Note that the following discussion refers also to cases where instead of oxygen we use any other gas that possesses electron affinity and is capable of forming compounds with metals (chlorine, fluorine, sulfur, and other substances). In the process, a film of a binary semiconductor is formed on the surface of the metal, and the thickness L of this film gradually increases with the passage of time t:

$$L = L(t).$$

This relation will be called the law of film growth. Experimental data shows that this law is different for different film thicknesses and in different temperature intervals. The more common laws of film growth are

(I) the logarithmic law $L = A \ln(t/\tau + 1)$,
(II) the parabolic law $L = B\sqrt{t}$,
(III) the linear law $L = Ct$.

A "collective" curve representing the experimental data can be found in [72]. The problem of finding the right law of film growth is central to the theory of corrosion. In this section we will discuss one possible mechanism of film growth. For different limiting cases this mechanism leads to different laws of film growth [73].

We will start from an assumption opposite to the one made in the well-known works of Mott [74], Cabrera and Mott [75], and Fromhold [76]. The authors of [74–76] assume that the limiting, i.e., slowest, oxidation stage is the reaction at the metal–oxide interface, or mass and charge transfer through the film. They completely ignore the processes on the oxide surface, e.g., oxygen adsorption. We assume that the rate-limiting reaction is that at the oxide–gas interface, i.e., the reaction in which a metallic ion is attached to a chemisorbed oxygen ion; obviously, this results in adding on new layers to the lattice. The authors of [77–80] used a similar approach. Here we will consider this problem from the standpoint of the electronic theory of chemisorption on semiconductors.

The assumption concerning the limiting role of the surface reaction at the oxide–gas interface does not mean, as it might seem at first glance, that surface coverage by chemisorbed oxygen must be low and, hence, that the partial oxygen pressure must be low. It is not the overall amount of the chemisorbed oxygen that must be small but only the fraction in the charged state, since the chemisorbed oxygen is partly in the charged state and partly in the neutral state and since only

---

*Section 4.7 was written in collaboration with E. V. Kulikova and A. I. Loskutova.

the charged atoms of the chemisorbed oxygen serve as the nucleation points of oxide growth. This condition is met not only for low pressures but for arbitrary pressures if the oxide is in a definite state, characterized by the position of the Fermi level.

We will assume, as many authors do, that the distribution of the electrons in the oxide is an equilibrium one and that there is equilibrium between the surface and the gaseous phase (adsorption equilibrium). Suppose that N^0 and N^- are equilibrium surface concentrations of chemisorbed oxygen atoms in the neutral and charged states, respectively, and N is the overall number of oxygen atoms (neutral and charged) per unit surface area of the oxide.

We will start from the equation

$$\frac{dL}{dt} = CN^-, \tag{4.132}$$

assuming that the film gets thicker due to addition of a metallic ion to a chemisorbed oxygen ion, the metallic ion being ejected from a lattice site onto the surface. The metal vacancies that form in this process leave for the oxide bulk, sooner or later reaching the metal surface or filling up en route with interstitial metallic ions, provided that such ions exist in the oxide. There is no need to go further into the mechanism of this process. Suffice it to note that the mechanism does not depend on whether the oxide film is an n- or p-type semiconductor.

The quantity N^- in Eq. (4.132) has the form (see Sections 3.1 and 3.2)

$$N^- = \eta^- N = N^* \left(\frac{1}{\eta^-} + \frac{a}{P} \frac{\eta^0}{\eta^-} \right)^{-1}, \tag{4.133}$$

where

$$a = a_0 \exp\left(-\frac{q}{kT} \right),$$

$$\eta^0 = \frac{N^0}{N} = \left[1 + \exp\left(-\frac{\epsilon - \upsilon + V_L}{kT} \right) \right]^{-1}, \tag{4.134}$$

$$\eta^- = \frac{N^-}{N} = \left[1 + \exp\left(\frac{\epsilon - \upsilon + V_L}{kT} \right) \right]^{-1}.$$

Here N^* is the surface concentration of the adsorption centers, P the partial oxygen pressure, T the absolute temperature, q the adsorption heat (the energy of the "weak" bond), a_0 a factor that slowly varies with temperature, and the meaning of the other symbols is clear from Fig. 4.16: υ and ϵ are the depths at which the local level of the chemisorbed oxygen atom and the Fermi level lie below the conduction band of an infinitely thick semiconductor, V_L is the potential energy of an electron in the plane of the surface, A the local level of chemisorbed oxygen, and FF the Fermi level. (Figure 4.16 repeats Fig. 4.13d; i.e., we restrict our discussion to a thin film and a negative charge on the outer surface.)

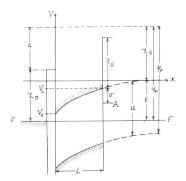

Fig. 4.16. Potential energy of the electron for negatively charged surface, positively charged metal, and thin film.

Obviously, as long as the film thickness is less than the screening length (Debye length) we will have

$$V_L = V(L). \tag{4.135}$$

For brevity we will think of V_L as the potential, as we did before, and measure it from the bottom of the conduction band of an infinitely thick semiconductor.

Substitution of (4.135) into (4.134) and then (4.134) into (4.133) yields

$$N^- = \frac{N^*}{1 + b \exp \dfrac{V(L) + \epsilon - \upsilon}{kT}} = N^-(L), \tag{4.136}$$

where

$$b = 1 + \frac{a_0}{P} \exp\left(-\frac{q}{kT}\right). \tag{4.137}$$

We see that the amount of chemisorbed oxygen in the charged state depends, for a sufficiently thin film and all other conditions remaining the same, on the film thickness L and, hence, varies with film growth. Substituting (4.136) into Eq. (4.132), we can find the law of film growth $L = L(t)$ if we know the function (4.135), i.e., the potential at the outer surface as a function of film thickness. Thus, the problem is reduced to finding the function (4.135).

4.7.2. The Electric Field in the Film

The function $V = V(x)$ was qualitatively studied in Section 4.6.1. For the case of a sufficiently thin film the behavior of V with x is sketched in Fig. 4.16, where the half space $x < 0$ is occupied by the metal, the half space $x > L$ by the gaseous phase, and the region $0 \leq x \leq L$ by the oxide film. The dashed curves correspond to an infinitely thick film. In Fig. 4.16 χ_M and χ_S are, respectively, the work function of the metal and semiconductor, φ_p and φ_T the photoelectric and thermionic work functions of the semiconductor, and ζ the contact potential difference between the semiconductor and the metal. We have (see Fig. 4.16)

$$\zeta = \epsilon - (\chi_M - \chi_S) = -V_0. \tag{4.138}$$

(Figure 4.16 refers to the case where ζ is positive.)

We can find the function $V = V(x)$ by solving Poisson's equation

$$\frac{d^2 V}{dx^2} = \gamma \rho(V), \tag{4.139}$$

where we have employed the notation $\gamma = 4\pi e^2/\kappa$, with ρ the space charge density in units of electron charge, e the absolute value of the electron charge, and κ the dielectric constant of the oxide. If the distributions of electrons and impurities in the oxide are equilibrium distributions (we will start with this assumption), we have (as shown in [61])

$$\rho(V) = 2n \sinh \frac{V}{kT}, \tag{4.140}$$

where n is the majority carrier concentration in the bulk of an infinitely thick crystal.

The boundary conditions have the form

$$V(0) = V_0 = -\zeta, \qquad \left(\frac{dV}{dx}\right)_{x=0} = \gamma \sigma_0, \tag{4.141a}$$

at the inner surface ($x = 0$), and

$$V(L) = V_L, \qquad \left(\frac{dV}{dx}\right)_{x=L} = -\gamma \sigma_L. \tag{4.141b}$$

at the outer surface ($x = L$). Here σ_L is the surface charge density on the outer side of the surface (this charge is generated by the adsorbed particles and the structural defects of the surface) and σ_0 the surface charge density on the inner surface (the charge acquired by the metal). Both σ_L and σ_0 are expressed in electron charge units. Obviously, here V_0 and σ_L are given while V_L and σ_0 must be found.

Substituting (4.140) into Eq. (4.139) and integrating the latter from $x = 0$ to $x = L$, we obtain [bearing in mind (4.141a) and (4.141b)]

$$2 \cosh \frac{V_0}{kT} - \left(\frac{\sigma_0}{\sigma}\right)^2 = 2 \cosh \frac{V_L}{kT} - \left(\frac{\sigma_L}{\sigma}\right)^2, \tag{4.142}$$

with

$$\sigma = \sqrt{\frac{\kappa kT}{2\pi e^2}} \, n. \tag{4.143}$$

If we integrate Poisson's equation (4.139) a second time from $x = 0$ to x, we find

$$\int_{V_0}^{V} \frac{dV}{\sqrt{2\left(\cosh\dfrac{V}{kT} - \cosh\dfrac{V_0}{kT}\right) + \left(\dfrac{\sigma_0}{\sigma}\right)^2}} = kT\frac{x}{l}, \tag{4.144}$$

with

$$l = \sqrt{\frac{\kappa kT}{8\pi e^2} \cdot \frac{1}{n}}. \tag{4.145}$$

the screening length.

We will restrict our discussion to the case of a surface charge so large that

$$2\left(\cosh\frac{V_0}{kT} - \cosh\frac{V}{kT}\right) \ll \left(\frac{\sigma_0}{\sigma}\right)^2. \tag{4.146}$$

This condition, which means that we neglect the space charge in comparison with the surface charge, is met, as it may be shown, only if the oxide film is thin, $L < l$. If L is greater than l, the surface charge is entirely screened by the space charge in the surface layer, i.e., the space charge is equal (in absolute value) to the surface charge.

If condition (4.146) is met, Eqs. (4.144) and (4.142) yield, respectively [using the notations (4.143) and (4.145)]

$$V_L - V_0 = \gamma\sigma_0 L, \tag{4.147}$$

$$\sigma_0 = -\sigma_L. \tag{4.148}$$

Thus, in the present approximation we are dealing with a homogeneous field in the film, which is a common assumption in theoretical works on corrosion.

Let us estimate the film thicknesses at which the condition is met. Noting that (see Fig. 4.16) $V_0 < V_L < 0$, we substitute for condition (4.146) the more stringent condition

$$2\left(\cosh\frac{V_0}{kT} - \cosh\frac{V_L}{kT}\right) \ll \left(\frac{\sigma_0}{\sigma}\right)^2. \tag{4.149}$$

Here is what this condition means. According to (4.138) and (4.148), we have

$$V_0 = -\zeta, \qquad V_L = \zeta + \gamma\sigma_0 L, \tag{4.150}$$

where, according to (4.143) and (4.145),

$$\gamma = \frac{kT}{\sigma l}. \tag{4.151}$$

Since V_L is negative, (4.150) implies that σ_0 and L must be such that

$$\gamma \sigma_0 L < \zeta \qquad \text{or} \qquad L < l \frac{\sigma}{\sigma_0} \frac{\zeta}{kT}. \tag{4.152}$$

Substituting (4.150) into (4.149), assuming that $\exp(V_L/kT) \ll 1$ and, hence, $\exp(V_0/kT) \ll 1$, and allowing for (4.151), instead of (4.149) we have

$$\exp\frac{\zeta}{kT} - \exp\left(\frac{\zeta}{kT} - \frac{\sigma_0}{\sigma}\frac{L}{l}\right) \ll \left(\frac{\sigma_0}{\sigma}\right)^2. \tag{4.153}$$

Obviously, the above condition is met at $L \ll L_0$, where

$$L_0 = l\frac{\sigma}{\sigma_0} \ln\left[1 - \left(\frac{\sigma_0}{\sigma}\right)^2 \exp\left(-\frac{\zeta}{kT}\right)\right]. \tag{4.154}$$

Now let us perform numerical estimates. We put $kT = 3 \times 10^{-2}$ eV, $\zeta = 0.15$ eV, $\kappa = 6$, $\sigma_0 = 0.5 \times 10^{12}$ cm^{-2}, and $n = 0.9 \times 10^{16}$ cm^{-3} and note that $e^2 = 14.4 \times 10^{-8}$ eV·cm. According to (4.152) and the above estimates, the condition $V_L < 0$ takes the form

$$L < 10^{-6} \text{ cm,}$$

while for L_0 (the limiting thickness of the film at which the field can still be considered homogeneous) we have, according to (4.154), the following estimate:

$$L_0 = 0.8 \cdot 10^{-6} \text{ cm.}$$

4.7.3. The Logarithmic Law of Film Growth

We know that the surface charge σ_L consists of two parts, the adsorption charge, which is produced by the chemisorbed oxygen, and the intrinsic charge, caused by the intrinsic defects of the surface. In our case, obviously, the adsorption charge is negative; let us assume, for definiteness, that the intrinsic charge, too, is negative. The absolute values of these charges (per unit surface area) expressed in units of electron charge are σ_A and σ_B. We have

$$\sigma_L = -(\sigma_A + \sigma_B), \tag{4.155}$$

where, obviously, $\sigma_A = N^-$. We will study the two limiting cases, namely,

$$\tag{4.156a}$$

$$\sigma_A \ll \sigma_B$$

and

$$\sigma_A \gg \sigma_B, \tag{4.156b}$$

for which the laws of film growth prove to be markedly different. We start with case (4.156a), or

$$N^- \ll \sigma_B. \tag{4.157}$$

We will also assume that

$$\sigma_B \leqslant N^*, \tag{4.158}$$

a condition we will use subsequently. Here N^- is a function of V_L, as shown by (4.136). Note that σ_B depends on the position of the Fermi level at the surface, i.e., is also a function of V_L. However, we will assume that σ_B is constant; i.e., the intrinsic charge of the surface does not vary in adsorption, which means that the intrinsic defects of the surface are assumed to be completely ionized.

What does condition (4.157) mean? According to (4.155) and (4.148),

$$V_L - V_0 = \gamma \sigma_B L. \tag{4.159}$$

Substituting (4.159) and (4.136) into (4.157) and allowing for (4.138), we can rewrite (4.157) as follows:

$$\frac{N^*}{\sigma_B} \ll 1 + b \exp\left\{\frac{\gamma \sigma_B L - [v - (\chi_M - \chi_S)]}{kT}\right\}. \tag{4.159a}$$

We can easily see that condition (4.159a) is met if

$$L_1 \ll L, \tag{4.160}$$

where

$$L_1 = \frac{v - (\chi_M - \chi_S)}{\gamma \sigma_B} + \frac{kT}{\gamma \sigma_B} \ln \frac{N^*}{b \sigma_B}. \tag{4.161}$$

Let us now determine the law of film growth. If we substitute (4.159) into (4.136), then (4.136) into (4.132), and allow for (4.138), we obtain

$$\frac{dL}{dt} = CN^* \left\{1 + b \exp \frac{\gamma \sigma_B L - [v - (\chi_M - \chi_S)]}{kT}\right\}^{-1},$$

or, on the basis of (4.159) and (4.158),

$$\frac{dL}{dt} = \frac{CN^*}{b} \exp\left\{-\frac{\gamma \sigma_B L - [v - (\chi_M - \chi_S)]}{kT}\right\}. \tag{4.162}$$

Integrating Eq. (4.162), we arrive at the logarithmic law of film growth:

$$L = A \ln\left(\frac{t}{\tau} + 1\right),$$ (4.163)

where

$$A = \frac{kT\kappa}{4\pi e^2 \sigma_B}, \quad \tau = \frac{A}{C}\frac{b}{N^*} \exp\left[-\frac{v - (\chi_M - \chi_S)}{kT}\right].$$ (4.164)

The applicability range for the logarithmic law (4.163) is determined by condition (4.160).

The law (4.163) is often observed in reality. It is valid for sufficiently thin films and in the low-temperature region (e.g., see [81, 82]). Various authors have suggested different mechanisms to explain this law; an analysis of these mechanisms is given in the work of Mott and Fehlner [80]. In our discussion we have considered another mechanism that leads to the logarithmic law.

4.7.4. The Parabolic and Linear Laws of Film Growth

Let us now take the case of (4.156b), i.e.,

$$\sigma_B \ll N^-,$$ (4.165)

which ensures that

$$\sigma_B \ll N^*.$$ (4.166)

We will discuss the meaning of (4.165) below.

According to (4.156b), (4.155), and (4.148), we have

$$V_L - V_0 = \gamma N^- L,$$ (4.167)

where N^- is a function of V_L [see (4.136)]. Equation (4.132) then assumes the form

$$\frac{dL}{dt} = \frac{C}{\gamma}\frac{V_L - V_0}{L},$$ (4.168)

where $V_L = V(L)$ can be found from Eq. (4.167).

How do we solve Eq. (4.167)? This equation, as (4.136) shows, is transcendental in V_L. Figure 4.17 depicts the N^- vs. V_L curve [see (4.136)] at $T > 0$ by a thin curve. We approximate this curve by a broken line (the heavy broken line in Fig. 4.17) tangent to the curve at points $V_L = \pm\infty$ and $V_L = V_L^0$. For V_L^0 we take the value of V_L corresponding to the point of inflection of the N^- vs. V_L curve, i.e.,

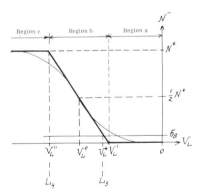

Fig. 4.17. Acceptor molecule concentration vs. potential energy of surface electrons.

$$V_L^0 = v - \epsilon - kT \ln b, \quad N^-(V_L^0) = \frac{1}{2} N^*.$$

The equation of the approximating line in region (b) (Fig. 4.17) is

$$N^- - \frac{1}{2} N^* = k(V_L - V_L^0),$$

where

$$k = \left(\frac{dN^-}{dV_L} \right)_{V_L = V_L^0} = -\frac{N^*}{4kT}. \tag{4.169}$$

Thus, we have [see (4.169) and Fig. 4.17)]

a) for $V_L' < V_L$ $N^- = 0,$

b) for $V_L'' \leqslant V_L \leqslant V_L'$ $N^- = \frac{1}{2} N^* \left(1 - \frac{V_L - V_L^0}{2kT} \right),$ (4.170)

c) for $V_L < V_L''$ $N^- = N^*,$

where, as we can easily see,

$$\begin{aligned} V_L^0 &= v - \epsilon - kT \ln b, \\ V_L' &= V_L^0 + 2kT, \\ V_L'' &= V_L^0 - 2kT. \end{aligned} \tag{4.171}$$

Let us estimate the relative error δ introduced by the approximation scheme (4.170). Obviously, the error is maximal at point $V_L = V_L''$ and, hence,

$$\delta \leqslant \frac{N_a^-(V_L'') - N^-(V_L'')}{N^-(V_L'')},$$

where $N_a^-(V_L'') = N^{**}$ is the value of N^- at point V_L'' [according to the approximation (4.170)], and $N^-(V_L'')$ is the exact value of N^- at this point given by (4.136). Substituting $V_L = V_L''$ from (4.171) into (4.136) and then (4.36) into the above estimates for δ, we obtain

$$\delta \leqslant \frac{1}{e^2} \approx 13.6\%$$

(here e is the base of the natural system of logarithms).

Let us return to condition (4.165). For this condition to be valid we must assume that V_L is restricted by the condition (see Fig. 4.17)

$$V_L < V_L^*. \tag{4.172}$$

We can find the expression for V_L^* if we substitute $N^- = \sigma_B$ and $V_L = V_L^*$ into Eq. (4.170b). This yields

$$\frac{1}{2} N^* \left(1 - \frac{V_L^* - V_L^0}{2kT}\right) = \sigma_B,$$

where, if we allow for (4.171),

$$V_L^* = V_L' - 4kT \frac{\sigma_B}{N^*}. \tag{4.173}$$

Now we go back to (4.170) and study region (b) (Fig. 4.17), assuming, in accordance with (4.170b) and (4.172), that

$$V_L'' \leqslant V_L < V_L^*. \tag{4.174}$$

We will start by explaining condition (4.174). Substituting (4.170b) into (4.167) and solving (4.167) for $(V_L - V_0)/L$, we obtain

$$\frac{V_L - V_0}{L} = \frac{B}{L_2 + L}, \tag{4.175}$$

where

$$L_2 = \frac{4kT}{\gamma N^*}, \quad B = \upsilon - (\chi_M - \chi_S) - kT (\ln b - 2). \tag{4.176}$$

If we substitute (4.175) into (4.174), we obtain

$$V_L'' - V_0 \leqslant \frac{BL}{L_2 + L} < V_L^* - V_0, \tag{4.177}$$

where, substituting (4.171) and (4.138) into (4.177), we find that

$$L_4 \leqslant L < L_3, \tag{4.178}$$

where

$$L_3 = \frac{v - (\chi_M - \chi_S)}{\gamma \sigma_B} - \frac{kT}{\gamma \sigma_B} \left(\ln b - 2 + 4 \frac{\sigma_B}{N^*} \right),$$

$$L_4 = \frac{v - (\chi_M - \chi_S)}{\gamma N^*} - \frac{kT}{\gamma N^*} (\ln b + 2), \tag{4.179}$$

and, as can easily be verified,

$$L_2 \ll L_3, \quad L_4 \ll L_3. \tag{4.180}$$

Indeed, on the basis of (4.176) and (4.179) we can write

$$L_3 = \left\{ \frac{v - (\chi_M - \chi_S)}{\gamma N^*} - \frac{kT}{\gamma N^*} (\ln b + 2) \right\} \frac{N^*}{\sigma_B} + \frac{4kT}{\gamma N^*} \left(\frac{N^*}{\sigma_B} - 1 \right) = L_4 \frac{N^*}{\sigma_B} + L_2 \left(\frac{N^*}{\sigma_B} - 1 \right),$$

where, if we allow for (4.166), for the temperature range in which $L_4 < 0$, we arrive at (4.180).

Thus, as we move along the horizontal axis in Fig. 4.17 from left to right, from point V_L'' to point V_L^*, the film thickness L increases from L_4 to L_3, and point L_2 may lie both to the left and to the right of point L_4 in Fig. 4.17.

We will now find the law of film growth assuming that

$$\begin{matrix} L_4 \ll L < L_3 & \text{for} & L_2 \leqslant L_4, \\ L_2 \ll L < L_3 & \text{for} & L_4 \leqslant L_2. \end{matrix} \tag{4.181}$$

Here, substituting (4.175) into (4.168) and integrating Eq. (4.168), we arrive at the parabolic law of growth:

$$L = \sqrt{2 \frac{BC}{\gamma} t}, \tag{4.182}$$

where $\gamma = 4\pi e^2/\kappa$, and B is defined in (4.176). Obviously, the applicability region of the parabolic law (4.182) is defined by conditions (4.181).

Now let us examine the cases where the laws of film growth are linear.

Note that, according to (4.176) and (4.179), L_2 monotonically increases and L_4 monotonically decreases as the temperature grows; i.e., as the temperature increases, the point $L = L_2$ moves to the right on the horizontal axis in Fig. 4.17 and the point $L = L_4$ to the left. We have $L_4 \leq L_2$ for $T' \leq T$ and $L_2 \leq L_4$ for $T \leq T'$, where T' is determined from the condition that $L_2(T') = L_4(T')$.

Now we turn to the high temperature region $T' \le T$ and put

$$L_4 \le L \ll L_2.$$ (4.183)

This places us in region (b) in Fig. 4.17. In this case, according to (4.175), we have

$$\frac{V_L - V_0}{L} = \frac{B}{L},$$ (4.184)

and Eq. (4.168), after we substitute (4.184) into it and integrate, yields the linear law of film growth:

$$L = \frac{B}{4kT} CN^* t,$$ (4.185)

where B is given in (4.176).

Let us now consider two cases:

1) the high-temperature region:

$$T' \le T, \quad L < L_4 \le L_2;$$ (4.186)

2) the low-temperature region:

$$T \le T', \quad L < L_2 \le L_4.$$

Both conditions in (4.186) are met if

$$V_L' < V_L'',$$ (4.187)

i.e., we have moved into region (c) in Fig. 4.17, which can easily be verified. Indeed, if condition (4.187) is met, then, according to [4.170, case (c)] and (4.167), we have

$$V_L = \gamma N^* L + V_0.$$ (4.188)

Substituting (4.188) into (4.187) and then (4.187) into (4.138) and (4.171), we arrive at $L < L_4$, or (4.186). Equation (4.132) again leads us to the linear law of growth, which, however, has a different coefficient:

$$L = CN^* t.$$ (4.189)

Obviously, the regions of applicability of linear laws (4.185) and (4.189) are determined, respectively, by conditions (4.183) and (4.186). We see that for small values of L and for any temperature the growth of film obeys a linear law.

This law is observed experimentally at early stages of metal oxidation [80].

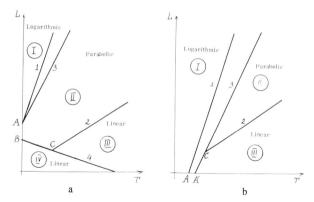

Fig. 4.18. Laws of film growth in different temperature and film thickness regions: a) α positive; b) α negative.

Note that there is a variety of reasons for the linear law to manifest itself. For instance, this law works when a porous, unprotecting layer is formed on a metal surface. In our case the linear law is due to an electric field in the film.

4.7.5. Succession of Laws of Film Growth with Temperature and Pressure Variation

Figures 4.18a and 4.18b clarify the question of regions of applicability for the laws of growth (4.163), (4.182), (4.185), and (4.189). (The abscissa is the absolute temperature T and the ordinate is the film thickness L.) The straight lines 1, 2, 3, and 4 depict the functions $L_1 = L_1(T)$, $L_2 = L_2(T)$, $L_3 = L_3(T)$, and $L_4 = L_4(T)$, according to (4.161), (4.176), and (4.179). For the sake of simplicity we put $b = 1$, which is the case, as shown by (4.137) for not too low pressures P. We have

$$\text{curve } 1 \quad L_1\,(T) = \alpha\,\frac{N^*}{\sigma_B} + kT\beta\,\frac{N^*}{\sigma_B}\,\ln\,\frac{N^*}{\sigma_B}\,,$$

$$\text{curve } 2 \quad L_2\,(T) = 4kT\beta,$$

$$\text{curve } 3 \quad L_3\,(T) = \alpha\,\frac{N^*}{\sigma_B} + 2kT\beta\left(\frac{N^*}{\sigma_B} - 2\right), \tag{4.190}$$

$$\text{curve } 4 \quad L_4\,(T) = \alpha - 2kT\beta,$$

where we have employed the notations

$$\alpha = \frac{\upsilon - (\chi_M - \chi_S)}{\gamma N^*}\,, \qquad \beta = \frac{1}{\gamma N^*}\,. \tag{4.191}$$

Figure 4.18a refers to the case where $\alpha > 0$ ($\nu > \chi_M - \chi_S$; see Fig. 4.16) and Fig. 4.18b to the case where $\alpha < 0$ ($\nu < \chi_M - \chi_S$). As the intrinsic disorder on the surface increases, i.e., as N^*/σ_B gets smaller [where, in accordance with (4.166) and (4.158), $N^*/\sigma_B \gg 1$], the slope of the curves 1 and 3 in Figs. 4.18a and 4.18b diminishes [see (4.190)], point A in Fig. 4.18a moves downward (and approaches point B, which remains fixed), and points A and A' in Fig. 4.18b move to the left.

Note that (4.190) and (4.166) imply

$$\frac{dL_2}{dT} < \frac{dL_3}{dT} ,$$

which ensures that there is an intersection point of the straight lines 2 and 3 in Fig. 4.18b and no such point in Fig. 4.18a. We can also show that

$$\frac{dL_3}{dT} < \frac{dL_1}{dT} . \tag{4.192}$$

Indeed, taking into account (4.190) and bearing in mind that, according to (4.166), $N^*/\sigma_B \geq 4$, we obtain

$$\frac{dL_1}{dt} - \frac{dL_3}{dt} = kT\beta \left[\frac{N^*}{\sigma_B} \ln \frac{N^*}{\sigma_B} - \left(2\frac{N^*}{\sigma_B} - 4 \right) \right]$$

$$= kT\beta \left[\frac{N^*}{\sigma_B} \left(\ln \frac{N^*}{\sigma_B} - 2 \right) + 4 \right] > kT\beta[4 (\ln 4 - 2) + 4] = 1.52\,kT\beta > 0.$$

Condition (4.192) means that the straight line 1 always lies above the straight line 3, as depicted in Figs. 4.18a and 4.18b.

Obviously, the region that satisfies condition (4.160) lies above the straight line 1 in Figs. 4.18a and 4.18b (region I), and in this region the logarithmic law of film growth (4.174) operates. In region II, which lies between the straight lines 2, 3, and 4 in Fig. 4.18a or between the straight lines 2 and in Fig. 4.18b, conditions (4.181) are met and the parabolic law (4.182) operates. The linear laws (4.185) and (4.189) operate in the region III and IV, respectively. Obviously, all the laws work only far from the straight lines 1, 2, 3, and 4 that separate these regions and close to the abscissa axis, i.e., for thin films that obey condition (4.146). Note that there is a single mechanism that produces these laws, and these laws represent various limiting cases.

Note further that as the pressure P increases, the value of parameter b diminishes and, hence, according to (4.161) and (4.179), the slope of the straight lines $L_1 = L_1(T)$ and $L_3 = L_3(T)$ increases. Thus, if for a given pressure we are at a certain point that belongs to region I, then an increase in pressure may shift us to region II or, in other words, as the pressure in the gaseous medium increases, the law of film growth ceases to be logarithmic and becomes parabolic. Another line of reasoning states that as P increases, so does σ_A, and the inequality $\sigma_A \ll \sigma_B$ becomes $\sigma_B \ll \sigma_A$, which means that the logarithmic law transforms into the parabolic, as shown in Sections 4.7.3. and 4.7.4. This result was observed experi-

Table 4.2

	Temperature, °C					
	0	200	400	600	800	1000
Mg	log.	par.	lin.			
Ca	log.	par.	lin.	lin.		
Th		par.	lin.	lin.		
Mo		par.		lin.	lin.	
Fe	log.	log.	par. par. par.	par.	par.	par.
Ni		log.	log. par.		par.	par.
Zn		log.	log. par. par.			

mentally by Loskutov *et al.* [83] in studies of the interaction of chlorine with an oxidized metal surface. At room temperature an increase in pressure leads to an increase in the work function (or the surface charge concentration) and to a transfer of the law of growth from the logarithmic law to the parabolic and then to the linear law.

Figure 4.18 clearly shows that the sequence that the laws of growth follow as the temperature T increases is from logarithmic to parabolic to linear, and, as can also be seen from the figure, any stage in this sequence may be omitted in certain conditions.

Table 4.2, taken from [84], illustrates which laws of oxidation operate at different temperatures and for different metals. We see that our results are in good agreement with the experimental data.

This section was devoted to a mechanism of film growth which, however (and this must be especially stressed), is only one of several possible mechanisms.

Chapter 5

THE CATALYTIC EFFECT OF A SEMICONDUCTOR

5.1. THE BASICS

5.1.1. Semiconductors as Catalysts of Chemical Reactions

The technical applications of semiconductors are extremely varied. However, until recently physics ignored the use of semiconductors as catalysts of chemical reactions. Such typical semiconductors as cuprous oxide, zinc oxide, and vanadium pentoxide are at the same time typical catalysts. Semiconductors serve as catalysts for many chemical reactions, including oxidation and hydrogenation.

Certain metals also act as catalysts. However, catalysis on semiconductors is widespread and, in fact, much more widespread than may appear at first glance. This is because most metals are usually covered with a semiconductor film. When oxygen, hydrogen, and nitrogen, come in contact with a clean metal surface, they are very rapidly and stably absorbed by it even at low temperatures. As a result the metal surface is covered with a film of a binary compound, the removal of which, as is well known, constitutes a difficult problem. Hence, in the majority of cases the metals prove to be covered with a semiconductor coating, so that the chemical processes that we regard as occurring on the surface of the metal are, in fact, proceeding on the surface of a semiconductor. For this reason, when we speak of the catalytic effect of a metal, we very often, though not always, are concerned with the catalytic effect of this semiconductor coating; the metal under the coating plays little or no part in the process.

The catalytic effect of semiconductors was discovered long before the concept of a semiconductor appeared. Chemists have long been interested in catalysis, but until recently they have not paid due attention to the fact that the majority of catalysts with which they are concerned are semiconductors. At the same time physicists concerned with semiconductors often do not know that semiconductors, among other things, possess catalytic activity and have long been used as catalysts in the chemical industry.

The problem of catalysis lies at the merging point of two sciences, physics and chemistry. It is now obvious that the catalytic properties of semiconductors are very closely connected with the electronic processes occurring inside and on the surface of semiconductors and are, in the final analysis, due to these processes. The problem of the catalytic activity of semiconductors, which is a chemical problem, can at the same time be considered quite rightly as a problem of semiconductor physics belonging to a broad group of problems that until recently have escaped the attention of physicists.

As examples note the influence of impurities introduced into the bulk of a semiconductor on its catalytic properties (well studied experimentally), the correlation between the electrical conductivity of a semiconductor and its catalytic action, the influence of radiation on the semiconductor's adsorption and catalytic activity (observed in a number of cases), which produces an internal photoelectric effect in the semiconductor, and finally, the variations that occur in the electrical conductivity and work function of a semiconductor when this acts as a catalyst.

In a catalytic process the semiconductor does not act as an inert substrate on which the chemical reaction proceeds but as an active participant in the process. It takes part in the intermediate stages of the reaction as one of the reaction components. The catalytic properties of a semiconductor are determined by its nature and electronic state. The mechanism of its catalytic action is hidden within the semiconductor, and a theory that attempts to disclose this mechanism cannot be constructed outside modern semiconductor theory. At the present stage of development of the theory of catalysis, the problem of catalysis on semiconductors is being included to an ever greater extent within the scope of semiconductor physics.

5.1.2. The Activity and Selectivity of a Catalyst

The basic concepts that concern us in catalysis are the activity and the selectivity of a catalyst. Let us dwell on these.

Let us imagine that a solid, the catalyst, is introduced into a mixture of reacting gases. The rate of reaction thereupon increases. The fractional increase in reaction rate characterizes what is called the activity of the catalyst. In some cases we are concerned with acceleration of a reaction by hundreds and thousands of times, so that the introduction of a catalyst may lead to violent development of a reaction that in the absence of a catalyst would hardly occur or would proceed very slowly.

The oxidation of carbon monoxide, which leads to the formation of carbon dioxide, may serve as an example:

$$2CO + O_2 \rightarrow 2CO_2.$$

This reaction proceeds very slowly in the absence of a catalyst and only at very high temperatures (of the order of several hundred degrees). In the presence of such catalysts as MnO_2, Ag_2O, and Co_2O_3 the reaction proceeds at a high rate even at temperatures below room temperature (right down to $-60°C$). At temperatures around $100°C$ and higher CuO and NiO are also catalysts for this reaction.

Note that the same catalyst has different activities in relation to different reactions. Although active in one reaction, it may prove inactive in another.

Usually a reaction consists of a number of intermediate steps that occur either in parallel or consecutively. A catalyst changes the rate of the individual steps and to different degrees.

When the reaction is a chain of consecutive steps, the rate of the resultant reaction is determined by the rate of its slowest step. In the presence of a catalyst this slowest (limiting) step may prove to be different from the step that is slowest in the absence of the catalyst. Moreover, the role of the catalyst may manifest itself in a change in the steps themselves. For instance, while in the absence of a catalyst a reaction breaks down into certain steps, in the presence of one the same reaction may prove to consist of different steps.

If the reacting substances participate in several reactions that occur in parallel, as often occurs, then the rate of each reaction may be changed by a given catalyst to different degrees; i.e., the activity of a catalyst may be different in relation to the different reactions occurring in parallel. This means that under the influence of a catalyst the direction of a reaction may be changed. This ability of a catalyst to control the direction of a reaction is known as its selectivity.

Thus, if as a result of some complicated reaction we obtain various products, a catalyst may change not only the absolute value of the yield of each product but also the relative yields of the various products. The decomposition reaction of isopropyl alcohol C_3H_7OH can serve as an example. This reaction may proceed generally in two directions, namely,

$$C_3H_7OH \rightarrow CO(CH_3)_2 + H_2$$

(dehydrogenation) and

$$C_3H_7OH \rightarrow C_3H_6 + H_2O$$

(dehydration). The catalyst ZnO promotes the reaction mainly in the first direction, while Al_2O_3 is a typical catalyst for the second direction. In the first case we obtain acetone, $CO(CH_3)_2$, and hydrogen, H_2, as the reaction products, and in the second case propylene, C_3H_6, and water H_2O.

The catalytic activity and selectivity of a semiconductor may change considerably as the result of some external influence on the semiconductor. For instance, the activity of a catalyst always markedly increases with temperature. Both the activity and selectivity of a semiconductor can be controlled by introducing minute quantities of an impurity. The catalytic properties of a semiconductor are determined not only by the chemical nature of the semiconductor but also by the prehistory of the given sample; i.e., they depend on the method of preparation and to a certain extent on the external influences to which the sample was subjected throughout its history. In other words, the catalytic properties of semiconductors belong to the category of structure-sensitive properties (see Section 1.1).

Note that a solid that catalyzes a reaction, i.e., changes its rate and course, is not a passive participant in the process but an active one. As we have already remarked, its role as a catalyst is due to the fact that it participates in the intermediate steps of the reaction as one of the reaction components. Characteristically, however, by the end of the reaction the catalyst is completely regenerated, i.e., it emerges from the reaction the same as when it entered it. This characteristic of catalysts can be regarded as a strict definition of the very concept of a catalyst.

In many cases that are met in practice, however, the composition of the catalyst does change to some extent during the reaction. As an example we can again cite the oxidation of CO catalyzed by solid oxides (see above), a reaction that often proceeds not only by means of the gaseous oxygen but also to a certain extent via the oxygen belonging to the crystal lattice of the catalyst. As a result the catalyst is gradually reduced in the reaction, and on completion of the reaction has a different stoichiometric composition.

Thus, in the ideal case the catalyst is completely regenerated at the end of the reaction. But in practice it gradually changes in the process of functioning and, finally, when it has functioned for a prolonged period, may fail. In both cases a catalyst, which accelerates a reaction, participates in the chemical process. It is precisely this that makes it a catalyst.

5.1.3. The Activation Energy

Every chemical process catalyzed by a solid involves adsorption and desorption as necessary steps. Adsorption and desorption are the initial and final steps of every heterogeneous catalytic reaction. Let us assume that a solid (the catalyst) is in contact with a gaseous phase. The gas molecules are first adsorbed on the surface of the solid, then, remaining in the adsorbed state, enter into reactions with each other or with molecules from the gaseous phase, after which the products of the reaction are desorbed. Thus, when we introduce a catalyst into a mixture of reacting gases, the reaction is transferred from the gaseous phase to the surface of the solid. A heterogeneous catalytic reaction is one that occurs at the interface between two phases.

Let us denote the chemical symbols of the substances participating in the reaction by $A_1, A_2, ..., A_n$, and the symbols of the reaction products by $A_1', A_2', ..., A_n'$. Any reaction can be written in the following form (the reaction equation):

$$\alpha_1 A_1 + \alpha_2 A_2 + ... + \alpha_n A_n \rightarrow \alpha_1' A_1' + \alpha_2' A_2' + ... + \alpha_n' A_n'.$$

Here α_i (with $i = 1, 2, ..., n$) and α_k (with $k = 1, 2, ..., n'$) are known as stoichiometric coefficients, which are integral numbers showing how many molecules of a given type participate in the reaction.

Suppose that the concentrations of molecules A_i and A_k' in the gaseous phase are N_i and N_k'. Obviously, the concentrations are functions of time:

$$N_i = N_i(t), \quad N_k' = N_k'(t).$$

If time is reckoned from the start of the reaction, then the quantities $N_i(0) - N_i(t)$ and $N_k'(t) - N_k'(0)$ are, respectively, the number of molecules of the initial substance A_i that have disappeared by time t and the number of molecules of the reaction product A_k' that have been created by time t. Obviously,

$$\frac{N_i(0) - N_i(t)}{\alpha_i} = \frac{N_k'(t) - N_k'(0)}{\alpha_k}$$

for all k's and i's. The quantity

$$g = -\frac{1}{\alpha_i}\frac{dN_i(t)}{dt} = \frac{1}{\alpha_k'}\frac{dN_k'(t)}{dt},$$

where $i = 1, 2, ..., n$ and $k = 1, 2, ..., n'$, is known as the reaction rate. Obviously, $g_i = \alpha_i g$ is the rate at which substance A_i disappears and $g_k' = \alpha_k' g$ is the rate at which the reaction product A_k' is created.

Generally speaking, g is a function of temperature and the concentrations of the molecules participating in the reaction. This fact is stated in the following form:

$$g = K(T) N_1^{\beta_1} N_2^{\beta_2} \dots N_n^{\beta_n}.$$

Here β_i (with $i = 1, 2, ..., n$) are the kinetic coefficients. The quantity β_i is the order of the reaction in component A_i. In the majority of cases the kinetic coefficients β_i coincide with the respective stoichiometric coefficients α_i:

$$\beta_i = \alpha_i.$$

Note that the for low surface coverages (the Henry region) the concentrations N_i are proportional to the corresponding partial pressures P_i. The coefficient $K(T)$ is known as the reaction rate constant and is often used to characterize the catalytic activity of a surface. Usually the reaction rate constant has the following form (the Arrhenius equation):

$$K(T) = K_0 \exp\left(-\frac{E}{kT}\right),$$

where E is the activation energy of the reaction, a quantity that plays an important role in catalysis. According to the Arrhenius law the $\ln K$ vs. T^{-1} dependence is a straight line, whose slope yields E and the initial ordinate yields K_0 (e.g., see Fig. 5.24). Note that the activation energy E may be different in different temperature intervals, a fact that is evident from breaks in the Arrhenius curve.

As a rule both the activation energy E and the pre-exponential factor K_0 depend on the preparation of the sample, for instance, on the nature and amount of impurity introduced into the sample.

Different samples that differ in their prehistory are characterized by different values of E and K_0. Figures 5.24 to 5.27 illustrate this fact (in each figure the Arrhenius curves are given for different impurity concentrations). Usually the transformation from sample to sample leads to a transformation of E and K_0 in the same direction, i.e.,

$$K_0 = K_0 (E),$$

with

$$\frac{dK_0}{dE} > 0.$$

Here we are encountering a compensation effect. Indeed, variations of E and K_0 to a certain extent compensate for each other; i.e., an increase in E leads to a decrease, and an increase in K_0 to an increase, in the reaction rate constant K (see Section 5.5.5).

5.1.4. The Electronic Theory of Catalysis

The mechanism of a heterogeneous catalytic reaction cannot be understood *in toto* if we fail to understand the mechanism of the intermediate stages of this process, the adsorption and desorption stages. The adsorption that we deal with in catalysis is chemisorption. This implies that the theory of heterogeneous catalysis is closely linked with chemisorption theory. While chemisorption theory may develop outside the scope of the theory of catalysis, the theory of heterogeneous catalysis is interwoven with chemisorption theory.

The ultimate problem confronting investigators in the field of catalysis is the selection of catalysts. The problem is to learn to control the activity and selectivity of catalysts, i.e., learn to vary their properties to the necessary extent and in the necessary direction.

This problem cannot be fully solved and cannot be taken beyond the bounds of crude empiricism until the mechanism of the action of a catalyst is understood. To solve the problem, we must elucidate the elemental (microscopic) mechanism of the catalytic act. Every heterogeneous catalytic process, like every chemical process in general, has basically an electronic mechanism. The elucidation of this mechanism constitutes a problem of the electronic theory of catalysis.

At present there is a vast body of experimental data showing that the electronic processes occurring in a semiconductor and determining its electrical, optical, and magnetic properties determine its catalytic properties. There exists a certain parallelism between the electronic properties of a semiconductor and its catalytic properties. To reveal the connection between these two groups is also a goal of the electronic theory of catalysis. This problem is closely connected to the first problem of electronic theory. The electronic theory of catalysis is being erected on the foundation of the modern theory of chemical bonding, on the one hand, and solid-state theory, on the other. Indeed, while the theory of chemical bonding is concerned with the transmutation of molecules, and solid-state theory with processes in crystals, the theories of chemisorption and heterogeneous catalysis are concerned with the transmutation of molecules situated on the surface of a crystal, bound to this crystal, and forming with it a single system that must be regarded as an integral whole.

The appearance of the electronic theory of catalysis marks the entry of modern semiconductor physics into the problem of catalysis. The electronic theory thereby introduces new concepts and ideas into the theory of catalysis. This, of course, does not mean that it excludes concepts and ideas used in other theories of catalysis. On the contrary, it makes use of them but attempts to elucidate their physical content.

The electronic theory of catalysis and other theories of catalysis, the latter having basically a phenomenological character, are not, as a rule, alternative nor do they compete with each other. They concern different aspects of catalysis and, therefore, differ form one another mainly in their approach to the problem. The electronic theory is interested in the elemental (electronic) mechanism of the phenomenon and approaches the problems of catalysis from precisely this point of view.

Between the existing phenomenological theories of catalysis and the electronic theory there is the same relationship as between the theory of the chemical bond of the last century, which made use of valence lines (and had at its disposal nothing other than these lines), and the modern quantum-mechanical theory of the chemical bond, which has filled the valence lines of the old chemistry with physical content and at the same time elucidated the physical nature of the chemical forces.

The founder of the electronic theory of catalysis was L. V. Pisarzhevskii (Kiev), whose work, begun as early as 1916, was part of his extensive project of investigations devoted to electronic phenomena in chemistry. He was the first to attempt to relate the catalytic properties of solids to their electronic properties. However, Pisarzhevskii formulated his electronic theory before the advent of quantum mechanics. It was based on Bohr's theory and, naturally, did not go beyond the bounds of this theory, which substantially limited its potential.

At the present time (starting from 1948) the electronic theory is developing along a more solid theoretical basis. The initiator of the rebirth of the electronic trend was S. Z. Roinskii (Moscow, USSR), from whose laboratory there has emerged a whole series of experimental and theoretical papers devoted to electronic phenomena in catalysis. We should note here the works of Soviet authors such as A. N. Terenin and his school (Leningrad), V. I. Lyashenko and L. V. Lyashenko and their collaborators (Kiev), I. A. Myasnikov and V. F. Kiselev and their collaborators (Moscow), and N. P. Keier (Novosibirsk). Among the works of non-Soviet authors we must note those of M. Boudart and P. B. Weisz (USA), J. E. Germain, P. Aigrain, S. J. Teichner, and B. Claudel (France), F. S. Stone (Great Britain), R. Coekelbergs and A. Crucq and their collaborators (Belgium), A. Bielanski and J. Haber and their collaborators (Poland), E. Segal and M. Teodorescu (Rumania), O. Peshev (Bulgaria), Y. Kwan (Japan), G. Rienäcker

(GDR) and, especially, the works of K. Hauffe and his collaborators (FRG) and S. R. Morrison (USA).

5.2. THE ROLE OF THE FERMI LEVEL IN CATALYSIS

5.2.1. Radical Mechanisms of Heterogeneous Reactions

We have seen (Section 2.4) that involvement of the free electrons and holes of a catalyst in participation in chemisorption bonds leads to the chemisorbed particle spending a certain fraction of its time in a radical state in the course of its life in the adsorbed condition. Thus, since radicals are always more reactive than saturated molecules, the very act of transferring molecules from the gaseous phase to the chemisorbed state leads to an increase in their reactivity.

A radical mechanism of heterogeneous reactions is provided by radical and ion radicals arising on the surface in chemisorption. Every heterogeneous reaction can be interpreted as proceeding according to a radical mechanism. This does not mean, of course, that nonradical mechanisms are completely excluded from heterogeneous catalysis. However, when radical and ion radical forms arise in sufficient concentration on the surface (and they do arise in certain conditions), the leading role in the heterogeneous catalytic process is transferred to them.

Here we will examine the various types of heterogeneous reactions and possible radical mechanisms for them [1, 2]. Let two molecules AB and CD, where A, B, C, and D are symbols for individual atoms or atomic groups, participate in the reaction.

First, let us assume that A and B and also C and D are bound by single bonds. Let us examine the exchange reaction

$$AB + CD \rightarrow AC + BD,$$

in the course of which two single bonds are disrupted and two are formed. The chlorination of ethane may serve as an example:

$$C_2H_6 + Cl_2 \rightarrow C_2H_5Cl + HCl,$$

which occurs, for example, on the catalyst $ZnCl_2$. A possible radical mechanism for a reaction that proceeds through the dissociation of both molecules entering the reaction is shown in Fig. 5.1. Here we have an example of a chain mechanism. The chain is started by a free catalyst valence (in our case an electron, which may be a free electron or an electron localized at a defect). The valence enters the reaction and is then recreated at the end of the reaction.

Let us assume that one of the molecules participating in the reaction, say molecule CD, contains a double bond, while A and B in molecule AB are joined by a single bond. We examine the reaction

$$AB + CD \rightarrow ACDB.$$

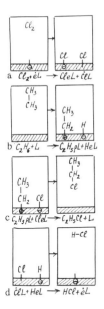

Fig. 5.1. Possible radical mechanism of a reaction proceeding through dissociation of both molecules entering the reaction: a–d) various stages in the chain reaction.

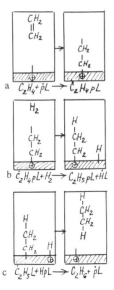

Fig. 5.2. Possible radical mechanism for the C_2H_4 + $H_2 \rightarrow C_2H_6$ reaction: a–c) various stages in the chain reaction.

A simple example is the hydrogenation of ethylene:

$$C_2H_4 + H_2 \rightarrow C_2H_6,$$

which occurs on MnO_3 and $ZnO + Cr_2O_3$ as catalysts. Figure 5.2 depicts a possible radical (chain) mechanism for this reaction, which is accompanied by the formation of surface radicals. In contrast to the previous case, the chain here is

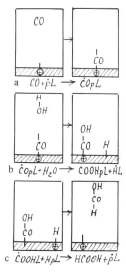

Fig. 5.3. Possible radical mechanism for the CO + $H_2O \to$ HCOOH reaction: a–c) various stages in the chain reaction.

started by a hole. A somewhat different mechanism, but also a chain, for heterogeneous hydrogenation was examined by Thon and Taylor [3].

Other examples of reactions of this type are provided by the addition reactions of hydrogen halides to olefins, e.g.,

$$C_2H_4 + HCl \to C_2H_5Cl.$$

In the same way reactions may occur in which an increase in the number of single bonds may arise not as a result of an opening up of a double bond but as a consequence of a change in the number of valances of one of the atoms in the system. For example, the mechanism of the reaction

$$CO + H_2O \to HCOOH,$$

which can take place on the surfaces of $CuCl_2$, CuI_2, and NaBr crystals, may be represented by a similar scheme (Fig. 5.3).

Finally, let us examine the case where both molecules, AB and CD, have double bonds. We investigate the reaction

$$AB + CD \to ABCD,$$

as a result of which a cyclic compound is formed. Diene synthesis may serve as an example:

$$(CH_2 = CH - CH = CH_2) + (CH_2 = CH_2) \to CH_2 \underset{CH \; = \; CH}{\overset{CH_2 \; - \; CH_2}{\diagup \diagdown}} CH_2,$$

which, however, proceeds quite easily homogeneously (without chains). On the

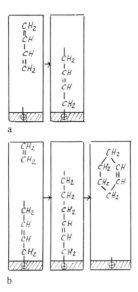

Fig. 5.4. Possible radical mechanism for diene synthesis: a, b) various stages in the chain reaction.

surface of a catalyst this reaction could proceed via a chain mechanism, as shown in Fig. 5.4. Here again the chain is started by a free catalyst valence (a hole).

Note that the radical schemes quoted here for a number of reactions (Figs. 5.1–5.4) must be regarded as only theoretically feasible. Generally speaking, the same reaction permits various radical mechanisms, depending on which bonds in the reacting molecules are broken and in what sequence and which forms of chemisorption for particles of a given type are to be considered active in the given reaction. For example, for the chlorination of ethane there is another possible mechanism (also a chain) depicted in Fig. 5.5 that can exist alongside the mechanism depicted in Fig. 5.1. Now a hole starts the reaction instead of an electron. The examples cited serve only to illustrate how a complicated reaction may break down on a surface into elemental acts, each of which requires a rupture of one and only one bond in a molecule and occurs with the participation of a surface radical.

The role of the catalyst amounts to creating such surface radicals. These are generated by free catalyst valences present on the surface or arising in the course of the reaction. Note that the supply of free valences on the surface is used up very slowly in the course of the reaction, since they are supplied to the surface from the bulk of the semiconductor, which is a practically inexhaustible reservoir of free valences (electrons and holes). The influx of valences from the bulk to the surface is limited only by the fact that the surface becomes charged during chemisorption. When the surface charge reaches a certain critical value, further arrival of free valences at the surface is stopped.

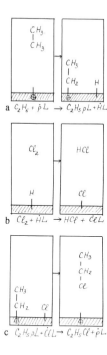

a $C_2H_6 + \dot{p}L \rightarrow C_2H_5pL + \dot{H}L$

b $Cl_2 + \dot{H}L \rightarrow HCl + \dot{C}lL$

c $C_2H_5pL + \dot{C}lL \rightarrow C_2H_5Cl + \dot{p}L$

Fig. 5.5. Alternative radical mechanism for chlorination of ethane (cf. Fig. 5.1): a–c) various stages in the chain reaction.

The catalyst thus emerges as a special kind of "poly-radical" and influences the course of the reaction for the same reason that the introduction of free radicals into a homogeneous medium influences the course of a homogeneous reaction. In both cases the acceleration of the reaction results from the free valences (electrons and holes) brought into the operation. In the case of heterogeneous catalysis these free valences are introduced by the catalyst. In the final analysis, it is these valences that effectively promote and control the reaction.

In Section 5.3 we will discuss the following interesting reactions from the viewpoint of the electronic theory:

(1) Oxidation of hydrogen:

$$2H_2 + O_2 \rightarrow 2H_2O.$$

(2) Dehydrogenation of ethyl alcohol:

$$C_2H_5OH \rightarrow CH_3CHO + H_2.$$

(3) Dehydration of ethyl alcohol:

$$C_2H_5OH \rightarrow C_2H_4 + H_2O.$$

(4) Oxidation of carbon monoxide:

$$2CO + O_2 \rightarrow 2CO_2.$$

(5) Hydrogen–deuterium exchange:

$$H_2 + D_2 \rightarrow 2\,HD.$$

In the same section we will calculate the rates of these reactions.

5.2.2. Acceptor and Donor Reactions

Of the total number of particles of a given kind chemisorbed on the surface of a catalyst only a certain fraction participates in the reaction, namely, those in the reactive state. In other words, among the various coexisting forms of chemisorption we must distinguish between active and nonactive forms (or more active and less active forms), from the viewpoint of the given reaction. The reaction rate for a given surface coverage, other conditions being the same, will obviously be determined by the relative fractions of such active forms on the surface.

Thus, the quantities η^0, η^-, and η^+, which are the relative fractions of the various chemisorption forms, will enter into the expression for the reaction rate. In the presence of electronic equilibrium the quantities η^0, η^-, and η^+ depend on the position of the Fermi level E_F in the surface plane. In this way g proves to be dependent on E_F, or

$$g = g(E_F). \tag{5.1}$$

The specific forms of function (5.1) for certain reactions will be given in Section 5.3.

We see that the reaction rate (and thereby the catalytic activity of a semiconductor in relation to a given reaction) is determined (all other things being equal) by the position of the Fermi level at the surface of the semiconductor. The position of the Fermi level determines the reaction rate not only at given partial pressures but the reaction rate at given surface coverages, too. (In the first case the catalytic activity of the semiconductor depends on its adsorptivity in relation to the reacting gases; in the second case the catalytic activity and the adsorptivity can be considered as two independent characteristics of the semiconductor.) The role of the Fermi level as regulator of catalytic activity was first investigated by the present author in 1950 [4]. Later this question was examined by Boudart [5] and Hauffe [6–8].

In accordance with the shape of the function (5.1) we must distinguish between two cases:

$$\frac{dg}{dE_F} > 0 \tag{5.2a}$$

and

$$\frac{dg}{dE_F} < 0. \tag{5.2b}$$

Thus, according to the nature of the variation of reaction rate with the position of the Fermi level, all heterogeneous reactions can be divided into two classes. To one of these belong all those reactions that proceed more rapidly the higher the Fermi level (all other conditions being equal). This case corresponds to (5.2a). Such reactions are accelerated by electrons, and we call them acceptor reactions or n-class reactions. To the other class belong all those reactions whose rate is greater the lower the Fermi level [case (5.2b)]. We call such reactions donor reactions or p-class reactions. These are accelerated by holes.

The dehydrogenation of alcohol, as we will see in Section 5.3.2, belongs to acceptor reactions, while the dehydration of alcohol belongs to donor reactions. The dependence of the reaction rate on the position of the Fermi level for these two reactions is shown in Fig. 5.9. We see that all the factors affecting the position of the Fermi level (all other conditions being equal) must affect the reaction rate g. For instance, an acceptor impurity, which always lowers the Fermi level, must lower the reaction rate, as shown in Fig. 5.9a, of an acceptor reaction and increase the rate, as shown in Fig. 5.9b, of a donor reaction. Donor impurities must act in the opposite manner. We will return to the question of impurities acting on the catalytic activity in Sections 5.5.3 and 5.5.4.

Whether a reaction belongs to the acceptor or donor class is determined above all by the concrete mechanism of the reaction in question. The same reaction, as a rule, can occur via different mechanisms and, depending on this, may fall into the acceptor or donor category. Moreover, for a given mechanism the reaction may be an acceptor reaction for certain values of E_F and a donor reaction for other values of E_F. In other words, it may be an acceptor or donor reaction depending on the position of the Fermi level on the semiconductor surface, i.e., on the history of the given sample acting as catalyst or, in other words, on the treatment to which the sample was subjected prior to the reaction.

We can illustrate this with the example of the oxidation of CO, which we will discuss in detail in Section 5.3.3. The dependence of g on E_F for this reaction is roughly depicted in Fig. 5.12; we see that as the Fermi level moves downward the reaction rate passes through a maximum. When the Fermi level lies sufficiently high (region a), the reaction is a donor reaction, but when the Fermi level is sufficient low (region b), the reaction is an acceptor reaction. Going from one sample to another (differently prepared) and changing the external conditions (temperature and partial pressures), we can easily transfer from the donor branch to the acceptor branch of the curve in Fig. 5.12 and back. On these two branches the same impurity introduced into the crystal and displacing the Fermi level upward or downward must exert, as we have seen, opposite influences on the reaction rate (decelerating the reaction in one case and accelerating it in the other). Moreover, introducing an impurity may transfer the reaction from one branch to the other. For instance, every acceptor impurity has a tendency to transfer the reaction from the donor class to the acceptor class. A negative charge on the surface due to the chemisorption of a foreign gas or to structural defects on the surface acts in the same direction. A donor impurity, on the other hand, has a tendency to transfer the reaction from the acceptor class to the donor class. A positive charge on the surface acts in the same manner, no matter what its origin is.

Note that in some cases the reaction rate may be independent of the position of the Fermi level. For instance, if the Fermi level lies high, in the case of alcohol dehydrogenation, or low, in the case of alcohol dehydration, the reaction rate g, as shown by Figs. 5.9a and 5.9b, does not depend on E_F, which, of course, does not contradict the electronic mechanism of the reaction, as is often thought, but follows from it. While the fact that catalytic activity depends on the position of the Fermi level (this is verified by experiments) speaks in favor of the electronic theory of catalysis, the absence of such a dependence, observed in special cases, cannot speak against the theory.

The question of whether a reaction belongs to an acceptor or donor class can be resolved experimentally on the basis of the data on the influence of acceptor and donor impurities on the reaction rate, provided that all the other conditions remain constant (See Section 5.5), on the basis of the data on the connection between the reaction rate and the electrical conductivity of the semiconductor, and, finally, on the basis of the data on the parallelism between the variations of the reaction rate and the work function of the semiconductor (see Section 5.4). The last data enable us to draw conclusions about the acceptor or donor character of the reaction in an unambiguous manner. Determining the nature of a reaction experimentally (to see whether it belongs to the acceptor or donor class) sheds some light on the reaction mechanism, which in many cases enables us to choose between the different mechanisms that are theoretically feasible for the given reaction.

5.3. ELECTRONIC MECHANISMS OF CATALYTIC REACTIONS

5.3.1. Oxidation of Hydrogen

As already noted in Section 5.2.1, the same reaction often admits of different theoretical mechanisms. The choice between these mechanisms requires additional experimental data. In this section we will discuss the possible mechanisms of reactions well known experimentally.

We start with the oxidation of hydrogen

$$2H_2 + O_2 \rightarrow 2H_2O.$$

Both n-type and p-type semiconductors, such as ZnO, V_2O_5, NiO, and CuO, may act as catalysts for this reaction.

Figure 5.6 shows one of the possible radical mechanisms of this reaction. We assume that the surface of the semiconductor contains chemisorbed atoms of oxygen that may be electrically neutral or negatively charged (i.e., they may be in a condition of "weak" or "strong" acceptor bonding with the surface). The first stage of the reaction is the adsorption of an H_2 molecule on the free valence of an O^- ion radical, accompanied by dissociation (Fig. 5.6a). As a result H atoms and OH molecules appear on the surface, and these atoms and molecules may be present in the form of neutral particles or as H^+ and OH^- ions. The second stage

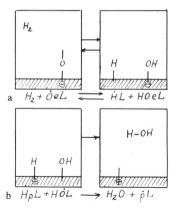

$$a \quad H_2 + \dot{O}eL \rightleftharpoons \dot{H}L + HOeL$$

$$b \quad HpL + H\dot{O}L \longrightarrow H_2O + \dot{p}L$$

Fig. 5.6. Possible radical mechanism for oxidation of hydrogen: a) adsorption of H_2 molecule on the free valence of an O^- ion radical accompanied by dissociation; b) recombination of an H atom with an OH molecule accompanied by desorption of an H_2O molecule.

is the recombination of an H atom with an OH molecule (e.g., as shown in Fig. 5.6b), accompanied by desorption of an H_2O molecule. In both stages of the reaction one of the two particles entering the reaction is in the radical state.

Assuming that for each type of particle (O, H, or OH) the various possible forms of chemisorption are in equilibrium (electronic equilibrium has been established) and that the surface coverage is low, and neglecting adsorption of the reaction products, we have

$$\frac{dN_{OH}}{dt} = \alpha_H P_{H_2} N_O^- - \beta_H N_H^0 N_{OH}^- - \gamma N_H^+ N_{OH}^0,$$

$$\frac{dN_O}{dt} = \alpha_O P_{O_2} - \beta_O (N_O^0)^2 - \alpha_H P_{H_2} N_O^- + \beta_H N_H^0 N_{OH}^-,$$

$$g = \gamma N_H^+ N_{OH}^0,$$

where g is the rate of the reaction (desorption of H_2O), P_{H_2} and P_{O_2} the partial pressures, N_O, N_H, and N_{OH} the surface concentrations of the corresponding particles (superior indices denote, as previously, the type of bonding between the chemisorbed particle and the surface), and α_H, β_H, α_O, β_O, and γ are coefficients whose form is irrelevant for us here. We assume at this point (but this can be proved rigorously) that only the O_2 molecules in the neutral state (i.e., are recombination products of two neutral O atoms) leave the surface.

Under equilibrium, i.e., with

$$\frac{dN_{OH}}{dt} = \frac{dN_O}{dt} = 0,$$

we have

$$\alpha_H P_{H_2} N_O^- = \beta_H N_H^0 N_{OH}^- + \gamma N_H^+ N_{OH}^0,$$

$$\alpha_O P_{O_2} = \beta_O (N_O^0)^2 + \gamma N_H^+ N_{OH}^0,$$

from which, assuming that

$$\gamma N_H^+ N_{OH}^0 \ll \alpha_O P_{O_2}, \qquad \gamma N_H^+ N_{OH}^0 \ll \alpha_H P_{H_2} N_O^-,$$

i.e., considering that desorption of the reaction product (H_2O) proceeds more slowly than adsorption of both reacting gases (O_2 and H_2), we obtain

$$g = \gamma N_H^+ N_{OH}^0 = K P_{H_2} \sqrt{P_{O_2}}, \tag{5.3}$$

where, using the notations (3.4),

$$K = \gamma \frac{\alpha_H}{\beta_H} \sqrt{\frac{\alpha_O}{\beta_O}} \left(\frac{\eta_H^+}{\eta_H^0} \frac{\eta_{OH}^0}{\eta_{OH}^-} \frac{\eta_O^-}{\eta_O^0} \right).$$

Here, according to (3.6a) and (3.6b) (see also Fig. 3.4a),

$$\frac{\eta_H^+}{\eta_H^0} = \exp\left(-\frac{E_F - E_H}{kT} \right),$$

$$\frac{\eta_{OH}^0}{\eta_{OH}^-} = \exp\left(-\frac{E_{OH} - E_F}{kT} \right), \qquad \frac{\eta_O^-}{\eta_O^0} = \exp\left(-\frac{E_O - E_F}{kT} \right)$$

and, consequently,

$$K = K_O \exp\left(-\frac{E + E_F}{kT} \right), \tag{5.4}$$

where

$$E = E_O - E_H - E_{OH}.$$

We see that the Fermi level E_F is a term in the activation energy of the reaction. To avoid misunderstanding, however, we must note that the sum $E + E_F$ in (5.4) cannot always be regarded as the activation energy in the strict sense of the word since E_F is generally a function of pressure and temperature (this may not be so in particular cases; see Section 4.1). For the same reason the order of the reaction with respect to hydrogen and oxygen cannot be regarded as completely elucidated in (5.3), generally speaking.

The reaction rate proves to be dependent on the position of the Fermi level at the surface of the crystal. Displacing the Fermi level (all other conditions being equal), we can control the reaction rate precisely: upward displacement (see Fig. 3.4a) slows down the reaction, and downward displacement speeds it up.

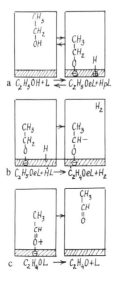

Fig. 5.7. Decomposition of alcohol (dehydrogenation): a) adsorption of an alcohol molecule; b) surface reaction; c) desorption of reaction products.

This result agrees with the experimental data of Boreskov and Popovskii [9], who investigated the oxidation of hydrogen on a number of semiconductors. Heating the specimens in an atmosphere of hydrogen, which raises the Fermi level, lowered their catalytic activity, while heating in an atmosphere of oxygen, which lowers the Fermi level, increased their activity.

5.3.2. Decomposition of Alcohol

Let us examine the decomposition of alcohol. For the sake of definiteness we will take ethyl alcohol, but any other alcohol can be considered in a similar manner. This reaction proceeds generally in two directions:

$$C_2H_5OH \Bigg\langle \begin{array}{ll} \longrightarrow CH_3CHO + H_2 & \text{dehydrogenation,} \\ \longrightarrow C_2H_4 + H_2O & \text{dehydration.} \end{array}$$

In the first case the reaction products are acetaldehyde CH_3CHO and hydrogen H_2, and in the second ethylene C_2H_4 and water H_2O.

Let us assume that the reaction proceeds via the mechanisms depicted in Fig. 5.7 (dehydrogenation) or in Fig. 5.8 (dehydration). In both cases the reaction breaks down into three stages: (a) adsorption of an alcohol molecule, (b) the surface reaction, and (c) desorption of the reaction products.

The course of the reaction (dehydrogenation or dehydration) is determined at the very first stage of the reaction and depends on which bond is broken in adsorption, the O–H bond (Fig. 5.7) or the C–OH bond (Fig. 5.8), which in turn depends on the nature of the catalyst. Generally speaking, both cases may be realized with the same catalyst, and the relative activity of the catalyst with respect to dehydrogenation and dehydration will depend, as we will subsequently see, on the position of the Fermi level.

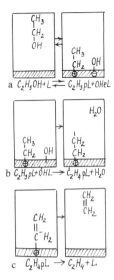

Fig. 5.8. Decomposition of alcohol (dehydration): a) adsorption of an alcohol molecule; b) surface reaction; c) desorption of reaction products.

Note that in an alcohol molecule the O–H and C–OH bonds are polarized, with the center of gravity of the electron cloud being shifted toward the O atom in the first case and toward the OH group in the second. Therefore, when these bonds are broken by the lattice field (dissociation during adsorption; see Section 2.4.4), structures will arise on the surface that are electrically charged as shown on the right-hand sides of Figs. 5.7a and 5.8a, respectively.

On the right-hand side of Figs. 5.7b and 5.8b we depict the "strong" (charged) forms of chemisorption of acetaldehyde and ethylene, respectively, while on the left-hand sides of Fig. 5.7c and 5.8c we depict the corresponding "weak" (electrically neutral) forms. The intermediate compound C_2H_5pL, depicted on the right-hand side of Fig. 5.8a or the left-hand side of Fig. 5.8b, which forms in the process of dehydration, is known as carbonium.

Neglecting the adsorption of the reaction products and by-product formation that may arise in the reaction, assuming that the surface coverage by alcohol molecules is low, and adopting for brevity the notation

$$A = CH_3CHO, \quad E = C_2H_4, \quad R = CH_3CH_2O, \quad Q = CH_3CH_2,$$

we have the following:

(1) In the case of dehydrogenation (Fig. 5.7),

$$\frac{dN_R}{dt} = \alpha P - \beta_1 N_R^- N_H^+ - \gamma_1 N_R^- N_H^0,$$

$$\frac{dN_A}{dt} = \gamma_1 N_R^- N_H^0 - \delta_1 N_A^0,$$

or in equilibrium,

$$\alpha P = \beta_1 N_R^- N_H^+ + \gamma_1 N_R^- N_H^0, \quad \gamma_1 N_R^- N_H^0 = \delta_1 N_A^0 = g_A, \tag{5.5}$$

where g_A is the reaction rate (the reaction of appearance of acetaldehyde in the gaseous phase), and P the pressure of alcohol.

(2) In the case of dehydration (Fig. 5.8),

$$\frac{dN_O}{dt} = \alpha P - \beta_2 N_Q^+ N_{OH}^- - \gamma_2 N_Q^+ N_{OH}^0,$$

$$\frac{dN_E}{dt} = \gamma_2 N_Q^+ N_{OH}^0 - \delta_2 N_E^0,$$

or in equilibrium,

$$\alpha P = \beta_2 N_Q^+ N_{OH}^- + \gamma_2 N_Q^+ N_{OH}^0, \quad \gamma_2 N_Q^+ N_{OH}^0 = \delta_2 N_E^0 = g_E, \tag{5.6}$$

where g_E is the reaction rate (the reaction of desorption of ethylene).

From (5.5) and using the notations (3.4) we obtain

$$g_A = \frac{\alpha P}{1 + (\beta_1 \eta_H^+)/(\gamma_1 \eta_H^0)} \tag{5.7}$$

or

$$g_A = \alpha \frac{\gamma_1}{\beta_1} \frac{\eta_H^0}{\eta_H^+} P, \qquad \text{if} \qquad \frac{\eta_H^0}{\eta_H^+} \ll \frac{\beta_1}{\gamma_1}, \tag{5.7a}$$

$$g_A = \alpha P, \qquad \text{if} \qquad \frac{\eta_H^0}{\eta_H^+} \gg \frac{\beta_1}{\gamma_1}. \tag{5.7b}$$

Similarly, from (5.6) we obtain

$$g_E = \frac{\alpha P}{1 + (\beta_2 \eta_{OH}^-)/(\gamma_2 \eta_{OH}^0)} \tag{5.8}$$

or

$$g_E = \alpha \frac{\gamma_2}{\beta_2} \frac{\eta_{OH}^0}{\eta_{OH}^-} P, \qquad \text{if} \qquad \frac{\eta_{OH}^0}{\eta_{OH}^-} \ll \frac{\beta_2}{\gamma_2}, \tag{5.8a}$$

$$g_E = \alpha P, \qquad \text{if} \qquad \frac{\eta_{OH}^0}{\eta_{OH}^-} \gg \frac{\beta_2}{\gamma_2}, \tag{5.8b}$$

where, according to (3.6a) and (3.6b) (see also Fig. 3.4a),

$$\frac{\eta_H^0}{\eta_H^+} = \exp\left(\frac{E_F - E_H}{kT}\right), \qquad \frac{\eta_{OH}^0}{\eta_{OH}^-} = \exp\left(\frac{E_{OH} - E_F}{kT}\right). \tag{5.9}$$

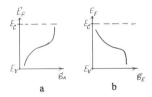

Fig. 5.9. Reaction rates vs. position of the Fermi level: a) g_A vs. E_F; b) g_E vs. E_F.

Thus, in cases (5.7a) and (5.8a) we have, respectively,

$$g_A = K_0' \exp\left(-\frac{E' - E_F}{kT}\right) P, \quad g_E = K_0'' \exp\left(-\frac{E'' + E_F}{kT}\right) P, \quad (5.10)$$

i.e., the Fermi level E_F enters the energy of activation in dehydrogenation and dehydration with opposite signs. In cases (5.7b) and (5.8b) the reaction rates (g_A and g_E) are independent of the position of the Fermi level.

The dependence of the reaction rates g_A and g_E on the position of the Fermi level is shown roughly in Figs. 5.9a and 5.9b, respectively [see (5.7), (5.8), and (5.9)], for the entire range of variation of E_F. We see that lowering the Fermi level slows down dehydrogenation and speeds up dehydration. Thus, dehydrogenation proves to belong to the class of acceptor reactions, and dehydration to the class of donor reactions. This provides a recipe for controlling the selectivity of a catalyst. The factors that lower the Fermi level (e.g., an acceptor impurity introduced into the crystal) poison the dehydrogenation reaction, and promote the dehydration reaction, while the factors that move the Fermi level upward (e.g., a donor impurity) promote the hydrogenation reaction and poison the dehydration reaction.

This agrees with the experimental data. For instance, decomposition of ethyl alcohol on zinc oxide proceeds mainly in the dehydrogenation direction and only to an insignificant degree in the dehydration direction, and the more zinc there is in the zinc oxide in excess of the stoichiometric proportion (the higher the Fermi level), the more pronounced is the first direction in comparison with the second [10–12]. Impurities that lower the Fermi level suppress dehydrogenation and simultaneously promote dehydration [13, 14]. Experimental data on dehydrogenation and dehydration is presented in Section 5.4.2.

Note that Hauffe [6] examined an electronic mechanism of catalytic decomposition of alcohol that radically differs from what we have just discussed. Hauffe's mechanism characteristically leads to a result diametrically opposite to that formulated above, namely that, according to Hauffe, lowering the Fermi level must promote dehydrogenation and suppress dehydration. The experimental data listed above (see also Section 5.4.2) contradicts Hauffe's mechanism.

Let us also note in conclusion that many authors believe that while dehydrogenation of alcohol does proceed via one or another electronic mechanism, dehy-

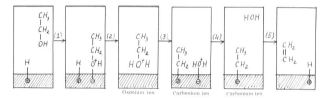

Fig. 5.10. Whitmore's mechanism of decomposition of alcohol.

dration proceeds via the acid–base mechanism, which is usually opposed to the electronic mechanism. In the acid–base mechanism the reaction is driven by the chemisorbed hydrogen atoms, which are assumed to be present on the surface. It is erroneously thought that acid–base heterogeneous reactions, in contrast to redox reactions, lie outside the scope of the electronic theory of catalysis. Actually the ionic mechanism of an acid–base reaction has an electronic mechanism at its foundation.

This can easily be seen from the example of the dehydration of ethyl alcohol, which is taken as a classical example of an acid–base reaction. According to Whitmore [15], this reaction proceeds via addition and detachment of a proton. By Whitmore's hypothesis the reaction consists of the following stages (here we have allowed for the modifications introduced by J. G. M. Bremner and D. A. Dowden; see [16]):

$$
\begin{aligned}
&1) \ (C_2H_5OH)_{gas} + (H^+)_{ads} \rightarrow (C_2H_5OH)_{ads} + (H^+)_{ads}, \\
&2) \ (C_2H_5OH)_{ads} + (H^+)_{ads} \rightarrow (C_2H_5O^+H_2)_{ads}, \\
&3) \ (C_2H_5O^+H_2)_{ads} \rightarrow (CH_3C^+H_2)_{ads} + (H_2O)_{ads}, \\
&4) \ (CH_3C^+H_2)_{ads} + (H_2O)_{ads} \rightarrow (CH_3C^+H_2)_{ads} + (H_2O)_{gas}, \\
&5) \ (CH_3C^+H_2)_{ads} + (H_2O)_{gas} \rightarrow (C_2H_4)_{gas} + (H_2O)_{gas} + (H^+)_{ads}.
\end{aligned}
\tag{5.11}
$$

In Fig. 5.10 Whitmore's mechanism is written without any changes in terms of the electronic theory. Such a scheme reveals the electronic mechanism of the reaction and thereby relates the reaction rate to the electronic state of the catalyst.

Both (5.11) and Fig. 5.10 show that the reaction proceeds on a surface that contains adsorbed ionized hydrogen. Stage (1) represents adsorption of alcohol combined with formation of "weak" bonding (the neutral form of chemisorption). Stage (2) leads to the formation on the surface of an oxonium ion $(C_2H_5O^+H_2)_{ads}$. Stage (3) leads to the formation of a carbonium ion $(CH_3C^+H_2)_{ads}$ and an H_2O molecule in the state of "weak" bonding with the surface. Stage (4) represents desorption of water. Stage (5) leads to restoration of ionized hydrogen, which drives the reaction.

We can easily see that reaction (5.11) shown in Fig. 5.10 belongs, just as the reaction depicted in Fig. 5.8, to the class of donor reactions; i.e., all other conditions being equal, the reaction is slowed down as the Fermi level moves upward in the energy spectrum. Indeed, if we assume that there is electronic equilibrium,

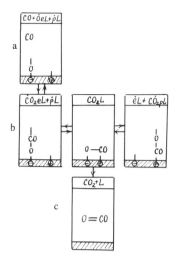

Fig. 5.11. Oxidation of carbon monoxide:
a–c) various stages in the reaction.

then when the Fermi level moves upward, a fraction of the chemisorbed hydrogen atoms that were present on the surface go over from the ionized state H^+ to the neutral state H and thereby leave the scene. As a result the reaction slows down since, according to (5.11) and Fig. 5.10, the reaction rate is proportional to the concentration of the H^+ ions on the surface. We see that the dependence of the reaction rate on the position of the Fermi level, explained in the electronic theory, in no way prohibits the acid–base mechanism from operating.

Note that Whitmore's dehydration mechanism cannot work if the catalyst surface contains no reactive hydrogen, as is sometime the case (e.g., see [17]).

5.3.3. Oxidation of Carbon Monoxide

Let us study the reaction of oxidation of CO:

$$2CO + O_2 \rightarrow 2CO_2 .$$

This reaction, which occurs both on n-type semiconductors (e.g., ZnO) and p-type conductors (e.g., NiO), has been studied from all aspects by many authors. A summary of the results is given in the paper by Takaishi [18] and in Germain's book [19]. Note that the results of different authors often contradict each other, e.g., Schwab and Block's results [20] (Hauffe and Schlosser [7] attempted an electronic interpretation of these results), on the one hand, and the results of Parravano [21] and Keier, Roginskii, and Sazonova [22], on the other (see below).

Let us examine one of the possible mechanisms of this reaction. Let us assume that the surface of the catalyst contains chemisorbed atomic oxygen and that these atoms, when in the ion-radical state, act as adsorption centers for the CO molecules. Then, on adsorption of CO molecules, surface CO_2^- ion radicals are formed as intermediate compounds. After being neutralized, these compounds are desorbed in the form of CO_2 molecules (see Fig. 5.11).

If we suppose that the various forms of chemisorption of CO_2 shown in Fig. 5.11 (see also Fig. 2.11) are in equilibrium, neglect the adsorption of CO_2 molecules, and assume the coverage of the surface by CO_2 molecules to be low, we have

$$\frac{dN_O}{dt} = \alpha_O P_{O_2}(1 - s_O N_O)^2 - \beta_O(N_O^0)^2 - \alpha_{CO} P_{CO} N_O^- + \beta_{CO} N_{CO_2}^-,$$

$$\frac{dN_{CO_2}}{dt} = \alpha_{CO} P_{CO} N_O^- - \beta_{CO} N_{CO_2}^- - \gamma N_{CO_2}^0, \tag{5.12}$$

where s_O is the effective area of a chemisorbed O atom (see Section 3.2.1), and the remaining symbols have the same meaning as in the previous sections. In equilibrium

$$\alpha_O P_{O_2}(1 - s_O N_O)^2 = \beta_O(N_O^0)^2 + \gamma N_{CO_2}^0,$$

$$\alpha_{CO} P_{CO} N_O^- = \beta_{CO} N_{CO_2}^- + \gamma N_{CO_2}^0.$$

Supposing that

$$\gamma N_{CO_2}^0 \ll \beta_O(N_O^0)^2$$

and adopting the notation (3.4), we have

$$N_O = \frac{\sqrt{\alpha_O P_{O_2}/\beta_O}}{\eta_O^0 + s_O \sqrt{\alpha_O P_{O_2}/\beta_O}},$$

$$\alpha_{CO} P_{CO} \eta_O^- N_O = \left(\frac{\beta_{CO}}{\gamma} \frac{\eta_{CO_2}^-}{\eta_{CO_2}^0} + 1\right) \gamma N_{CO_2}^0,$$

from which we obtain the expression for the reaction rate:

$$g = \gamma N_{CO_2}^0 = a \frac{\eta_O^-}{b + \eta_{CO_2}^-/\eta_{CO_2}^0} \frac{P_{CO}\sqrt{P_{O_2}}}{c\sqrt{P_{O_2}} + \eta_O^0}, \tag{5.13}$$

where we have used the following notations for brevity:

$$a = \gamma \frac{\alpha_{CO}}{\beta_{CO}} \sqrt{\frac{\alpha_O}{\beta_O}}, \qquad b = \frac{\gamma}{\beta_{CO}}, \qquad c = s_O \sqrt{\frac{\alpha_O}{\beta_O}}$$

and where, according to (3.6b) and Fig. 3.4a,

$$\eta_O^0 = \frac{1}{1 + \exp\left(-\dfrac{E_O - E_F}{kT}\right)},$$

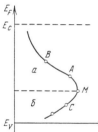

Fig. 5.12. Reaction rate of oxidation of carbon monoxide vs. position of the Fermi level.

$$\eta_O^- = \frac{1}{1 + \exp\left(\dfrac{E_O - E_F}{kT}\right)}, \qquad \frac{\eta_{CO_2}^-}{\eta_{CO_2}^0} = \exp\left(-\frac{E_{CO_2} - E_F}{kT}\right). \qquad (5.14)$$

Figure 5.12 shows the reaction rate g [in accordance with (5.13) and (5.14)] as a function of the position of the Fermi level E_F. As the Fermi level moves downward, the reaction rate increases, reaches a maximum for a certain position of the Fermi level (sufficiently low), and then falls off. In region a in Fig. 5.12 we have

$$\frac{\eta_{CO_2}^-}{\eta_{CO_2}^0} \gg b \quad \text{if} \quad \gamma N_{CO}^0 \ll \beta_{CO} N_{CO_2}^-,$$

i.e., the reaction is limited by desorption of CO_2 [see (5.12)] (the donor stage). In region b we have

$$\frac{\eta_{CO_2}^-}{\eta_{CO_2}^0} \ll b \quad \text{if} \quad \beta_{CO} N_{CO_2}^- \ll \gamma N_{CO_2}^0,$$

i.e., adsorption of CO is the limiting stage (the acceptor stage). When the Fermi level E_F lies far above the maximum point on the g vs. E_F curve (point M in Fig. 5.12) or far below this point, Eq. (5.13) yields, respectively,

$$g = K_a \exp\left(-\frac{E_a + E_F}{kT}\right) P_{CO}, \qquad g = K_b \exp\left(-\frac{E_b - E_F}{kT}\right) \frac{P_{CO}\sqrt{P_{O_2}}}{1 + c\sqrt{P_{O_2}}}. \qquad (5.15)$$

Let us study Eq. (5.13). We denote the values of E_F and g at point M in Fig. 5.12 by $(E_F)_M$ and g_M. By finding the maximum of function (5.13), we obtain

$$\exp\frac{(E_F)_M}{kT} = \sqrt{bx\left(1 + \frac{1}{c\sqrt{P_{O_2}}}\right)},$$

$$g_M = \frac{ay P_{CO}}{c\left(\sqrt{by} + \sqrt{1 + \dfrac{1}{c\sqrt{P_{O_2}}}}\right)^2}, \qquad (5.16)$$

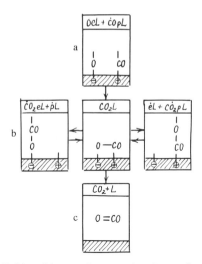

Fig. 5.13. Alternative mechanism of oxidation of carbon monoxide: a–c) various stages in the reaction.

where we used for brevity the following notations:

$$x = \exp\frac{E_{CO_2} + E_O}{kT}, \qquad y = \exp\frac{E_{CO_2} - E_O}{kT}.$$

Thus, from the analysis of Eq. (5.13) it follows that the position of point M in Fig. 5.12 depends on the conditions in which the reaction occurs, i.e., on P_{CO}, P_{O_2}, and T. As P_{O_2} increases, point M, as (5.16) shows, moves downward and for a sufficiently high P_{O_2} may become situated in the valence band. In this case throughout the range of variation of E_F from E_V to E_C the reaction is of a donor nature. On the other hand, as P_{O_2} drops, point M may be raised into the conduction band. In this case in the entire range of variation of E_F the reaction is of an acceptor nature.

We see that whether the reaction of oxidation of CO is an acceptor or donor reaction depends not only on the position of the Fermi level but also on the conditions in which the reaction occurs (for a fixed position of the Fermi level), i.e., on the temperature and partial pressures of the gases in the reaction mixture. When these quantities vary, the very curve that expresses the dependence of the reaction rate on the position of the Fermi level (the curve in Fig. 5.12) moves, with the result that a fixed Fermi level may prove to lie on the acceptor or donor branch of this curve.

Together with the mechanism examined and shown in Fig. 5.11 other mechanisms are possible for oxidation of CO. For example, the reaction may proceed via the mechanism shown in Fig. 5.13, in which the CO and O_2 molecules, in contrast to the preceding case, are adsorbed independently of each other (adsorption of O_2 is accompanied by dissociation). It can be shown that here the reaction rate g, as a function of the position of the Fermi level E_F, is represented by a curve similar to that shown in Fig. 5.12 (a curve with a maximum); i.e., lowering the Fermi level replaces a donor stage by an acceptor stage.

One explanation of the discrepancy between the experimental data on oxidation of CO obtained from different sources is that different authors were concerned with different branches of the curve in Fig. 5.12. For example, according to Schwab and Block's data [20], Li_2O in NiO promotes the oxidation of CO, while, according to Parravano [21] and Keier, Roginskii, and Sazonova [22], Li_2O in the same semiconductor acts as poison in relation to the same reaction. Direct measurements have shown that Li on the surface of NiO in the experiments of Keier, Roginskii, and Sazonova acted as donor (see [23] and Section 5.4.2). This implies that Parravano, Keier, Roginskii, and Sazonova dealt with the donor branch of the g vs. E_F curve in Fig. 5.12 (e.g., at point A or B). The results of Schwab and Block can be fitted to the above-mentioned data if we assume that Li acts as donor but that Schwab and Block dealt with the acceptor branch of the curve (e.g., at point C).

We note in passing that the discrepancy between [20], on the one hand, and [21, 22], on the other, can be resolved in another manner. We could have assumed that all the data refer to the donor branch of the curve in Fig. 5.12, but that Li introduced into NiO acts as donor (forms an interstitial solution), when we are dealing with the data of [21, 22], and hence moves us from point A to point B in Fig. 5.12, while in relation to the data of [20] Li acts as acceptor (forms a substitutional solution) and moves us from point B to point A. Indeed, Bielanski and Deren [24] have shown (see also Section 4.2.1) that with an increase in the concentration of Li in NiO the nature of the solution changes, namely, the substitutional solution at low concentrations transfers into an interstitial solution at high concentrations. In the process the reaction rate passes through a maximum as the amount of Li increases (transition $B \to A \to B$ in Fig. 5.12).

We will return to the reaction of oxidation of CO in the section devoted to the mechanism of the promoting and poisoning action of an impurity (Section 5.5.3) and to the relation between the catalytic activity and the electrical conductivity of a semiconductor (Section 5.4.2; see also Section 6.1.2).

5.3.4. Hydrogen–Deuterium Exchange

Finally, let us study the reaction of hydrogen–deuterium exchange

$$H_2 + D_2 \to 2HD.$$

This is one of the simplest heterogeneous reactions that occurs at the surface of a semiconductor. At present there is a vast body of experimental data concerning this reaction. (Below we give a summary of the experimental results.) This reaction was also studied theoretically. Hauffe [25] considered it within the framework of the boundary layer theory, and Dowden *et al.* [26] undertook a theoretical study of this reaction from the viewpoint of the crystal field theory. We will employ the electronic theory of catalysis [27].

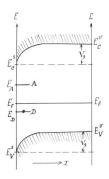

Fig. 5.14. Band bending in the hydrogen–deuterium exchange reaction.

Let us assume that the H_2 and D_2 molecules dissociate into atoms in the adsorption process. We will also assume that the adsorption centers for deuterium and hydrogen atoms coincide. (We will ignore the nature of these centers.) We will denote the surface concentration of these centers by N^* and the surface concentrations of the chemisorbed hydrogen and deuterium atoms by N_H and N_D. The surface is assumed to be saturated by hydrogen and deuterium atoms, i.e.,

$$N_H + N_D = N^+. \tag{5.17}$$

We know that chemisorbed hydrogen atoms, as well as chemisorbed deuterium atoms, may be in an electrically neutral or (positively or negatively) charged state, with the electrically neutral form being a radical (reactive) form and the charged forms being valence-saturated forms (see Section 2.4.3). We will assume that these forms exist on the surface in concentrations that correspond to the case of electronic equilibrium. Let N_H^0 and N_D^0 be the surface concentrations of H and D atoms in the reactive state (state of "weak" bonding).

As a first approximation we may assume that the chemisorbed H and D atoms have the same energies of affinity for a free lattice electron ($E_C - E_A$) and the same ionization energies ($E_C - E_D$). Then, as follows from (3.5), we have

$$\eta^0 = \frac{N_H^0}{N_H} = \frac{N_D^0}{N_D} = \left[1 + \exp\left(-\frac{E_A - E_F}{kT} \right) + \exp\left(-\frac{E_F - E_D}{kT} \right) \right]^{-1}, \tag{5.18}$$

where E_F is the Fermi level, and E_A and E_D are the acceptor and donor levels representing the hydrogen or deuterium atom (see Fig. 5.14).

In Fig. 5.14, V_s is the surface potential (which gives the bending of the energy bands), with

$$V_s = E_C^s - E_C^v = E_V^s - E_V^v, \tag{5.19}$$

where E_C^s and E_V^s are the bottom of the conduction band and the top of the valence band in the plane of the surface, and E_C^v and E_V^v are the same quantities in the crystal's bulk. Obviously, V_s is positive if the bands are bent upward and negative if they are bent downward.

We will assume that only the chemisorbed H and D atoms in the radical state participate in the exchange reaction and that the reaction occurs via the following equations:

$$H_2 + \dot{D}L \rightarrow HD + \dot{H}L, \quad D_2 + \dot{H}L \rightarrow HD + \dot{D}L, \tag{5.20}$$

where L is the symbol of the lattice, and the dot above a symbol stands for a free valence ($\dot{H}L$ and $\dot{D}L$ stand for the chemisorbed H and D atoms in a state of "weak" bonding with the surface).

According to (5.20), the reaction rate is given by the formula

$$g = \alpha_D N_D^0 P_H + \alpha_H N_H^0 P_D,$$

or, according to (5.18),

$$g = \eta^0 (\alpha_D N_D P_H + \alpha_H N_H P_D), \tag{5.21}$$

with

$$\frac{dN_H}{dt} = -\frac{dN_D}{dt} = \eta^0 (\alpha_D N_D P_H - \alpha_H N_H P_D), \tag{5.22}$$

where P_H and P_D are the partial pressures of H_2 and D_2.

In a steady-state equilibrium process

$$\frac{dN_H}{dt} = -\frac{dN_D}{dt} = 0$$

and, hence, according to (5.22),

$$\alpha_D N_D P_H = \alpha_H N_H P_D, \tag{5.23}$$

where, on the basis of (5.17)

$$N_H = \frac{\alpha_D P_H}{\alpha_H P_D + \alpha_D P_H} N^*. \tag{5.24}$$

According to (5.21), (5.23), and (5.24), we have

$$g = 2\eta^0 \alpha_H N_H P_D = 2\eta^0 \frac{\alpha_H \alpha_D P_H P_D}{\alpha_H P_D + \alpha_D P_H} N^*. \tag{5.25}$$

If we assume that $P_H = P_D = P$, then, instead of (5.25), we have

$$g = \eta^0 \alpha P N^*, \tag{5.26}$$

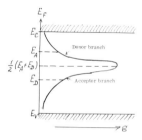

Fig. 5.15. Reaction rate of hydrogen–deuterium exchange vs. position of the Fermi level.

where

$$\alpha = 2\,\frac{\alpha_H \alpha_D}{\alpha_H + \alpha_D}$$

and η^0 is given by (5.18).

The dependence of the reaction rate g on the position of the Fermi level E_F at the surface of the crystal (for a given pressure P and temperature T) is depicted in Fig. 5.15 in accordance with (5.26) and (5.18). In the region where E_F is higher than $(E_A + E_D)/2$ the reaction is of the donor type, i.e., speeds up as the Fermi level moves downward, while in the region where E_F is lower than $(E_A + E_D)/2$ the reaction is of the acceptor type, i.e., slows down as the Fermi level moves downward.

We will now relate the results obtained here with the experimental data. Let us see what effect the various factors in (5.26) have on the rate of hydrogen–deuterium exchange.

5.3.4.1. Pressure

Pressure P enters into (5.26) not only explicitly but through the parameter η^0, as seen from (5.18), since both $E_A - E_F$ and $E_F - E_D$ depend on V_s and, hence, are generally functions of P. In the present model, in which the surface is assumed to be saturated with hydrogen and deuterium atoms [all adsorption centers are assumed to be occupied; see (5.17)], V_s may be taken as being independent of P. Thus, according to (5.26), the hydrogen–deuterium exchange reaction is a first-order reaction with respect to hydrogen (deuterium).

A number of papers are devoted to investigations of the dependence of the rate of this reaction on pressure, and in almost all (e.g., see [28–30]) the reaction proved to be first-order with respect to both hydrogen and deuterium.

5.3.4.2. Impurities

Introducing an impurity into the crystal shifts the Fermi level E_F, which leads to a variation in g, according to (5.26). A donor impurity, we know, moves the

Fermi level upward, while an acceptor impurity moves it downward. The effect of the same impurity on the catalytic activity of the crystal in the case of an acceptor reaction will be opposite to the effect in the case of a donor reaction.

Most experimental data point to the fact that the reaction of hydrogen–deuterium exchange belongs to the acceptor class (i.e., is accelerated by electrons and slowed down by holes). This means that the experimenter, as a rule, deals with the acceptor branch of the g vs. E_F curve (see Fig. 5.15), on which branch chemisorbed hydrogen and deuterium atoms act as donors. Here a donor impurity increases the catalytic activity and an acceptor impurity reduces it.

For example, Heckelsberg, Clark, and Bailey [31] found that introduction into ZnO of a donor impurity (Al_2O_3) increases the reaction rate, while doping with an acceptor impurity (Li_2O) reduces the rate. Molinari and Parravano [28] noted that a donor impurity (Al_2O_3, Ga_2O_3) added to a ZnO sample promotes the reaction, while an acceptor impurity (Li_2O) slows it down. An increase in the catalytic activity of SiO_2 in relation to the hydrogen–deuterium exchange reaction when the sample is doped with a donor impurity was observed by Kohn and Taylor [29]. We especially note the work of Holm and Clark [32], who found that the reaction rate g passes through a maximum as the fraction of donor impurity (sample of Al_2O_3 with SiO_2 as impurity) steadily increases. Apparently this maximum is due, as shown by Fig. 5.15, to a transition from the acceptor branch of the g vs. E_F curve to the donor branch as E_F steadily increases.

5.3.4.3. The State of the Surface

Surface treatment of any kind that changes V_s, e.g., adsorption of foreign gases on it (i.e., bends the energy bands), must lead, according to (5.18) and (5.26), to a change in g. As a result of adsorption of a donor gas we move upward on the g vs. E_F curve in Fig. 5.15, while adsorption of an acceptor gas shifts us downward. If we remain on the acceptor branch of this curve, adsorption of a donor gas must lead to higher catalytic activity, and adsorption of an acceptor branch to lower activity. For instance, heating the sample in a hydrogen atmosphere (which leads to adsorption and absorption of hydrogen) does indeed, as follows from a multitude of papers (e.g., see [28, 31–36]), increase catalytic activity. At the same time, as discovered by Sundler and Gazith [37], adsorption of oxygen has a poisoning effect.

At this point the experiments of Volt and Weller [33] must be noted. In these a drop in catalytic activity was observed as a result of adsorption of water (water usually acts as donor). To understand this result, we must assume that the two researchers dealt with the donor branch of the curve in Fig. 5.15, or with the acceptor branch, in which case water molecules acted as acceptors. We note in this connection that the acceptor functions of water (negative charging of a surface under adsorption of water) were observed earlier (e.g., see [38]).

5.4. THE RELATIONSHIP BETWEEN THE CATALYTIC ACTIVITY OF A SEMICONDUCTOR AND ITS ELECTRONIC PROPERTIES

5.4.1. The Origin of the Relationship between Catalytic Activity, Work Function, and Electrical Conductivity

In Sections 5.2 and 5.3 we found that the catalytic activity of a semiconductor is determined, among other things, by the position of the Fermi level at the surface of the crystal. There are two important consequences of this.

The first consequence is a correlation between catalytic activity of a semiconductor's surface and the work function of an electron emerging from the semiconductor. We will characterize the position of the Fermi level by the distance between the Fermi level and the bottom of the conduction band. This distance at the surface and in the bulk will be denoted by ϵ_s and ϵ_v (see Fig. 5.14):

$$\epsilon_s = E_C^s - E_F^s, \quad \epsilon_v = E_C^v - E_F^v .$$

As usual, we denote the reaction rate by g and the thermionic work function by φ_T. According to (5.1) and (4.38),

$$\begin{cases} g = g(\epsilon_s), & (5.27) \\ \varphi_T = \epsilon_s + \delta + V_D, & (5.28) \end{cases}$$

where δ is the energy of affinity of a free electron for the lattice, and V_D the dipole component of the work function. Combining (5.27), (5.28), (5.2a), and (5.2b), we obtain

$$\frac{dg}{d\varphi_T} = \frac{dg}{d\epsilon_s} \begin{cases} > 0 & \text{for a donor reaction,} \\ < 0 & \text{for an acceptor reaction.} \end{cases} \quad (5.29)$$

Thus, the catalytic activity and work function vary in the same direction if the reaction is of the donor type and in the opposite manner if the reaction is of the acceptor type. The type of dependence of the reaction rate on the work function may therefore serve as an experimental criterion for determining unambiguously whether a reaction belongs to the donor type or the acceptor type.

The second consequence of the fact that catalytic activity is determined by the position of the Fermi level at the surface of the semiconductor is the relationship between catalytic activity g and electrical conductivity κ. In general we have (see Sections 1.6.2 and 4.2.2)

$$\kappa = \kappa(\epsilon_s, \epsilon_v), \quad (5.30)$$

with

$$\frac{\partial \kappa}{\partial \epsilon_s} < 0, \quad \frac{\partial \kappa}{\partial \epsilon_v} < 0, \quad (5.31a)$$

Table 5.1

	n-Type semiconductor	p-Type semiconductor
n-Reaction	$\dfrac{dg}{d\kappa} > 0$	$\dfrac{dg}{d\kappa} < 0$
p-Reaction	$\dfrac{dg}{d\kappa} < 0$	$\dfrac{dg}{d\kappa} > 0$

for n-type conduction and

$$\frac{\partial \kappa}{\partial \epsilon_s} > 0, \qquad \frac{\partial \kappa}{\partial \epsilon_v} > 0. \qquad (5.31b)$$

for p-type conduction. Combining (5.27) and (5.30) yields

$$\frac{dg}{d\kappa} = \frac{dg/d\epsilon_s}{\partial \kappa/\partial \epsilon_s + (\partial \kappa/\partial \epsilon_v)(d\epsilon_v/d\epsilon_s)}. \qquad (5.32)$$

Bearing in mind (5.2a), (5.2b), (5.31a), and (5.31b) and recalling that, according to (4.13),

$$\frac{d\epsilon_v}{d\epsilon_s} \geqslant 0, \qquad (5.33)$$

we can construct, on the basis of (5.32), Table 5.1.

Thus, the catalytic activity and electrical conductivity vary together or in the opposite manner depending on the type of reaction (acceptor or donor) and the type of conduction of the semiconductor at which the reaction takes place (n-type or p-type conduction). Knowing the dependence of activity on electrical conductivity (established experimentally) and the type of semiconductor, we can establish the type of reaction.

The parallelism between activity and electrical conductivity has a simple physical meaning; namely, the electrical conductivity is determined by the concentrations of free electrons and holes in the semiconductor, and these free electrons and holes, as we have seen, take part in reactions (as reaction components) and, hence determine its rate.

We can therefore conclude that the factors influencing the work function or electrical conductivity of a semiconductor must also influence its catalytic activity. One such factor is the treatment of the sample, for instance, the introduction of impurities (of one or another nature or concentration) into the semiconductor's bulk or onto its surface. Consequently, different samples of the same semiconductor, prepared differently and differing in work function or electrical conductivity, must also differ in catalytic activity. In going from sample to sample, the work function and electrical conductivity, on the one hand, and the catalytic activity, on the other, must vary in parallel.

The established correlations must manifest themselves not only in the case of single crystals, as might seem at first glance, but also in polycrystalline samples, with which specialists in catalysis usually have to deal. Indeed, according to Petritz's theory [39], the electrical conductivity of a sample consisting of grains separated by contact barriers (according to Petritz, practically all the voltage applied to the sample drops across these barriers) depends, of course, on the shape and height of the barriers but at the same time remains proportional to the electrical conductivity of a grain. This leads to (5.33) and, naturally, to (5.29).

The existence of a correlation between catalytic activity, on the one hand, and work function and electrical conductivity, on the other, was pointed out in the 1950s (see [4, 40]), when there was no experimental work disproving or confirming this theoretical prediction. At present there exists a broad spectrum of papers in which such a correlation was observed experimentally; several authors have measured the work function and electrical conductivity simultaneously with the catalytic activity of samples that differ in their history, and the variations in these characteristics were found to be related in the same or opposite manner in going from sample to sample (See Section 5.4.2).

Note that the parallelism between electrical conductivity and catalytic activity and between work function and catalytic activity may sometimes be violated. This happens in the interval of values E_F for which the reaction rate is independent of E_F. For example, for alcohol dehydrogenation this occurs when E_F is high and for alcohol dehydration when E_F is low (see Section 5.3.2 and Fig. 5.9). We also note that the above-mentioned connection between electrical conductivity and catalytic activity may also be destroyed when the semiconductor has a quasiisolated surface, i.e., when ϵ_s is independent of ϵ_v (see Section 4.3), if the dimensions of the crystal are not too small.

It is important to emphasize that the parallelism between the electronic parameters (such as the electrical conductivity and work function) has meaning only in cases where variations in electronic parameters and activity are produced by the action of the same factors on the semiconductor, other conditions being unaltered. There is no reason to expect any parallelism in variations in the work function and catalytic activity or the electrical conductivity and catalytic activity when different semiconductors (i.e., of different chemical nature) are compared. Indeed, the reaction rate is determined, as we have seen in Section 5.3, not by the position of the Fermi level relative to the energy bands but by its position relative to those local surface levels that correspond to the chemisorbed particles participating in the reaction. The position of these levels in the energy spectrum is generally different for semiconductors of different kinds. For this reason the parallelism between the electrical conductivity and catalytic activity or the work function and catalytic activity may in these cases be completely masked. For the same reason there are no grounds for seeking a connection between the catalytic activity of semiconductors and their conduction type (n or p) if we are dealing with semiconductors of different chemical origin.

In conclusion we note that along with the connection between electrical conductivity and catalytic activity there must exist, generally speaking, a relationship between electrical conductivity and adsorptivity of a semiconductor. Indeed, as demonstrated in Section 3.2.1, adsorptivity depends on the position of the Fermi level, ϵ_s, at the surface of the crystal. For an acceptor gas it is higher the smaller ϵ_s is (i.e., the higher the Fermi level), while for a donor gas it is higher the greater ϵ_s is (i.e., the lower the Fermi level). From this we may conclude that for an acceptor the relation between the electrical conductivity and the adsorptivity on an n-type semiconductor must be symbatic, while on a p-type semiconductor it must be the opposite. For a donor gas the situation is reversed. This effect can easily be demonstrated by using the family of isotherms taken for samples with different electrical conductivities. The isotherms must lie below one another in order of increasing electrical conductivity (inverse relation) or decreasing conductivity (direct relation).

5.4.2. Experimental Results

Different authors investigating the connection between electronic parameters and catalytic activity often attach a different meaning to this concept and interpret their task differently. For this reason experimental papers devoted to this question must be divided into two groups.

The first group comprises papers in which the relation between the electronic parameters and catalytic activity is interpreted as the relation between two characteristics of the sample; e.g., its thermionic work function φ_T is related to its reaction rate g, or its electrical conductivity κ related to g (the reaction rate characterizes the catalytic activity of the given sample in relation to the reaction). Some of the papers of this group are considered below.

The second group comprises papers in which the change in work function or electrical conductivity occurring due to the catalytic action is measured. In these papers the rate of the catalytic reaction occurring at the surface of the semiconductor is measured simultaneously with the work function or electrical conductivity, which changes in the course of the reaction. Some of the papers of this group are considered in Section 5.4.3.

Let us consider the first group of papers.

(a) We start with the reaction of decomposition of alcohol. Here we must note the work of Myasnikov and Pshezhetskii [10], who investigated the dehydrogenation of isopropyl alcohol C_3H_7OH on ZnO with oxygen chemisorbed at the surface of the latter, the oxygen acting as a surface impurity. Under the action of oxygen both the electrical conductivity and the catalytic activity were lowered simultaneously. For zinc oxide, which is an n-type semiconductor, this direct relation between the variation in electrical conductivity and activity was further evidence (see Section 5.3.2) of the acceptor nature of alcohol dehydrogenation.

Another important paper is that of Zhabrova and his collaborators [14], who studied the decomposition of isopropyl alcohol on zinc oxide with various impurities and found that the greater the work function, the lower the activity of the sample in relation to dehydrogenation but the higher its activity in relation to dehydration.

Similar results were obtained by Keier and his collaborators[41], who studied the decomposition of isopropyl alcohol on TiO_2 with W and Fe as impurities. Varying the fraction of the impurity enabled them to vary the work function of the catalyst and thereby control its selectivity. A decrease in the work function (i.e., a shift of the Fermi level upward) leads to a speeding up of dehydrogenation and a slowing down of dehydration, while an increase in the work function (i.e., a shift of the Fermi level downward) produces the opposite effect, as predicted by theory.

The findings of Spitsyn *et al.* [42] agree with the above results. Here the researchers studied the correlation between variations in the work function and those in the selectivity of the catalyst caused by the introduction of radioactive impurities. They studied the decomposition of isopropyl alcohol on Y_2O_3 and found that an increase in the work function leads to speeding up of dehydrogenation and slowing down of dehydration. In another paper [17] the same group of researchers studied the dehydration of decyl alcohol $C_{10}H_{21}OH$ on samples of tungsten disulfide WS_2, which either possesses a sulfur deficiency compared to its stoichiometric ratio (*n*-samples) or an excess over stoichiometry (*p*-samples). They observed a correlation between catalytic activity and electrical conductivity. On *p*-samples the activity increased with electrical conductivity (i.e., as the Fermi level moved downward), while on *n*-samples the activity decreased as electrical conductivity grew (i.e., as the Fermi level moved upward). This points once more to the donor nature of dehydration.

(b) We now turn to oxidation of carbon monoxide. Here we note the work of Keier, Roginskii, and Sazonova [22], cited earlier. The researchers studied oxidation of CO on NiO samples that differed in impurity content. They introduced Li_2O as the impurity, which forms a substitutional solution with the NiO lattice; i.e., the doubly charged ions Ni^{++} at the lattice sites were partially replaced by singly charged ions Li^+, which consequently acted as acceptor defects. Introduction of Li into NiO must lead to a lowering of the Fermi level (an increase in ϵ_v) and, since NiO is a *p*-type semiconductor, to an increase in conductivity, which was indeed observed. The catalytic activity in relation to oxidation of CO was measured on samples with different Li content and consequently with different electrical conductivities. A clear inverse relation was established between activity and electrical conductivity, i.e., the higher the initial conductivity of the sample the lower the activity. Lithium introduced into the crystal thus acted as a poison.

On the basis of these results the introduction of lithium could be expected to slow down oxygen adsorption (oxygen acting as an acceptor gas) and stimulate the adsorption of CO (carbon monoxide acting as a donor gas) (see Section 3.2.1). Measurements made by the same authors, however, produced a directly opposite result, which thus contradicted the theory. This contradiction was resolved when the work function of the sample was measured [23]. It was established that introduction of lithium decreases the work function instead of increasing it (as was expected), which is unambiguous evidence of positive charging of the surface on introduction of lithium. This also clearly pointed to the fact that the authors

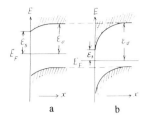

Fig. 5.16. Band bending in the oxidation of carbon monoxide: a) low lithium content; b) high lithium content.

were dealing with the donor branch of the g versus E_F curve (Fig. 5.12, the region where dg/dE_F is negative).

Comparison of the data on electrical conductivity and the work function compels us to conclude, on the basis of (5.32), that in the given case an increase in ϵ_v is accompanied by a decrease in ϵ_s rather than by an increase, i.e., condition (5.3) or (4.13) proves to be violated (this is illustrated by Fig. 5.16, where Fig. 5.16a corresponds to a sample with low lithium content and Fig. 5.16b to a sample with high lithium content). This result can be understood if we assume that on the surface of the sample and in the layer adjacent to it and enriched by lithium the excess lithium atoms are situated in the interstices and act as donors.

The work of Keier, Roginskii, and Sazonova is an example of the case where even knowing whether the relation between catalytic activity and electrical conductivity is symbatic or the opposite, we cannot make an unambiguous conclusion concerning the class (acceptor or donor) to which the given reaction belongs, since condition (5.33) is not always met.

(c) Let us now turn to the reaction of hydrogen–deuterium exchange. The dependence of the reaction rate on the position of the Fermi level for this case is shown in Fig. 5.15. On the acceptor branch of the curve in Fig. 5.15 the catalytic activity is directly related to the electrical conductivity for an n-type semiconductor and the opposite for a p-type semiconductor. This pattern was observed in many experiments. For instance, the heating of a sample in a hydrogen atmosphere leads, as noted in Section 5.3.4, to an increase in its catalytic activity and at the same time to an increase in the electron component of conduction and a decrease in the hole component. Heating in an oxygen atmosphere acts in the opposite manner (see [28, 31, 33, 37]). At the same time, the introduction into ZnO of donor impurities (Al_2O_3, Ga_2O_3), which increase conductivity, leads (as we have seen) to an increase in activity, while an acceptor impurity (such as Li_2O), which reduces conductivity, causes a lowering of activity [28, 31].

It is worthwhile to note in conclusion that the list of experimental papers devoted to the connection between catalytic activity and conductivity and work function is not exhausted by the works cited here. This connection has been studied by many authors for different catalysts and reactions. A summary of the results obtained (also not exhaustive) is given in Table 5.2.

Table 5.2

Reaction	Catalyst	Conduction type	Nature of relation between conductivity and activity	Source
$H_2 + D_2 \rightarrow 2HD$	ZnO with different contents of added Li_2O, Al_2O_3, Ga_2O_3	n	symbatic	[31, 28]
$H_2 + D_2 \rightarrow 2HD$	Cr_2O_3 treated differently in H_2 and O_2 atmospheres	p	in reverse order	[33, 37]
$CH_3OH \rightarrow CO + 2H_2$	ZnO with different contents of added Zn in excess of stoichiometry	n	symbatic	[43]
$C_3H_7OH \rightarrow C_3H_6O + H_2$	ZnO with different contents of chemisorbed O_2	n	symbatic	[10]

Reaction	Semiconductor	Type	Effect	Ref.
$C_{10}H_{21}OH \rightarrow C_{10}H_{20} + H_2O$	WS_2 with different contents of added W in excess of stoichiometry	p	symbatic	[17]
$C_{10}H_{21}OH \rightarrow C_{10}H_{20} + H_2O$	WS_2 with different contents of added W in excess of stoichiometry	n	in reverse order	[17]
$2CO + O_2 \rightarrow 2CO_2$	NiO with different contents of added Li_2O	p	in reverse order	[22]
$2N_2O \rightarrow 2N_2 + O_2$	NiO with different contents of added Al_2O_3	p	symbatic	[44]
$C_7H_{16} \rightarrow C_6H_5CH_3 + 4H_2$	$Cr_2O_3 \cdot Al_2O_3$ samples of different composition	?	symbatic	[45]

5.4.3. Variations in Electrical Conductivity and Work Function in the Course of a Reaction

We will now turn to the second group of experimental papers that study the relation of catalytic activity to work function and conductivity, understood as the variation of these parameters in the course of the reaction. This effect (the variation of work function and electrical conductivity under the influence of the reaction) may have two causes.

In the reaction process the relative fractions of the molecules adsorbed on the surface and entering into the reaction and of the molecules that are reaction products change. The coverage of the surface by the reagents gradually decreases, while the coverage by the products increases. In other words, the nature and concentration of the adsorbate change, which, as we know, may influence the electrical conductivity and work function of the sample. Here variations in conductivity and work function occurring in the reaction process are due to the influence of adsorption, whose mechanisms we examined in Sections 4.2.1 and 4.2.3.

With suitable experimental conditions this effect may be avoided. If the reaction products are removed and the pressures of the gases participating in the reaction are kept constant so that the coverage of the surface by adsorbed molecules (of each kind) remains unchanged in the course of the reaction, variations in the electrical conductivity and the work function due to this factor will be reduced to zero. These two quantities will remain as established at the beginning of the reaction.

Another factor capable of producing variations in the work function and electrical conductivity under the influence of the reaction is the change in the chemical composition of the catalyst during the reaction. While in the previous case variations in the electronic parameters were due to the variations in adsorbate composition, in the present case they are due to changes in the adsorbent. For instance, many reactions catalyzed by solid oxides are usually accompanied by oxidation or reduction of the catalyst. As the reaction proceeds, the degree (and sometimes the very nature) of stoichiometric disturbances in the semiconductor gradually changes, which, of course, is reflected in the work function and conductivity of the semiconductor. Variations in these two quantities observed during a reaction may thus be produced by this fairly trivial reason.

This effect can be removed by a suitable choice of system and experimental conditions in which stability of the catalyst during its operation (i.e., constancy of structure and composition) is guaranteed.

Among the papers of this group we note the one by Boreskov and Matveev [43]. The researchers investigated the decomposition of methyl alcohol on ZnO:

$$CH_3OH \rightarrow CO + 2H_2 .$$

The electrical conductivity increased in the course of the reaction and acquired a metallic character; i.e., the exponential nature of the temperature dependence of electrical conductivity disappeared. The increase in conductivity in these experiments is caused by the reduction of zinc oxide under the influence of methyl alcohol vapor. In the course of the reaction the zinc oxide becomes enriched by zinc in excess of the stoichiometric measure, and this leads to degeneracy of the electron gas, or metallic conduction.

Different samples, i.e., prepared differently and differing considerably in the magnitude of their conductivity, gradually become equal in the course of the reaction (i.e., in the course of the reduction) as regards both the value of their conductivity and their catalytic activity. This means that the process of sample preparation cannot, strictly speaking, be considered as finished at the moment when the reaction starts. The final preparation of a sample occurs in the reaction process.

The increase in conductivity in the process of reaction is accompanied by a simultaneous increase in catalytic activity. There was a clear direct relation between variations in catalytic activity and electrical conductivity.

Let us also note a series of papers by the Polish investigators Bielanski, Deren, and Haber [46], who studied the change in electrical conductivity of a catalyst during its operation and the connection between electrical conductivity and catalytic activity for different catalysts (both n- and p-type semiconductors) in dehydrogenation and oxidation of ethyl alcohol.

When the reacting gases were introduced, the electrical conductivity of the catalyst changed sharply and a new value was established. This value remained constant during the further course of the reaction. The magnitude of this change proved to depend on the temperature and the composition of the catalyst. In all cases a strict parallelism between variations in conductivity and catalytic activity was observed, with the activity being characterized by the yield of the reaction products.

Finally, we note the work of Lyashenko, Romanova, and Stepko [47–50]. In these experiments a mixture of carbon dioxide and oxygen was directed onto the surface of CuO at various temperatures, and then the electrical conductivity and work function of CuO were measured. At temperatures at which CO was oxidized a sudden drop in the value of the work function, and a sudden change in the activation energy were observed. The same results were obtained for other semiconductors, such as MnO_2, NiO, and Ge.

The same authors studied the decomposition of formic acid and nitrogen monoxide on Ge. In this case, too, a sharp drop in the value of the work function was observed at temperatures at which the reaction starts.

Since in the course of a reaction on the surface of a semiconductor the electrical conductivity and work function of the semiconductor change, generally speaking, we can use these changes to judge the course of the reaction. This problem was studied by the Rumanian researchers Nicolescu, Spinzi, and Suceveanu [89, 90], who worked with ammonia synthesis on iron oxides. Direct adsorption measurements showed that in this reaction hydrogen acts as donor and nitrogen as acceptor. According to these authors, the reaction consists of the stages shown on the next page.

$$N_2 + 2eL \to 2NeL,$$
$$H_2 \to 2HpL + 2eL,$$

$$NeL + HpL \to NHL,$$
$$NHL + HpL + eL \to NH_2L,$$
$$NH_2L + HpL + eL \to NH_3L,$$

$$NH_3L \to NH_3 + L.$$

The adsorption of nitrogen is the limiting stage here.

5.4.4. Correlation between Catalytic Activity and the Forbidden Gap Width in the Energy Spectrum of a Semiconductor

In 1967, while comparing the catalytic and electronic properties of semiconductors, Krylow [51] noticed a certain correlation between the rate g of a reaction on their surfaces and the width u of the forbidden gap between the valence and conduction bands ($u = E_C - E_V$). This correlation is clearly seen in Fig. 5.17 (taken from [51]). Let us establish whether this correlation between the catalytic activity and the forbidden gap width of a semiconductor (established empirically) can be derived from the principles of the electronic theory of catalysis [52].

We will consider a reaction occurring at the surface of a semiconductor and consisting of a sequence of stages. We will assume that a certain stage in this sequence has a reaction rate proportional to the concentration n of free electrons on the crystal's surface or to the concentration p of free holes on the surface.

A simple example is the reaction of dissociation of a diatomic molecule AB that proceeds according to the equation

$$AB + eL \to A + BeL.$$

Here we used notations common in the electronic theory of catalysis: L is the

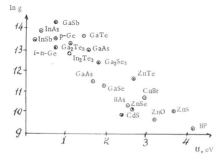

Fig. 5.17. Correlation between catalytic activity and forbidden gap width.

lattice symbol, eL the symbol of a free lattice electron, and BeL the symbol of a chemisorbed B molecule in a state of "strong" bonding with the lattice. This reaction was calculated quantum-mechanically in [53].

Many reactions of this type are considered in the electronic theory of catalysis, e.g., chemisorption of O_2 (see Fig. 2.18a) and ethylene hydrogenation (see Fig. 58a in [53]); see also Figs. 5.4a and 5.5a. We will assume that there is a limiting stage in such reactions, i.e., the reaction rate is determined by the rate of this stage.

In the first case, where the reaction rate (denoted by g_n) is proportional to the free hole concentration, we are dealing with an acceptor reaction, i.e., a reaction whose rate is higher the higher the Fermi level in the energy spectrum. In the second case, where the reaction rate (denoted by g_p) is proportional to the free hole concentration, we are dealing with a donor reaction, whose rate is higher the lower the Fermi level in the energy spectrum. Thus,

$$g_n = a_n\, n, \qquad g_p = a_p\, p, \tag{5.34}$$

where a_n and a_p are factors that are of no interest to us. In electronic equilibrium

$$n = b_n \exp\left(-\frac{\epsilon}{kT}\right), \qquad p = b_p \exp\left(-\frac{u - \epsilon}{kT}\right), \tag{5.35}$$

where $\epsilon = E_C - E_F$ is the distance from the bottom of the conduction band to the Fermi level (at the surface), and b_n and b_p are factors that depend on the statistical weights of the conduction and valence bands and carrier effective masses and weakly depend (in comparison with an exponential factor) on temperature. In the region of intrinsic conduction we may put

$$\epsilon = u/2 \tag{5.36}$$

(to within a constant term that is of no interest).

Substituting (5.26) into (5.35) and then (5.35) into (5.34), we obtain

$$g = C \exp\left(-\frac{u}{kT}\right), \tag{5.37}$$

where $C = C_n$ and $g = g_n$ or $C = C_p$ and $g = g_p$ depending on whether the reaction is of an acceptor or donor type. We have thus obtained the sought relation between the reaction rate g and the forbidden gap width u. If the factor C in (5.37) was the same for all catalysts, then (5.37) would serve as a universal relationship between g and u. Actually, however, C depends on a number of parameters characterizing the specific catalysts. This results in a spread of points in Fig. 5.17, and (5.37) is a correlation dependence instead of a functional dependence. We see that the origin of the relation between g and u in the electronic theory is quite simple.

5.5. THE EFFECT OF VARIOUS FACTORS ON CATALYTIC ACTIVITY

5.5.1. The Effect of an External Electric Field

By acting on a semiconductor with various agents that shift the Fermi level or, in other words, change the concentrations of the electron and hole gases at the surface, we can control the catalytic activity of the semiconductor. Here are some of these agents.

(1) An impurity introduced into the bulk of the crystal or onto its surface. We will study its effect in Sections 5.5.3 and 5.5.4.

(2) Illumination of the crystal by a photoelectrically active light. The study of this factor is beyond the scope of the present book. We will study this factor below.

(3) An external electric field applied perpendicularly to the surface of the catalyst. We will study this factor below.

(4) The thickness of the thin semiconductor film on a metal if the catalyst is this film. By varying the film's thickness we can change the position of the Fermi level at its outer surface. We will study this effect in Section 5.5.2.

Let us now turn our attention to the effect that an external electric field has on the catalytic activity (the electrocatalytic effect; see [54, 55]). In Section 4.4.2 we examined the effect of a field on the adsorptivity of a semiconductor (the electro-adsorptive effect). The electrocatalytic effect is characterized by the fractional change in the reaction rate in the field,

$$\frac{\Delta g}{g_0} = \frac{g - g_0}{g_0}, \tag{5.38}$$

where g and g_0 are the reaction rates with and without field, with $g = g(\epsilon_s)$ and $g = g(\epsilon_{s0})$ (here ϵ_s and ϵ_{s0} are the distances between the Fermi level and the bottom of the conduction band in the surface plane with and without field). When a positive potential is applied to the sample, the bands bend upward and the above-mentioned distance increases, i.e., $\epsilon_s > \epsilon_{s0}$, and, hence the effect is positive for a donor reaction ($g > g_0$), i.e., the reaction rate increases, and negative for an acceptor reaction ($g < g_0$), i.e., the reaction rate drops. A negative potential produces an opposite effect. Thus, under the action of a field the catalytic reaction speeds up or slows down, depending on the direction of the field and type of reaction.

Let us examine the particular (but widespread) case where g has the simple form

$$g(\epsilon_s) = C \exp\left(\mp \frac{\epsilon_s}{kT} \right). \tag{5.39}$$

(Reactions considered in Section 5.3 are examples.) The upper sign in (5.39) corresponds to an acceptor reaction and the lower sign to a donor reaction. Combining (5.39) with (5.38) yields

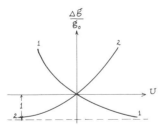

Fig. 5.18. Effect of an external electric field on catalytic activity. Curve 1 corresponds to an acceptor reaction, and curve 2 to a donor reaction.

$$\frac{\Delta g}{g_0} = \exp\left(\mp \frac{\Delta V_s}{kT}\right) - 1, \tag{5.40}$$

where

$$\Delta V_s = \epsilon_s - \epsilon_{s0} = V_s - V_{s0}$$

is the change in the bending of the bands initiated by the field.

We will now employ the dependence of ΔV_s on U, where U is the potential difference across the sample (see Section 4.4.2). Substituting (4.81) and (4.76) into (4.80) and then (4.80) into (5.40), we obtain $\Delta g/g_0$ as a function of U. The function is depicted in Fig. 5.18; curve 1 corresponds to an acceptor reaction and curve 2 to a donor reaction. We see that the effect is unsymmetric with respect to the sign of the field, namely, for the same absolute value of U the positive effect is greater than the negative.

Until recently the electrocatalytic effect was not observed experimentally and remained only a theoretical "forecast." There now exists several works, however, in which the effect has been established experimentally. Ivankiv, Miliyanchuk, and Filatova [56] studied ethyl alcohol dehydrogenation on ZnO. When an electric field was applied, the reaction rate changed. Depending on the direction of this field, the reaction rate either drastically increased or insignificantly decreased (for the same absolute value of field strength). Fentsik and Stadnik [57] studied the effect of an electric field on the rate of catalytic oxidation of methanol and formaldehyde and some alcohols on silver oxide. Their results, the authors believe, agree with the predictions of the electronic theory. Keier, Mikheeva, and Usol'tseva [58] observed the effect of an electric field on the rate of isopropyl alcohol dehydration on TiO_2.

5.5.2. Catalytic Properties of a Semiconductor Film on a Metal

Let us consider a thin semiconductor film that covers the surface of a metal. The adsorptive properties of such a film were studied in Section 4.6. We will assume that all the conditions stated in Section 4.6.1 hold. Let σ_0 and σ_L be the charge densities at the inner and outer surfaces of the film, and ϵ_0 and ϵ_L the distances between the Fermi level and the bottom of the conduction band at the corresponding surfaces (see Fig. 4.13). Obviously, ϵ_0 and ϵ_L are (to within constant terms) the work function of an electron leaving the metal and leaving the film. Now let us assume that a catalytic reaction proceeds on the outer surface of the film. We wish to find g as a function of film thickness L and work function ϵ_0 of the substrate metal [59]. We have

$$g = g(\epsilon_L),$$

where

$$\epsilon_L = \epsilon_L(L, \epsilon_0).$$

We consider the derivatives

$$\frac{dg}{dL} = \frac{dg}{d\epsilon_L} \frac{d\epsilon_L}{dL} = \frac{dg}{d\epsilon_L} \frac{dV_L}{dL},$$

$$\frac{dg}{d\epsilon_0} = \frac{dg}{d\epsilon_L} \frac{d\epsilon_L}{d\epsilon_0} = \frac{dg}{d\epsilon_L} \frac{dV_L}{dV_0}, \tag{5.41}$$

where [see (4.118)]

$$\begin{cases} V_L = \epsilon_L - \epsilon_v, \\ V_0 = \epsilon_0 - \epsilon_v. \end{cases} \tag{5.42}$$

Here ϵ_v is the distance between the Fermi level and the bottom of the conduction band of a thick semiconductor (Fig. 4.14), and V_L and V_0 are the potential energies of an electron at the outer and inner surface of the film reckoned from ϵ_v (see Section 4.6.1). The dependence of V_L on L and V_0 was studied in Section 4.6 and is depicted in Fig. 4.15. Figure 4.15a refers to the case where the outer surface of the film (if it is thick) is positively charged ($\sigma_L > 0$), while Fig. 4.15b refers to the case where this surface is negatively charged ($\sigma_L < 0$). In Figure 4.15a the potential energy V_L steadily increases; i.e., the Fermi level at the (outer) surface of the film steadily moves downward, as shown by (5.42), as the film gets thinner. In Figure 4.15b the pattern is the same as long as $V_0 \geq V_L{}^0$, where $V_L{}^0$ is the critical value of V_L; at $V_0 < V_L{}^0$ there is a maximum on the V_L versus L curve. The critical value $V_L{}^0$ is the value of the potential energy (corresponding to a certain position of the Fermi level) at which the charge density at the outer surface of the film vanishes ($\sigma_L = 0$; see Section 4.6.1).

Noting that (by definition)

$$\frac{dg}{d\epsilon_L} > 0 \text{ for a donor reaction,}$$

$$\frac{dg}{d\epsilon_L} < 0 \text{ for an acceptor reaction,} \tag{5.43}$$

and employing (5.41), (5.43), and Fig. 4.15, we arrive at the following conclusions:

(1) If for a large L the (outer) surface of the film is positively charged ($\sigma_L > 0$) or if it is charged negatively ($\sigma_L < 0$) but $V_0 \geq V_L{}^0$, the catalytic activity of the film in relation to a donor reaction steadily increases and in relation to an acceptor reaction steadily decreases as the film gets thinner. but if L is large and the (outer) surface is negatively charged ($\sigma_L < 0$) and $V_0 < V_L{}^0$, the catalytic activity of the film in relation to a donor reaction passes through a maximum and in relation to an acceptor reaction through a minimum as L decreases. Note that the surface charge σ_L is determined by the entire collection of molecules entering into the reaction, the reaction products, and the intrinsic surface defects.

(2) If the catalytic reaction consists of two stages, one of which is a donor reaction and the other an acceptor reaction, then, as the film thickness changes, the limiting role may be transferred from one stage to the other. For example, if for a film of fixed thickness the limiting stage is the donor reaction, then for a thinner film (all other conditions being equal) the acceptor stage may prove to be limiting. This may change the order of the reaction.

(3) Let us assume that the reaction proceeds in two parallel directions, of which one is a donor reaction (reaction rate g_D) and the other an acceptor reaction (reaction rate g_A). The ratio g_D / g_A characterizes the selectivity of the catalyst. Obviously, the selectivity of the film will vary with thickness. For instance, if for a film of fixed thickness the reaction proceeds mainly in the acceptor direction ($g_A \gg g_D$), a thinner film may lead the reaction (all other conditions being equal) in the donor direction ($g_A \ll g_D$). Thus, the very direction of a reaction may change as the film thickness changes.

(4) For a given film of a given thickness L that is less than the screening length l, the catalytic activity of the film in relation to a donor reaction is higher, and in relation to an acceptor reaction is lower, the greater ϵ_0 is, i.e., the higher the work function of the substrate metal.

In the majority of cases encountered in practice the semiconductor jacket covering the metal appears as a result of oxidation of the metal surface. This jacket is an oxide film whose thickness can to a certain extent be controlled. An example is the layer of cuprous oxide grown on the parent copper.

We see that by varying the thickness of the jacket we can to a certain extent control the catalytic activity and selectivity of the sample. It is important, however, that the thickness of the jacket be less than the screening length ($L < 10^{-4}$–10^{-5} cm) and yet not too small, so that the jacket can be considered as a separate phase ($L > 10^{-6}$ cm).

In conclusion we refer to the work of Deren, Russer, and Haber [60, 61], who gave an experimental confirmation of the theory in a study of the decomposition of hydrogen peroxide on a chromium oxide film covering metallic chromium. Depending on the temperature at which the metal oxidized, films of different thicknesses were produced. As the thickness grew, the reaction rate increased, reached a maximum, and then decreased.

5.5.3. The Mechanism of the Action of an Impurity

Let us now examine qualitatively the mechanism by which an impurity acts on the catalytic activity and selectivity of a semiconductor.

We have seen that by introducing impurities of one type or another in definite concentrations onto the surface of a crystal or into its bulk we can, to some extent, control the position of the Fermi level at the surface of the crystal. The mechanism by which an impurity affects the adsorptivity of the surface (i.e., the surface coverage at a given pressure and temperature) and the catalytic activity of the surface (for a given surface coverage) consists in this displacement of the Fermi level produced by the impurity. It enables us to understand how minute amounts of an impurity can produce a perceptible acceleration or deceleration of the reaction without entering into direct contact with the reacting particles. The physical meaning of the mechanism is that the impurity controls the concentrations of the electron and hole gases at the semiconductor surface, which, in turn, control the reaction rate. Thus, the influence of an impurity on the catalytic activity, on the one hand, and the parallelism between catalytic activity and electrical conductivity, on the other, are in fact two aspects of the same effect.

Here the word "impurity" does not necessarily mean chemically foreign atoms introduced into the lattice. As in semiconductor physics (see Sections 1.1 and 1.2), this concept has a broader meaning. Any local imperfections in the strict periodicity of the lattice are assumed to be impurities. These may be vacant sites, foreign atoms substituting for lattice atoms proper, or foreign or native atoms ejected into interstices or onto the surface of the crystal. Thus, stoichiometric disturbances and, in general, any deviations from ideal periodicity of the crystal lattice are, in this sense, impurities. The role of an impurity (in our case a surface impurity) is also fulfilled by chemisorbed particles not participating in the reaction, by reacting chemisorbed particles, and by chemisorbed particles that are reaction products. Consequently, foreign gases in whose atmosphere the catalyst operates can lower or increase the catalytic activity. As a result of accumulation of reaction products on the surface the catalytic activity may be raised or lowered in the process of the reaction.

As we know, two types of impurities must be distinguished, i.e., acceptor and donor impurities, which act as traps, or localization centers, for free lattice electrons and holes, respectively. It is important to stress that foreign particles dissolved in the crystal act as acceptors or donors depending not only on their nature but also on the manner of their inclusion in the crystal, i.e., whether they form an interstitial solution or a substitutional solution with the semiconductor. For example, Li atoms introduced into the NiO lattice, as noted earlier (see Section 5.3.3), act as donors if they are situated in the interstices (interstitial solution) or as acceptors if they are situated at the lattice sites, replacing nickel atoms (substitutional solution). In the case of a substitutional solution the same foreign particles may act as acceptors and donors depending on the lattice in which they are dissolved. For instance, Ga atoms are donors in the ZnO lattice and acceptors in the Ge lattice. Thus, if foreign atoms of any definite sort act as, say, acceptors in the adsorbed state, these same particles, on being dissolved in the bulk of the crystal, may act as donors, and vice versa.

Let us take a crystal whose dimensions are much greater than the screening length. Suppose that ϵ_s and ϵ_v are, as was previously the case, the distances from the bottom of the conduction band and the Fermi level at the surface and in the bulk of the crystal, respectively, Z the (surface or bulk) concentration of impurity centers of a certain kind, and T the absolute temperature.

Introducing an impurity (irrespective of whether we are speaking of a surface impurity or an impurity introduced into the bulk) generally leads to a displacement of the Fermi level at the surface of the crystal. Surface impurities act directly on ϵ_s, leaving ϵ_v unchanged, while bulk impurities effect ϵ_v and thereby produce changes in ϵ_s (see Section 4.1.1).

For the rate g of a reaction occurring at the surface we have

$$g = g(T, \epsilon_s),$$

where

$$\epsilon_s = \epsilon_s(T, Z).$$

This means that

$$\frac{\partial g}{\partial Z} = \frac{\partial g}{\partial \epsilon_s} \frac{\partial \epsilon_s}{\partial Z}, \tag{5.44}$$

and an impurity is called a promoter if

$$\frac{\partial g}{\partial Z} > 0, \tag{5.45a}$$

and a poison if

$$\frac{\partial g}{\partial Z} < 0. \tag{5.45b}$$

By definition, in (5.44)

$$\frac{\partial g}{\partial \epsilon_s} > 0 \text{ for a donor reaction,}$$

$$\frac{\partial g}{\partial \epsilon_s} < 0. \text{ for an acceptor reaction.} \tag{5.46}$$

At the same time we always have

$$\frac{\partial \epsilon_s}{\partial Z} \leqslant 0, \quad \frac{\partial \epsilon_s}{\partial T} \geqslant 0 \text{ for donor impurities,}$$

$$\frac{\partial \epsilon_s}{\partial Z} \geqslant 0, \quad \frac{\partial \epsilon_s}{\partial T} \leqslant 0 \quad \text{for acceptor impurities},\tag{5.47}$$

i.e., donor impurities always shift the Fermi level upward, while acceptor impurities always shift it downward. If the temperature is increased or the impurity content lowered, the Fermi level is always drawn to the center of the forbidden gap between the energy bands.

We see from (5.44), (5.46), and (5.47) that acceptor reactions are accelerated by a donor impurity and slowed down by an acceptor impurity. In the case of donor reactions the situation is reversed. Thus, the same impurity in the same catalyst may act as promoter for one reaction and poison for another.

If the reaction consists of two (or more) consecutive stages of which one belongs to the donor class and the other to the acceptor class, then, as the impurity content (or Z) increases, i.e., as the Fermi level is steadily displaced [see (5.47)], the limiting role may be transferred from one stage, say, the donor stage, to the other, the acceptor stage.

This is illustrated by Fig. 5.12, in which the dependence of the reaction rate g on the position of the Fermi level E_F (we recall that $\epsilon_s = E_C - E_F$) is shown for such a two-stage reaction. As a result of enrichment of the crystal by an impurity we may be transferred from a point A on the upper branch of the curve to a point C on the lower branch. At points A and C the derivative $\partial g/\partial \epsilon_s$ has different signs and the derivative $\partial \epsilon_s/\partial Z$ (for given nature of the impurity) has the same sign. Thus, from (5.44) we see that an impurity which at one concentration assumes the role of promoter for a given reaction may at another concentration assume the role of poison for the same reaction.

We note further that if the reaction rate as a function of temperature (in a given range of temperatures and partial pressures) obeys the Arrhenius equation, then the displacement of the Fermi level (which enters as a separate term into the energy of activation with a plus or minus sign) is generally reflected in both the activation energy and the preexponential factor in the reaction rate constant, since the Fermi level is itself a function of temperature. In other words, both the activation energy and the preexponential factor may prove to be functions of Z.

The example of oxidation of CO (see Section 5.3.3) shows that the curve expressing the dependence of reaction rate g on the position of the Fermi level E_F (Fig. 5.12) shifts when the external conditions (temperature T and partial pressures) vary. Figure 5.19 (cf. Fig. 5.12) depicts a set of g vs. E_F curves corresponding to different T's for oxidation of CO (see Section 5.3.3). If we fix T and vary Z, we move along a specific curve, while if we fix Z and vary T, we go from one curve to another. Consequently, as a result of variation of temperature (Z constant) we may be transferred from point A in Fig. 5.19 to point B or back. From (5.44) we see that the same impurity in the same concentration may act as promoter at one temperature and as a poison at another.

Thus, the promoting and poisoning effect of an impurity is determined not only and not so much by the nature of the impurity and the character of the reaction as by the position of the Fermi level at the surface of the crystal, i.e., the state of the system as a whole. While at some values of T and Z condition (5.45a)

Fig. 5.19. Reaction rate of oxidation of carbon monoxide at different temperatures vs. position of the Fermi level.

is met, at other values of T and Z it may be replaced by condition (5.45b). The concepts of promoter and poison lose their universal meaning and are often replaced by a more general concept of a modifier.

By regulating the amount of impurity in a semiconductor we can control not only the activity but also the selectivity of the catalyst. Indeed, if the reaction proceeds along two parallel routes, one an acceptor reaction and the other a donor reaction, then, as the Fermi level is steadily shifted downward or upward (i.e., as Z changes monotonically), one reaction route will be accelerated and the other slowed down, as is evident in the case of alcohol decomposition in Figs. 5.9a and 5.9b. The introduction of an impurity may accelerate one route and slow down the other.

Note that in some cases it is possible to reduce the effect of impurities on the activity and selectivity of a catalyst to zero. The reaction rate proves to be insensitive to a bulk impurity (dissolved in the interior of the crystal) when the reaction occurs on a quasiisolated surface and also when the temperature is so high that the semiconductor is in the region of intrinsic conduction, i.e., the Fermi level in the bulk is stabilized at the middle of the forbidden gap between the bands and no impurity can displace it from this position. In such cases only the surface impurities continue to have an effect.

Thus, if the reaction does not lose its sensitivity to impurities at intrinsic conduction temperatures, this may be regarded as indicative of the fact that impurities that exhibit a promoting or poisoning effect are situated on the surface (completely or at least partially).

In those ranges of ϵ_s where the reaction rate is independent of ϵ_s, the action of both bulk and surface impurities ceases. The electronic mechanism of the promoting and poisoning actions of an impurity is completely suppressed.

Let us note in conclusion that the electronic mechanism examined does not, of course, exclude more trivial mechanisms (e.g., poisoning as a result of surface blocking) and, generally speaking, coexists with them.

5.5.4. The Experimental Data on the Effect of Impurities

The phenomenon of catalyst poisoning and promotion by impurities has long been well known and made use of in the chemical industry. Numerous experimental papers are devoted to this question. The phenomenon of modification (poisoning by promoters and promotion by poisons) was discovered in 1940 in the laboratory of S. Z. Roginskii. A theoretical interpretation of the phenomenon was given (within the framework of the concept of a quasiisolated surface) in the first papers on the electronic theory of catalysis [62, 4, 40, 63]. A review of the experimental data is given in [65] (see also [64]). Here we will illustrate with the experimental results the main qualitative results obtained in Section 5.5.3.

(1) From the viewpoint of the electronic mechanism, as we have seen, there is an inherent difference between chemically foreign impurities introduced into the crystal, stoichiometric disturbances in the crystal, and structural defects in the lattice. We may expect that changes in the degree of deviation from stoichiometry and, more than that, changes in the degree of disorder in the semiconductor with its chemical composition remaining unchanged must lead to changes in its catalytic activity.

This is often observed in reality. As an example we refer to the already cited paper of Boreskov and Matveev [43], who found that Zn in ZnO in excess of stoichiometry promotes the decomposition of methyl alcohol. Another example is the work of Élement [66], who found that the catalytic activity of NiO in relation to complete oxidation of isooctane depends greatly on the content of oxygen in NiO in excess of stoichiometry. Finally, we note the paper of Shekhter and Mashkovskii [67], who produced a change in the catalytic activity of ZnO not by introducing a chemically foreign impurity into the sample but by treating it thermally ("thermal modification").

(2) We have seen, furthermore, that the promoting and poisoning effects of an impurity in a given reaction is determined not only by the nature of the impurity but also by the character of the reaction (i.e., whether it belongs to the donor or acceptor class). We can expect an impurity that poisons one reaction to promote another.

This is a well-known fact. An example is provided by ZnO, which is a catalyst for oxidation of CO and for decomposition of N_2O. Minute quantities of Li_2O introduced into ZnO as an impurity poison the oxidation of CO, according to the results of Keier and Chizhikova [68], but promote the decomposition of N_2O, according to the results of Schwab and Block [69].

(3) We have also seen that the same impurity in the same catalyst in the same reaction can act as promoter or poison at a given temperature depending on its concentration. We can therefore expect that the introduction of an impurity into a sample will produce either an increase or a decrease in its catalytic activity in a given reaction, depending on the amount of impurity introduced.

This too is often observed in reality. Figures 5.20–5.23 show the experimental curves for the variation of activity (rate constant K) with the impurity content Z for various temperatures. Figure 5.20 is taken from the paper of Zhabrova and Fokina [70], who studied the decomposition of hydrogen peroxide on MgO crys-

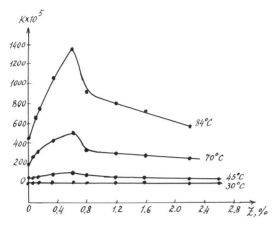

Fig. 5.20. Rate constant (activity) vs. impurity content for the decomposition of hydrogen peroxide on MgO crystals with Sb_2O_3 as impurity (Zhabrova and Fokina [70]).

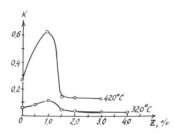

Fig. 5.21. Rate constant (activity) vs. impurity content for complete oxidation of isooctane on WO_3 crystals with NaOH as impurity (Roginskii [71]).

tals with Sb_2O_3 as the impurity. Figure 5.21 is taken from the paper of Roginskii [71] and corresponds to the results obtained by Krylov (complete oxidation of isooctane on WO_3 crystals with NaOH as the impurity). Figures 5.22 and 5.23, taken from the work of Élement [66], belong to the same reaction on WO_3 crystals with nitric acid (Fig. 5.22) or B_2O_3 (Fig. 5.23) as the impurity.

In Figure 5.23 we are concerned with normal poisoning. In this case, according to Section 5.5.3, we remain on the upper or lower branches of the curves $g = g(T, E_F)$ depicted in Fig. 5.19, i.e., within the limits of the same limiting stage. The transition through a maximum in Figs. 5.20–5.22 (cf. Fig. 5.19) indicates, according to Section 5.5.3, a replacement of the limiting stage. Note that the presence of shallow minima in Fig. 5.22 (modification of the second kind) cannot

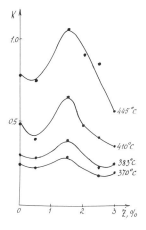

Fig. 5.22. Rate constant (activity) vs. impurity content for complete oxidation of isooctane on WO_3 crystals with nitric acid as impurity (Élement [66]).

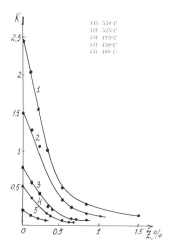

Fig. 5.23. Rate constant (activity) vs. impurity content for complete oxidation of isooctane on WO_3 crystals with B_2O_3 as impurity (Élement [66]).

be interpreted within the framework of Section 5.5.3. Indeed, transfer of the limiting role from one stage to another cannot lead to a minimum. A minimum is produced via another mechanism, not discussed in Section 5.5.3. In Section 3.1.3 we saw that when the Fermi level is steadily displaced (while all the other conditions remain the same), the quantity η^0, which gives the relative fraction of the "weak" form of chemisorption on the surfaces, passes through a maximum (see Fig. 3.4b). This fact in certain cases can lead to a maximum or a minimum on the g vs. ϵ_s curve within the limits of the same limiting stage [72].

We see from Figs. 5.20–5.22 that an impurity that promotes a given reaction at low concentrations can act as a poison for the same reaction at high concentra-

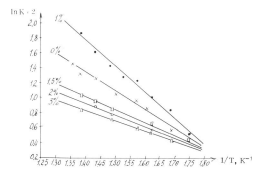

Fig. 5.24. Arrhenius straight lines corresponding to complete oxidation of isooctane on WO$_3$ crystals with NaOH as impurity (Roginskii [71]).

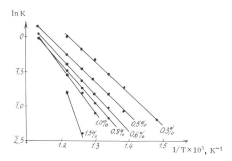

Fig. 5.25. Arrhenius straight lines corresponding to complete oxidation of isooctane on WO$_3$ crystals with B$_2$O$_3$ as impurity (Élement [66]).

tions. The same effect can be observed in Figs. 5.24–5.27, which display the experimentally obtained Arrhenius straight lines (ln K as a function of $1/T$) for samples varying in impurity content. Figure 5.24, taken from the paper by Roginskii [71], is constructed according to the results obtained by Krylov (complete oxidation of isooctane on WO$_3$ with NaOH as the impurity). Figure 5.25 belongs to the same reaction on NiO with B$_2$O$_3$ as the impurity (according to the data of Élement [66]). The figures at the curves in Figs. 5.24 and 5.25 give the amount of additives in weight percent. Figure 5.26 is constructed using the data in the paper by Margolis and Todes [73], who studied the oxidation of isooctane on a magnesium–chromium oxide catalyst with H$_3$PO$_4$ as the impurity. Finally, Figure 5.27 is taken from the paper by Krylov and Margolis [74] and refers to the oxidation of ethylene on a MgO·Cr$_2$O$_3$ catalyst with Na$_2$SiO$_3$ as the impurity. The straight lines in Figs. 5.26 and 5.27 are numbered in order of increasing impurity

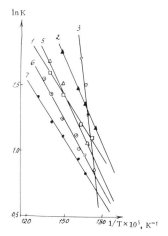

Fig. 5.26. Arrhenius straight lines corresponding to oxidation of isooctane on a magnesium–chromium oxide catalyst with H_3PO_4 as impurity (Margolis and Todes [73]).

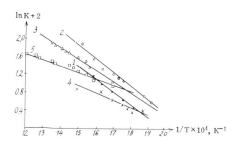

Fig. 5.27. Arrhenius straight lines corresponding to oxidation of ethylene on a $MgO \cdot Cr_2O_3$ catalyst with Na_2SiO_3 as impurity (Krylov and Margolis [74]).

content. Note that if the temperature is fixed and we go from sample to sample (Figs. 5.24, 5.26, and 5.27), the activity of a sample is not a monotonic function of the sample's impurity content.

(4) Finally, we have seen that the same impurity at the same concentration in the same catalyst and in the same reaction may act as a promoter or poison depending on the temperature of the experiment. Thus, we can expect that of two samples differing in impurity content the more active sample at a given temperature may be the less active one at another temperature.

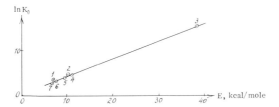

Fig. 5.28. Constable's law representing the compensation effect and corresponding to the data of Fig. 5.26.

This is indeed observed in a number of cases. Figures 5.26 and 5.27 offer examples, namely, samples 2 and 3 in Fig. 5.26 or samples 1 and 2 in Fig. 5.27, as is evident, change places as regards their activity as the temperature varies.

5.5.5. The Compensation Effect

As we have noted in Section 5.1.3, when Arrhenius's equation

$$K = K_0 \exp\left(- \frac{E}{kT} \right)$$

or

$$\ln K = \ln K_0 - \frac{E}{kT} \tag{5.48}$$

is valid, the so-called compensation effect takes place; i.e., the quantities E and K_0 vary symbatically as we go from one sample to another prepared differently. These samples may differ in the impurity concentration. For an example we can take Fig. 5.24, where the initial ordinate $\ln K_0$ increases with the slope E of the Arrhenius curve (straight line).

The compensation effect is often expressed by a linear dependence of $\ln K_0$ on E:

$$\ln K_0 = A + BE. \tag{5.49}$$

This law was first found experimentally by Constable [75] as early as 1923. It was later confirmed by many authors for various catalysts in different reactions. Constable's law is illustrated in Fig. 5.28, which is constructed from the data of Fig. 5.26, namely, each point in Fig. 5.28 corresponds to an Arrhenius straight line in Fig. 5.26, i.e., a pair of values (K_0, E) (the numbers of the points in Fig. 5.28 correspond to the numbers of the straight lines in Fig. 5.26).

Constable's law is reminiscent of the well-known empirical law of Meyer and Neldel [76] for the electrical conductivity of semiconductors, according to which

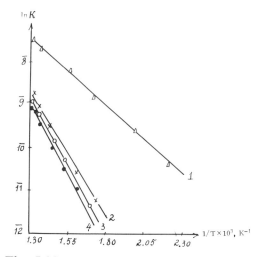

Fig. 5.29. Meyer–Neldel's law in ln K vs. $1/T$ coordinates.

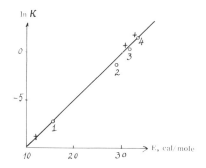

Fig. 5.30. Meyer–Neldel's law in ln K vs. E coordinates.

the parameters κ_0 and u in the expression for the temperature dependence of the electrical conductivity of a semiconductor,

$$\kappa = \kappa_0 \exp\left(-\frac{u}{kT}\right)$$

or

$$\ln \kappa = \ln \kappa_0 - \frac{u}{kT}, \tag{5.50}$$

are connected by the relationship

$$\ln \kappa_0 = a + bu. \tag{5.51}$$

As illustration of Meyer–Neldel's law we present Figs. 5.29 and 5.30, taken from the paper of Claudel and Veron [77]. Figure 5.29 shows the dependence of $\ln \kappa$ on $1/T$ according to (5.50) for various samples of thorium oxide differing in their preparation (the duration and temperature of preheating in various temperatures and at various pressures). Figure 5.30 is constructed from the data of Fig. 5.29; i.e., each point in Fig. 5.30 corresponds to a straight line in Fig. 5.29, i.e., a pair of values (κ_0, u) (the numbers of the points in Fig. 5.30 correspond to the numbers of the straight lines in Fig. 5.29).

Note that Constable's law can be expressed in another form. We can easily show that if the relation between $\ln K_0$ and E is linear, i.e., is given by (5.49) (Constable's law), the corresponding Arrhenius straight lines (5.48) must intersect at one point. To verify this we will consider Arrhenius straight lines (5.48) for which the parameters K_0, E have the values K_0', E' and K_0'', E''. The point of intersection of these two lines can be found from the condition

$$\ln K_0' - \frac{E'}{kT} = \ln K_0'' - \frac{E''}{kT} . \tag{5.52}$$

Substituting (5.49) into (5.52) and introducing the notation

$$B = \frac{1}{kT^*} , \tag{5.53}$$

we can rewrite (5.51) as follows:

$$A + \left(\frac{1}{kT^*} - \frac{1}{kT} \right) E' = A + \left(\frac{1}{kT^*} - \frac{1}{kT} \right) E''. \tag{5.54}$$

From (5.54), (5.48), and (5.49) we have

$$T = T^*, \quad \ln K = A$$

for the point of intersection. Thus, for each pair of Arrhenius straight lines we have the same intersection point; in other words, all Arrhenius straight lines (5.48) that satisfy condition (5.49) must have a common intersection point.

The above reasoning can be applied to the Meyer–Neldel law. The straight lines (5.50) that satisfy (5.51) intersect at one point. At this point

$$T = T^{**}, \quad \ln \kappa = a,$$

where T^{**} is defined by a condition similar to (5.53):

$$b = \frac{1}{kT^{**}}$$

Let us assume that the catalytic and electrical conductivity measurements are done on the same samples. Obviously, at $T = T^{**}$ all samples have the same catalytic activity. According to the electronic theory of catalysis, these samples must have the same conductivity, i.e., $T^* = T^{**}$, or $B = b$.

For the semiconductors investigated by Meyer and Neldel [76], $b = 0.15$–0.30 mole/kcal, while the values of B in Constable's law, obtained by different authors from catalytic measurements [78], vary from 0.2 to 0.5 mole/kcal. Unfortunately, the measurements of electrical conductivity and catalytic activity were performed on different samples. The only two semiconductors for which both the coefficient b in the Meyer–Neldel law and the coefficient B in Constable's law are known are zinc oxide and thorium oxide. As could be expected, $b = B$ in these two cases, i.e., the electronic theory is valid. Indeed, for zinc oxide, according to Meyer and Neldel's results [76], $b = 0.29$ mole/kcal, while according to the measurements of Schwab [79] and Shekhter and Moshkovskii [78] (methanol decomposition) $B = 0.35$ mole/kcal; for thorium oxide, according to Claudel and Veron's results [77], $b = 0.48$ mole/kcal and $B = 0.49$ mole/kcal (oxidation of CO).

Note that the compensation effect (K_0 increases with E) may be expressed by a dependence of $\ln K_0$ on E that differs from the linear dependence, or Constable's law. Often Constable's law (5.49) is no more than a rough approximation. To elucidate this fact we refer the reader to Figs. 5.26 and 5.28. The straight lines in Fig. 5.26 intersect at more than one point, while the points in Fig. 5.28 (which is constructed from the data of Fig. 5.26) fall onto a straight line sufficiently well. This straight line, therefore, only approximately reflects the dependence of $\ln K_0$ on E.

The origin of the compensation effect has for many years been the subject of theoretical discussion (e.g., see [80]). A brief summary of the papers published before 1960 is given in [81]. Here we will name the papers that have appeared since 1960. In the paper by Roginskii and Khait [81] the compensation effect appears as the result of a statistical study of an elemental reaction act. Lebedev [82] has shown experimentally that the $\ln K$ in Arrhenius's equation (5.48) varies little in comparison with E/kT (the rate constant K varies within one to two orders of magnitude), which means that the term $\ln K$ may be replaced by its average. Then (5.48) becomes (5.49) and we arrive at Constable's law for each given temperature. Likhtenshtein [83] studied the compensation effect for catalysts with an inhomogeneous surface (i.e., different sections of the surface are independent of each other and their activation energy distribution is Gaussian). Exner [84] has given an exhaustive analysis of the credibility of the experimental methods of establishing the correlation between activation energy and the preexponential factor in Arrhenius's equation.

To conclude this chapter, let us see how the compensation effect is interpreted in the electronic theory of catalysis. Kuznetsov [85], in a study of doped semiconductors, showed that the E and K_0 in Arrhenius's equation are functions of one parameter, the impurity concentration. In this way E and K_0 prove to be interdependent, and the manner of this dependence is such that their variations compensate each other to a certain extent. In his calculations Kuznetsov ignored the bending of the energy bands at the surface, i.e., the surface is assumed to be electrically neutral, which we know is generally not the case. Peshev and Bliznyakov [86–88] studied microgranular semiconductors (the size of grains L is of the order of the screening length l) and showed that the activation energy E and preexponential factor K_0 are monotonic functions of L, which leads to the direct relation between E and K_0 (the compensation effect).

Chapter 6

PROCESSES ON A REAL SURFACE

6.1. DEVIATIONS FROM LANGMUIR'S THEORY ON A REAL SURFACE

6.1.1. The Concept of an Inhomogeneous Surface

As we know, the regularities in adsorption observed experimentally often deviate from those predicted by the classical Langmuir theory of adsorption. These deviations, which also arise both in adsorption kinetics and in adsorption equilibrium (non-Langmuir isotherms and the dependence of the differential adsorption heat on surface coverage), indicate the violation of one or more of the initial assumptions unnderlying Langmuir's theory. These assumptions were discussed in detail in Section 2.1.1. Here we will name them once more.

1. We assume that the surface contains adsorption centers of one kind.
2. We assume that the adsorbed molecules do not interact with each other.
3. We assume that the number of adsorption centers remains constant.
4. Finally, we assume that the energy of the bond of an adsorbed molecule with an adsorption center does not vary with time or with the surface coverage.

The deviations of the experimental laws from the Langmuir laws are explained in two ways.

(a) The first assumption enumerated above is dropped while all the others are retained. In this way we introduce the concept of an inhomogeneous surface, i.e., a surface whose various sections have different adsorptivity or, in other words, a surface with adsorption centers of different kinds, which are characterized by different values of the heat of adsorption.

261

(b) The second assumption is dropped while all the others are retained. In this way we introduce the concept of intersection between adsorbed molecules.

Both the concept of an inhomogeneous surface and the concept of interaction of adsorbed molecules are flexible tools in the hands of the theoretician, making it possible to explain many experimental facts that fail to fit into the elementary Langmuir theory.

Let us start with the concept of an inhomogeneous surface. We assume, for the sake of definiteness, that the surface is inhomogeneous in adsorption heat q. Let the surface be characterized by a discrete set of values q. The adsorption centers are of different types differing (in relation to molecules of a given type) in values of q. Let q_i be the adsorption heat for centers of the ith type and n_i the number of such centers per unit area, with

$$n = \sum_i n_i$$

the total number of adsorption centers. Let us further assume that centers of each kind are partly occupied by adsorbed molecules (of the same sort). Let N_i be the number of molecules adsorbed on centers of the ith type ($N_i \leq n_i$), with

$$N = \sum_i N_i$$

the total number of the adsorbed molecules per unit surface area. We have

$$\frac{N}{n} = \frac{1}{n} \sum_i N_i = \sum_i \frac{N_i}{n_i} \frac{n_i}{n} . \tag{6.1}$$

Adopting the notations $\Theta = N/n$, $\Theta_i = N_i/n_i$, and $F_i = n_i/n$, we can rewrite (6.1) as follows:

$$\Theta = \frac{N}{n}, \qquad \Theta_i = \frac{N_i}{n_i}, \qquad F_i = \frac{n_i}{n} , \tag{6.2}$$

Here Θ is the total surface coverage, Θ_i the surface coverage of centers of the ith type, and $F_i = F(q_i)$ the relative fraction of this type of centers on the surface. The function $F(q_i)$ is known as the integral adsorption-heat distribution function.

We will now assume that q varies continuously and the surface consists of sections each of which has different values of q. Let dS be the part of the surface in which q lies in the interval from $q + q + dq$. We denote the concentration of the adsorbed particles on dS by $dN(q, P)$, where P is the pressure, and assume $N(P)$ to be the average concentration over the entire surface. Obviously,

$$N(P) = \int dN(q, P)$$

or

$$\frac{N(P)}{S} = \int \frac{dN(q, P)}{dS} \frac{dS}{S} , \tag{6.3}$$

where integration is carried out over the entire surface S. Adopting the notations

$$\Theta(P) = \frac{sN(P)}{S},$$

$$\Theta(q, P) = s \frac{dN(q, P)}{dS}, \qquad f(q)dq = \frac{dS}{S},$$

where s is the effective area of an adsorbed molecule, we can rewrite (6.3) as follows:

$$\Theta(P) = \int \Theta(q, P)f(q)dq, \tag{6.4}$$

where integration with respect to q is carried out over all values of q from q_{min} to q_{max}. The function

$$f(q) = \frac{1}{S} \frac{dS}{dq} \tag{6.5}$$

is known as the differential distribution function. Obviously, $f(q)$ is normalized, since

$$\int f(q)dq = \frac{1}{S} \int dS = 1,$$

where integration is over all values of q and over the entire surface S.

Thus, in the theory of adsorption on inhomogeneous surfaces the surface is built from separate homogeneous sections, and it is assumed that the Langmuir theory works (which is important) for each of these sections. The adsorption regularities, observed in experiments, that correspond to the surface as a whole are the result of averaging over the separate sections. Note that it is not necessary to consider these homogeneous sections as geometrically isolated. On the contrary, it is more correct to think of them as sets of points distributed at random geometrically but having the same adsorptive properties.

If we allow for the electronic processes that occur on the surface in adsorption, then even on a homogeneous surface and in the absence of interaction between the adsorbed particles the adsorption regularities, as shown in Section 3.5.2, may prove to be non-Langmuir. This is due, as we know, to the fact that each chemisorbed particle "feels" the presence of all the other particles through the adsorbent electron gas. This somewhat modifies the theory of adsorption on inhomogeneous surfaces. There is no more need for the separate homogeneous sections, which constitute a given inhomogeneous surface, to obey the Langmuir regularities (although in certain conditions the opposite may become possible). What changes is the elementary law over which we must average, i.e., the shape of the functions $\Theta(q, P)$ in (6.4) and Θ_i in (6.2).

In the theory of adsorption on inhomogeneous surfaces the inhomogeneous surface is characterized by the distribution function $F(q)$ or $f(q)$, which is regarded as the "identity card" of the surface. From a given distribution function the adsorption isotherm (and other adsorption characteristics) can be determined from (6.2) or (6.4); conversely, the distribution function can be determined from the experimentally observed adsorption behavior.

The theory of adsorption on inhomogeneous surfaces, developed chiefly by Roginskii and his collaborators [1, 2], sets itself this second task (rather than the first) and solves it; i.e., the distribution function is not fixed by any physical considerations but is constructed in such a way that the isotherm calculated from (6.5) coincides with the isotherm observed experimentally. The physical origin and meaning of the distribution function determined in this manner remain obscure. Moveover, the distribution function thus determined is often different for the same adsorbent for different adsorbates, i.e., the surface changes its identity card for each new adsorbate. Consequently, the distribution function thus determined cannot be regarded as a characteristic of the surface of the adsorbent as such, since it is a characteristic of the system (adsorbent and adsorbate) as a whole. We see that though the theory of adsorption on inhomogeneous surfaces gives a convenient means of describing the behavior of inhomogeneous surfaces, it remains devoid of physical content.

It will acquire physical meaning only when it is given the task opposite to that which it has been set up to the present. The task consists in revealing the physical meaning of one or another distribution function and determining the isotherm from a given distribution function fixed by the physical conditions instead of determining the distribution function from an experimentally observed isotherm.

6.1.2. The Concept of Interaction

Let us now turn to the concept of interaction. Note that the same experimental regularities can often be obtained both from a certain choice of the inhomogeneity of the surface (ignoring the interaction of the molecules) and from a certain choice of the law of interaction between the molecules (ignoring the inhomogeneity of the surface). Therefore, we can establish a correspondence between the nature of the inhomogeneity, on the one hand, and the law of interaction, on the other. It is often possible to determine to what interaction law a certain type of inhomogeneity belongs (from the viewpoint of the experimental data) and, conversely, to what type of inhomogeneity a certain interaction law belongs [3]. Note, however, that the inhomogeneity factor and interaction factor are not alternatives. Deviations from the Langmuir regularities may be caused by the simultaneous action of both factors. The only question is which of them must be considered dominant under the given conditions.

The interaction between adsorbed molecules in the case of physisorption is usually interpreted as the interaction between dipoles oriented in parallel. The moment of each individual dipole is determined not only by the nature of the lattice and the nature of the adsorbed molecules but also by the interaction of a given dipole with all the other dipoles, and in the case of a metal with their electrical images as well. The magnitude of the dipole moment of every adsorbed molecule is thus dependent on the total number of adsorbed molecules, i.e., on the surface coverage. This dipole interaction and the special features in adsorption behavior resulting from it have been examined by a number of authors. Detailed calculations are given in Robert's book [4].

In the case of chemisorption three types of interaction come into play:

(1) Between particles in a state of "strong" bonding there arises electrostatic interaction, which (for not too small distances between the particles) can be regarded as a Coulomb interaction between point charges. Particles in a state of "strong" acceptor or "strong" donor bonding repel each other as charges of like sign, but if one is in a state of acceptor bonding and the other in the state of donor bonding, such chemisorbed particles are attracted to one another as two charges of opposite sign. An interaction of this type and the resulting non-Langmuir behavior as well as the spectrum of surface states caused by this interaction have been discussed in the paper by Ainbinder and Enikeev [49].

(2) Between particles in a state of "weak" bonding with the surface there is dipole–dipole interaction. However, we must bear in mind that the dipole moments arising in chemisorption have an essentially different origin (purely quantum mechanical) than the dipole moments in physisorption and may exceed them considerably (see Section 2.5.3).

(3) Moreover, for particles in a state of "weak" bonding with the surface a third type of interaction is possible when these particles approach each other closely enough. This was pointed out by Koutecky [5]. To clarify the nature of this interaction we will examine two isolated nonovalent (electropositive) atoms separated by a distance such that their electron clouds can be considered nonoverlapping (i.e., there is no exchange interaction between the atoms). Let us now imagine that these two atoms are transferred to the adsorbed state while the distance between them remains unaltered. The valence electron of each atom is now drawn into the lattice and is spread out over a certain region in the lattice (see Sections 2.5.2 and 2.5.3). This spreading out may be such that the electron cloud of one of the atoms overlaps the electron cloud of the other inside the lattice, which means that there is now an exchange interaction between the two atoms. Thus, we must now deal with a special long-range exchange interaction brought about through the agency of the crystal lattice, which in this case acts as a medium extending the region of such interaction to a lesser or greater degree. An interaction of this type may introduce a certain correction term (at small distances between the chemisorbed particles) to the dipole–dipole interaction noted above.

The inhomogeneity of the surface, on the one hand, and the interaction between the adsorbed particles, on the other, are not, as is often considered, the only possible factors leading to deviations from the regular Langmuir behavior. Indeed, the Langmuir theory contains (see Section 2.4.1), along with Assumption 1 (which is removed in the concept of an inhomogeneous surface) and Assumption 2 (which is removed in concept of interaction), Assumptions 3 and 4, which may also be subject to scrutiny. Whether Assumptions 3 and 4 may be retained or removed depends on the physical nature of the adsorption centers and the properties of these centers.

If we drop Assumption 3 (the number of adsorption centers is constant) and retain all the other assumptions, we obtain a typical non-Langmuir behavior. This has been done in Section 1.2.1, where we examined the adsorption on centers having a thermal origin (the concentration of which increases with temperature), and in Section 1.2.2, where we considered the adsorption on localized carriers (the concentration of which increases with surface coverage). At the same time the surface remained homogeneous (adsorption centers of only one type) and interaction between the adsorbed particles was ignored.

If we drop Assumption 4 and retain all the other assumptions, we again obtain a typical non-Langmuir behavior. This has been done in Section 3.5.2, where we studied the adsorption on centers whose nature and bonding strength with the particles on them varied in the course of the life of the particles in the adsorbed state. Here we are concerned with the manifestation of a special kind of interaction between the adsorbed particles, which find expression in the dependence of the properties and nature of the bonding of each individual adsorbed particle with the surface on the state of the system as a whole (adsorbent and adsorbates). At the same time we are concerned with a special kind of inhomogeneity since at any given moment three different forms of bonding with the surface are brought into being for particles of the same kind. In this case the difference between the concepts of surface inhomogeneity and interaction is eradicated, so that both these concepts in a certain sense merge, expressing only two different aspects of the same mechanism [6].

The removal of Assumption 4, which together with the other assumptions forms the foundation of the Langmuir theory, is, as we have seen, characteristic of the electronic theory of chemisorption. This does not, of course, mean that the electronic theory fails to recognize the usual surface inhomogeneity (which is an experimental fact) or the usual interaction between adsorbed molecules. It only indicates that the logical consideration of the chemisorbed particles and the adsorbent lattice as a single quantum-mechanical system leads to the concept of a special type of inhomogeneity or the concept of a special type of interaction that cannot be expressed as a manifestation of some force, concepts that manifest themselves both for a homogeneous surface (in the usual sense of the word) and in the case of a zero force between the adsorbed particles. An important consequence is that in the theory of adsorption on inhomogeneous surfaces the individual homogeneous sections of the surfaces over which summation is carried out prove to be generally non-Langmuir (with the exception of special cases).

6.2. THE ADSORPTION-HEAT DISTRIBUTION FUNCTION

6.2.1. Inhomogeneity Due to Irregularities in the Impurity Distribution

Let us consider adsorption on a surface with irregularly distributed defects. Such a surface, as we have seen, acts as an inhomogeneous surface even when adsorption centers of only one type act in the adsorption process. Let us see how the inhomogeneity, which is characterized by a distribution function, is related to distribution of the defects on the surface.

We recall that for the concentration N of chemisorbed particles on a homogeneous surface we have the following formula (see Section 3.2.1):

$$N(q, P) = \frac{N^*}{1 + (b/P) \exp(-q/kT)}, \tag{6.6}$$

where q is the differential heat of adsorption, P the pressure, N^* the surface concentration of adsorption centers, and b a factor of no interest to us at the moment. We must assume that

$$q = q^0 - kT \ln(1 - \eta), \tag{6.7}$$

where q^0 is the energy of the bonding of a chemisorbed particle with the surface (the "weak" bonding energy), and η the fraction of particles on the surface in a charged state (in he state of "strong" bonding). According to (3.6a) and (3.6b),

$$\eta = \frac{1}{1 + \exp\left(\mp \dfrac{E_F - E_A}{kT}\right)}, \tag{6.8}$$

where E_F and E_A are the position of the Fermi level and the local level of the adsorbed particle. Obviously,

$$E_F - E_A = v - \epsilon_s.$$

where ϵ_s and v are the distances from the bottom of the conduction band to the Fermi level at the surface and to the local level of the chemisorbed particle, respectively. In (6.8) and in what follows the upper sign corresponds to the case where the chemisorbed particle is an acceptor and the lower sign a donor.

At $\eta = 0$, when all the chemisorbed particles are electrically neutral, we have [according to (6.7)]

$$q = q^0,$$

which brings us back into the realm of Langmuir's theory. In the other limiting case, at $1 - \eta \ll 1$, i.e., when the charged form of chemisorption is predominant over the neutral, we have [on the basis of (6.7) and (6.8)]

$$q = q^0 \mp (E_A - E_F) = q^0 \mp (\epsilon_s - v). \tag{6.9}$$

In what follows we will confine ourselves to this last case.

If ϵ_s remains constant in adsorption, then Eq. (6.6) represents an adsorption isotherm. But if the position of the Fermi level at the surface of the adsorbent depends on the concentration of chemisorbed particles, as is often the case, i.e.,

$$\epsilon_s = \epsilon_s(N),$$

then Eq. (6.6) yields the dependence of N on P in implicit form and, therefore, is not the equation of an isotherm. Solving Eq. (6.6) for N, we can obtain different non-Langmuir isotherms (for different approximations) (See Section 3.5.2).

In what follows we will assume that the surface charge is essentially of a nonadsorption origin (i.e., is intrinsic), so that the chemisorbed particles contribute very little to the charge (this, however, does not mean that $\eta \ll 1$). In this case we can assume that ϵ_s is practically independent of N, or

$$\epsilon_s = \text{const.}$$

This means that the initial adsorption regularities at separate homogeneous sections of the surface are considered of Langmuir form.

Substituting (6.8) into (6.7) and then (6.7) into (6.6), we see that for given external conditions (i.e., at given pressure and temperature) the surface coverage of each homogeneous section depends not only on q^0, i.e., the energy of the bonding of a particle to the surface, but on the position of the Fermi level, ϵ_s, for this section, i.e., the electronic state of the system as a whole. As long as we remain within the scope of Langmuir's theory ($\eta = 0$), then inhomogeneity of the surface with respect to the adsorption heats q is due solely to the difference in the sections with respect to the binding energy q^0. In the electronic theory ($\eta \neq 0$) q may be different for different sections even if q^0 is the same for all sections, provided that ϵ_s is different for different sections.

Thus, all mechanisms that ensure that different sections of the surface have different Fermi levels ϵ_s may be the physical cause of the inhomogeneity of the surface. For one, inhomogeneity may be due to the irregularities in the impurity distribution in the bulk or on the surface of the semiconductor. Indeed,

$$\epsilon_s = \epsilon_s(x, y), \qquad \text{or} \qquad n = n(x, y),$$

where n is the impurity concentration in the semiconductor, and x and y are coordinates on the adsorbent surface (the adsorbent occupies the half-space $z \geq 0$). Thus, a gradient in the impurity distribution shapes the topography of the Fermi level ϵ_s at the surface and thereby the topography of the inhomogeneity. In the final analysis the inhomogeneity of the surface is determined (in this case) by the distribution of the impurity in the semiconductor.

Note that this may be an impurity of a definite type and one that takes no direct part in the chemisorption act (e.g., is distributed in the bulk). The inhomogeneity in impurity distribution is a natural feature of real samples, while the concept of a uniform distribution, which is used in theoretical work, is no more than an idealization.

Below we will show that different adsorption-heat distribution functions correspond to different laws governing the distribution of the impurity at the surface or in the bulk of the adsorbent. We note, in passing, that in the case where $1 - \eta \ll 1$ (to which we will restrict our discussion, as noted earlier) the adsorption-heat distribution function $f(q) = (1/S)dS/dq$ [see (6.5)] coincides with the distribution function $\Upsilon(\epsilon_s) = (1/S)dS/d\epsilon_s$. Indeed, according to (6.9),

$$f(q) = \mp \varphi(\epsilon_s).$$

6.2.2. The Relation between the Impurity Concentration Gradient and the Adsorption-Heat Distribution Function

Let us assume that the crystal has two types of impurities, a surface impurity and a bulk impurity, whose concentrations we denote by n_s and n_v, respectively. We will discuss the following cases of impurity distribution:

$$n_s = n_s(x, y), \quad n_v = \text{const}, \tag{6.10a}$$

$$n_s = \text{const}, \quad n_v = n_v(x, y). \tag{6.10b}$$

In the case (6.10b) we will assume that n_v does not vary in the direction normal to the surface (n_v is independent of z; any irregularities in the impurity distribution along the z axis cannot by themselves lead to a dependence of ϵ_s on x or y).

Note that in our problem the distance from the bottom of the conduction band to the Fermi level, ϵ, is generally a function of three coordinates,

$$\epsilon = \epsilon(x, y, z), \tag{6.11}$$

where, by definition,

$$\epsilon_s(x, y) = \epsilon(x, y, 0),$$

$$\epsilon_v(x, y) = \epsilon(x, y, \infty).$$

In the general case the function (6.11) can be determined from the three-dimensional Poisson equation

$$\frac{\partial^2 \epsilon}{\partial x^2} + \frac{\partial^2 \epsilon}{\partial y^2} + \frac{\partial^2 \epsilon}{\partial z^2} = \frac{4 \pi e}{\chi} \rho \tag{6.12}$$

with the boundary condition

$$\left(\frac{\partial \epsilon}{\partial z} \right)_{z=0} = \frac{4 \pi e}{\chi} \sigma, \tag{6.13}$$

where χ is the dielectric constant of the adsorbent, e the absolute magnitude of the electron charge, and ρ and σ the space and surface charge densities, respectively:

$$\rho = \rho(n_v), \quad \sigma = \sigma(n_s). \tag{6.14}$$

If we assume that the impurity concentration varies smoothly along the x and y axes (we take this for granted), so that

$$\left| \frac{\partial n}{\partial x} \right| \ll \frac{n}{l}, \quad \left| \frac{\partial n}{\partial y} \right| \ll \frac{n}{l}, \tag{6.15}$$

where l is the Debye length, and

$$\begin{aligned} n &= n_s \quad \text{in the case of (6.10a),} \\ n &= n_v \quad \text{in the case of (6.10b),} \end{aligned} \tag{6.16}$$

then we can show (see [7], pp. 110–117, 131–135) that

$$\frac{\partial^2 \epsilon}{\partial x^2} \ll \frac{\partial^2 \epsilon}{\partial z^2}, \qquad \frac{\partial^2 \epsilon}{\partial y^2} \ll \frac{\partial^2 \epsilon}{\partial z^2}$$

and therefore the three-dimensional problem (6.12), (6.13) is reduced to the one-dimensional problem

$$\frac{\partial^2 \epsilon}{\partial z^2} = \frac{4\pi e}{\chi} \rho, \quad \left(\frac{\partial \epsilon}{\partial z}\right)_{z=0} = \frac{4\pi e}{\chi} \sigma, \tag{6.17}$$

in which the coordinates along the surface (x and y) are parameters [since n_s and n_v in (6.14) are assumed to be functions of x and y]. In this case ϵ_s depends on x and y through n_s or n_v, so that the value of ϵ_s at each given point of the surface is determined solely by the values of n_s and n_v at the point:

$$\epsilon_s = \epsilon_s(n_s, n_v). \tag{6.18}$$

We will now return to the adsorption-heat distribution function (6.5). We can rewrite (6.5) as follows:

$$f(q) = \frac{1}{S} \frac{dS}{dq} = \frac{1}{S} \frac{dS}{dn} \frac{dn}{d\epsilon_s} \frac{d\epsilon_s}{d\eta} \frac{d\eta}{dq}, \tag{6.19}$$

where dS is the fraction of the surface for which q lies between q and $q + dq$ or, in other words, on which the impurity concentration n lies between n and $n + dn$. [Here n has the meaning defined in (6.160).] Let us consider the derivatives in (6.19).

The derivatives $d\eta/dq$ and $d\epsilon_s/d\eta$ in (6.19) may easily be calculated. From (6.7) and (6.8) we obtain

$$\frac{dq}{d\eta} = \frac{kT}{1 - \eta}, \tag{6.20}$$

$$\frac{d\eta}{d\epsilon_s} = \frac{\eta(1 - \eta)}{kT}. \tag{6.21}$$

The derivative $dn/d\epsilon_s$ can be found from (6.18). The function (6.18), which can be obtained by solving Poisson's equation (6.17), depends on the nature of the semiconductor and the nature of the impurity in the bulk of this semiconductor and on its surface.

As for the derivative dS/dn, it may be written in the following form if we take into account the fact that the distance between two curves of equal concentration is inversely proportional to the concentration gradient:

$$\frac{dS}{dn} = \Sigma \oint \frac{dl}{|\operatorname{grad} n|}. \tag{6.22}$$

Integration is carried out along a curve of equal concentration, and summation is over all the curves of equal concentration corresponding to one value of n. We can illustrate formula (6.22) with Fig. 6.1. We have

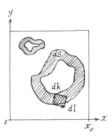

Fig. 6.1. Illustration of formula (6.22).

$$\frac{dS}{dn} = \Sigma \oint \frac{dl \cdot dx}{dn} = \Sigma \oint \frac{dl}{dn/dx} = \Sigma \oint \frac{dl}{|\operatorname{grad} n|}.$$

Thus, if we keep to the case when $1 - \eta \ll 1$ and note that

$$|\operatorname{grad} n| = \sqrt{\left(\frac{\partial n}{\partial x}\right)^2 + \left(\frac{\partial n}{\partial y}\right)^2},$$

we can write the adsorption-heat distribution function as follows:

$$f(q) = \mp \varphi(\epsilon_s) = \mp \frac{1}{S}\left(\frac{d\epsilon_s}{dn}\right)^{-1} \Sigma \oint \left[\left(\frac{\partial n}{\partial x}\right)^2 + \left(\frac{\partial n}{\partial y}\right)^2\right]^{-1/2} dl, \qquad (6.23)$$

where we have employed (6.19)–(6.22). This formula expresses the adsorption-heat distribution function or $\varphi(\epsilon_s)$ in terms of the impurity concentration gradient in the semiconductor.

6.2.3. Examples of Inhomogeneous Surfaces

We start with the example of adsorption of an acceptor gas on an n-type semiconductor that has both surface and bulk impurities of the donor type ($\rho \leq 0$, $\sigma > 0$). Suppose that $n_v = \mathrm{const}$ and $n_s = n_s(x, y)$ [the case (6.10a)], with $0 \leq x \leq x_0$ and $0 \leq y \leq y_0$. Let us assume that the surface impurity is completely ionized, i.e.,

$$\sigma = e n_s, \qquad (6.24)$$

and the bending of the energy bands at the surface is considerable compared with kT, so that

$$\frac{\epsilon_v - \epsilon_s}{kT} \gg 1. \qquad (6.25)$$

We will start by calculating the derivative $d\epsilon_s/dn$ in (6.23). On the basis of (4.7) combined with (6.17) we can write

$$\int_{\epsilon_v}^{\epsilon_s} \rho(\epsilon)d\epsilon = \frac{2\pi e}{\chi}\sigma^2. \tag{6.26}$$

We will restrict ourselves to the case where all the donors in the bulk of the semiconductor are completely ionized and the electron gas in the crystal is nondegenerate. Then

$$\rho(\epsilon) = e\left[n_v - n_0\exp\left(-\frac{\epsilon}{kT}\right)\right], \tag{6.27}$$

where n_0 is the effective electron concentration in the conduction band. The condition of electron neutrality in the bulk of the crystal yields

$$n_v = n_0\exp\left(-\frac{\epsilon_v}{kT}\right), \tag{6.28}$$

where, substituting (6.28) into (6.27), we obtain

$$\rho(\epsilon) = en_0\left[\exp\left(-\frac{\epsilon_v}{kT}\right) - \exp\left(-\frac{\epsilon}{kT}\right)\right]. \tag{6.29}$$

Substituting (6.29) into (6.26), integrating, and allowing for (6.24) and (6.25), we obtain

$$\exp\left(\frac{\epsilon_s}{kT}\right) = \frac{\chi kT}{2\pi e^2}\frac{n_0}{n_s^2}, \tag{6.30}$$

where

$$\frac{d\epsilon_s}{dn} = -\frac{2kT}{n_s}. \tag{6.31}$$

Now let us calculate dS/dn_s. We will restrict our discussion to the simple case where n_s depends only on one of the coordinates (x or y), so that the curves of equal concentration are parallel lines and the impurity concentration n_s changes in the direction normal to these lines. Here are two examples:

(1) Suppose that the impurity is distributed according to the law

$$n_s(x) = (x_0 - x)^{\frac{m}{2}},$$

where m is an arbitrary constant ($m > 0$). Then

$$\frac{dn_s}{dx} = \frac{1}{2} m n_s^{-\frac{m+2}{m}} \qquad (6.32)$$

Substituting (6.31) and (6.32) into (6.23) and recalling that $S = x_0 y_0$, we have

$$f(q) = \frac{1}{kTmx_0} n_s^{-\frac{2}{m}},$$

where, expressing n_s in terms of ϵ_s from (6.30) and then expressing ϵ_s in terms of q according to (6.9), we arrive at an exponential adsorption-heat distribution function,

$$f(q) = C \exp\left(-\frac{q}{mkT}\right), \qquad (6.33)$$

where

$$C = \frac{1}{kTmx_0} \left(\frac{2\pi e^2}{\chi kTn_0}\right)^{1/m} \exp\left(\frac{q_0 + v}{kT}\right).$$

(2) Suppose that the impurity is distributed according to the law

$$n_s(x) = a \exp(bx),$$

where a and b are arbitrary constants. Then

$$\frac{dn_s}{dx} = b n_s. \qquad (6.34)$$

Substituting (6.31) and (6.34) into (6.23), we arrive at a uniform adsorption-heat distribution,

$$f(q) = \frac{1}{2kTbx_0} = \text{const.} \qquad (6.35)$$

Note that according to the theory of adsorption at an inhomogeneous surface (see [1], p. 94) the exponential distribution function (6.33) leads (at $m > 1$) to the Freundlich isotherm

$$\Theta = AP^{1/m},$$

while the distribution function (6.35) corresponds to the logarithmic isotherm

$$\Theta = B \ln \frac{P}{P_{\min}}.$$

We see that different non-Langmuir isotherms are obtained as a result of an irregular distribution of the impurity in the semiconductor rather than as a result of the fact that different sections of the surface have different energies of bonding, q^0, with a chemisorbed particle (this quantity is everywhere the same). In the final analysis it is the gradient in the impurity distribution that determines the shape of the adsorption isotherm. We note once more that we are speaking of an impurity that takes no direct part in the chemisorption act.

The adsorption-heat distribution function (and thus the adsorption isotherm) is determined solely by the impurity concentration gradient. Note, however, that the inverse problem is not so straightforward, i.e., different impurity distributions may lead to the same distribution function.

Of course, we do not believe (and this must be stressed) that in all cases the inhomogeneity of the surface has this origin. The mechanism studied here is no more than one possible physical cause of inhomogeneity. It provides a certain physical meaning for the distribution function, which, however, is not the only possible one.

6.3. THE ROLE OF SURFACE STRUCTURAL DEFECTS IN ADSORPTION

6.3.1. Adsorption on a Structural Defect

Let us now turn to another mechanism of the influence of defects on the adsorption and catalytic properties of the surface due to their participation in the adsorption process as adsorption centers. The problem of chemisorption on defects was examined quantum mechanically by Bonch-Bruevich [8], and from the point of view of the boundary layer theory by Hauffe [9]. It was also investigated in detail by Kogan and Sandomirskii [10]. To elucidate the special features of adsorption we will start with the adsorption of a monovalent electropositive atom C on an F-center in an MR lattice consisting of M^+ and R^- ions. Such an F-center, formed by a vacant metalloid site with an electron localized in its neighborhood, is depicted in Fig. 6.2a. We will denote it by DL. From the chemical viewpoint it is a localized free vacancy capable of accepting a foreign particle. Figure 6.2b depicts the same F-center but without the electron belonging to it. We will denote such an ionized F-center by DpL. Figures 6.2c and 6.2d show two forms of chemisorption for atom C: the strong (two-electron) acceptor bonding (Fig. 6.2c) and the weak (one-electron) bonding (Fig. 6.2d). We will denote these two forms by CDL and CDpL. In the first case (Fig. 6.2c) we are concerned with adsorption on an F-center and in the second (Fig. 6.2d) with adsorption on an ionized F-center. Figure 6.3, which shows the energy band structure of the surface of the crystal (the y axis is parallel to the surface), depicts F-centers (Fig. 6.2a) as represented by the local donor levels D, and F-centers with the atoms C adsorbed on them (Fig. 6.2b) by the local donor levels CD.

Note that in the adsorption on an F-center the strong form of chemisorption is electrically neutral and the weak form is charged (in contrast to what occurs on an

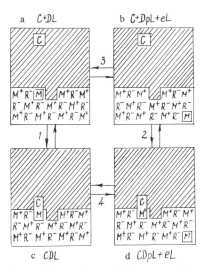

Fig. 6.2. Adsorption of a monovalent electropositive atom C on an F-center in an MR lattice consisting of M^+ and R^- ions: a) an F-center formed by a vacant metalloid site with an electron localized in its neighborhood; b) the same F-center but without the electron; c, d) two forms of chemisorption for atom C.

ideal surface). Indeed, the chemisorbed particle in the case at hand is bound to the vacant metalloid site equivalent to a positive point charge equal in magnitude to the electron charge. In the case of the strong bond the charge of this site is compensated for by the charge of the electron participating in the bond, while in the case of the weak bond this charge remains uncompensated.

Moreover, in this case the strong and weak bonds are stronger than the same types of bond on an ideal surface. Indeed, these bonds are formed in the field of a vacant metalloid site (i.e., in the field of a positive charge), which, as can be shown, strengthens the bond. But since the weak and strong bonds are not strengthened to the same extent, the weak (one-electron) bond may become stronger than the strong (two-electron) bond.

The structural forms represented in Figs. 6.2a–6.2d are capable of changing into one another. This may be expressed in the following way (arrows directed to the right correspond to exothermic reaction; the heats of reactions are entered at the left):

$$
\begin{aligned}
&1)\ \ C + DL \rightleftarrows CDL && \text{for } \ q^-, \\
&2)\ \ C + DpL \rightleftarrows CDpL && \text{for } \ q^0, \\
&3)\ \ DpL + eL \rightleftarrows DL && \text{for } \ v_D, \\
&4)\ \ CDpL + eL \rightleftarrows CDL && \text{for } \ v_{CD}.
\end{aligned}
\qquad (6.36)
$$

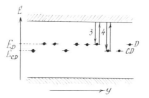

Fig. 6.3. Energy band structure of the surface of the crystal corresponding to Fig. 6.3.

Reactions (3) and (4) are electron transitions shown in Fig. 6.3 by arrows 3 and 4. Equations (6.36) yield

$$q^- = q^0 + (E_D - E_{CD}),\qquad(6.37)$$

which implies that (see Fig. 6.3)

$$\text{if}\quad q^- > q^0,\quad \text{then}\ E_{CD} < E_D,\qquad(6.38a)$$
$$\text{if}\quad q^- < q^0,\quad \text{then}\ E_D < E_{CD}.\qquad(6.38b)$$

As adsorption proceeds, the levels D in Fig. 6.3 disappear and are replaced by the same number of levels CD, which leads to a displacement of the Fermi level: in the case (6.38b) this is displaced upward, i.e., the atoms C behave as donors, while in the case (6.38a) it is displaced downward, i.e., the atoms C act as acceptors, although they are represented in Fig. 6.3 by local donor levels.

If the levels D and CD in Fig. 6.3 lie deeply enough inside the conduction band (compared with kT), as occurs, for instance, in the alkali-halide crystals, then transitions 3 and 4 can be considered as excluded (provided that there are no other additional ionizing agents apart from temperature), and in this case adsorption on neutral and ionized F-centers will occur independently. In this case the adsorptivity (at low pressures) of a "colored" crystal (i.e., one containing neutral F-centers) will be

$$\gamma = \exp\left(\frac{q^- - q^0}{kT}\right) = \exp\left(\frac{E_D - E_{CD}}{kT}\right)\qquad(6.39)$$

times lower [see (6.37)] than that of a "colorless" crystal (i.e., one containing the same quantity of F-centers but ionized). For example, at room temperature and $E_D - E_{CD} = 0.2$ eV we have $\gamma = 10^3$.

The effect was observed by Bauer and Staude [11], who investigated the adsorption of quinine on AgBr crystals. They found that in the case of "colorless" crystals the adsorptivity was practically zero, but after preparatory "coloration" of the crystal it increased sizably. The authors were undoubtedly dealing with adsorption on F-centers.

Fig. 6.4. F-Center consisting of a vacant oxygen site with two electrons localized in its neighborhood.

F-centers can act as adsorption centers not only on alkali-halide crystals but on any crystals as well. For instance, on a ZnO crystal, which to a first approximation can be considered as made up of Zn^{++} and O^{--} ions, an F-center is a vacant oxygen site with two electrons (instead of one) localized in its neighborhood, as shown in Fig. 6.4. From the chemical viewpoint such a center consists of two free vacancies of the same sign localized together and which (this must be emphasized) cannot appear on an ideal surface because of the Coulomb repulsive force between them. As a result of this property such an F-center can play a specific role in catalysis, acting as an active center in a number of reactions.

Besides F-centers, there are, of course, other surface structural defects that can act as adsorption centers. For one, V-centers, which in MR crystals, built of M^+ and R^- ions, are vacant metallic sites with holes localized at them, can serve as adsorption centers. In addition, with respect to molecules of a given gas the role of adsorption centers may be played by particles of another gas chemisorbed on the surface. For instance, chemisorbed oxygen atoms can act as adsorption centers for CO molecules. We used this model as one of the possible models when examining the oxidation of CO (see Section 5.3.3). In this model every adsorption bond formed is accompanied by the disappearance of one OL acceptor level and the appearance of the CO_2L acceptor level in place of it. If the CO_2L levels lie above the OL levels, then adsorption of CO molecules moves the Fermi level at the surface upward, i.e., the surface acquires a positive charge; in other words, the CO molecules behave like donors, although no donor levels arise on the surface in this process.

Here we should note the papers of Rzhanov and his collaborators [12, 13], who showed that the adsorption of H_2O molecules at the surface structural defects of germanium leads to creation of entirely new structural defects, which act as recombination centers. Later Novototskii-Vlasov with his collaborators [14, 15] obtained similar results for surface recombination centers on silicon.

Since surface defects are adsorption centers and are also localization centers for free surface valences, they can serve as active centers in catalysis. Ensembles made up of such defects may assume the same role. They may be treated as groups of localized ruptured valences.

Note in conclusion that the idea of surface defects, and ensembles of them, being active centers in catalysis was developed in a number of papers by Kobozev [16, 17]. In these papers, however, such ensembles are regarded (which is characteristic of this approach) as structures independent of the lattice. Actually, as we have seen, individual defects on the surface, and ensembles of them, form an integral whole with the crystal lattice, and their properties are to a considerable degree determined by the state of the lattice as a whole. By examining ensembles in isolation from the lattice (and thus ascribing the role of an inert substrate to the lattice) we lose the specific features of the heterogeneous case, transferred only from three-dimensional space to two-dimensional space.

6.3.2. Adsorption on Defects of Thermal Origin

At the very beginning of this book (see Section 1.1.2) we remarked that on every real surface the defects may be of either intrinsic or thermal origin. If thermal defects act as adsorption centers for molecules of a given type, then adsorption on such defects has a number of specific features and is characterized by essentially non-Langmuir regularities [18]. To clarify the role of thermal disorder we will examine the following example.

Let us suppose that a surface contains two types of defects, which we denote by AL and BL, and that only the defects of the AL type act as adsorption centers for the given molecules, which we denote by C. Let us also suppose that the defects participate in the following reactions:

$$C + AL \rightleftarrows CAL \ldots q_C, \quad BL + AL \rightleftarrows BAL \ldots q_A. \tag{6.40}$$

The first of these constitutes the adsorption and desorption reaction, while the second represents blocking and unblocking of the adsorption center AL (an arrow directed from left to right corresponds to an exothermic reaction). Denoting by N_A, N_B, N_{CA}, and N_{BA} the concentrations of the corresponding defects, where, obviously, N_A is the concentration of the adsorption centers, N_{CA} the concentration of occupied centers, i.e., the concentration of chemisorbed particles, and N_{BA} the concentration of blocked centers, we will have (in equilibrium)

$$N_A = \alpha N_{CA}, \quad \text{where} \quad \alpha = \frac{\alpha_0}{P} \exp\left(-\frac{q_C}{kT}\right), \tag{6.41a}$$

$$N_A N_B = \beta N_{BA}, \quad \text{where} \quad \beta = \beta_0 \exp\left(-\frac{q_A}{kT}\right), \tag{6.41b}$$

with

$$N_A + N_{BA} + N_{CA} = N_A^*, \quad N_B + N_{BA} = N_B^*, \tag{6.42}$$

where N_A^* and N_B^* are given constants (we will assume that $N_A^* \geq N_B^*$).

If the overall disorder at temperature T is characterized by a concentration $N = N_A + N_{CA}$, i.e., the concentration of unblocked adsorption centers, then $N^* = N_A^* - N_B^*$ is the intrinsic part of the disorder, $N - N^* = N_B$ the thermal part, and N_A^* the maximum disorder that can occur at the surface (corresponding to $T = \infty$).

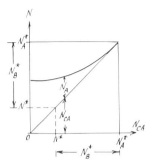

Fig. 6.5. Variation of the number of vacant
adsorption centers with surface coverage.

Indeed, at $T = 0$ all the defects B are used to block the defects A, i.e., $N_B = 0$, and, hence, $N_B = N_{BA}$, which implies $N = N_A + N_{CA} = N_A{}^* - N_B{}^*$. This is disorder at $T = 0$ (intrinsic disorder). As the temperature increases, unblocked defects A are added to intrinsic disorder, and the number of such defects N_B increases with temperature. This is the thermal fraction of disorder. At $T = \infty$ all defects A are unblocked, i.e., $N_{BA} = 0$, and, hence, disorder becomes maximal, i.e., equal to $N = N_A + N_{CA} = N_A{}^*$.

From Eq. (6.41b) combined with (6.42) we obtain

$$N_A = \frac{1}{2} \{ N^* - N_{CA} - \beta + \sqrt{(N^* - N_{CA} + \beta)^2 + 4N_B^*\beta} \} . \tag{6.43}$$

This equation gives us the law by which the number of vacant adsorption centers decreases as the surface coverage grows. Figure 6.5 illustrates this dependence (at $T = \text{const}$) by a thick curve. We see that the total number of unblocked adsorption centers $N = N_A + N_{CA}$ does not remain constant in the adsorption process (i.e., as N_{CA} increases), but increases. The very adsorption process thus leads to the appearance of new adsorption centers on the surface.

Substituting (6.43) into (6.41a) and then solving Eq. (6.41a) for N_{CA}, we obtain the equation of an isotherm. We can easily prove that this is a non-Langmuir isotherm, which, however, at $N_B{}^* = 0$ (i.e., if we ignore the thermal fraction of the disorder) becomes the usual Langmuir isotherm. Thus, although the surface has only adsorption centers of one kind on it (characterized by a single heat of adsorption q_C) and, therefore, is homogeneous (in the usual sense of the word), it behaves as an inhomogeneous surface. This effective inhomogeneity, which reveals itself in the violation of Langmuir regularities, is the result of the varying of concentration of adsorption centers in the adsorption process, which in the final analysis is the result of allowing for thermal disorder.

If the entire disorder has a thermal origin ($N^* = 0$) and the maximum disorder for the given surface $N_A{}^*$ is sufficiently large ($N_A{}^* \gg \beta$), then at low coverages (for $N_{CA} \ll N_A$) Eq. (6.43) yields

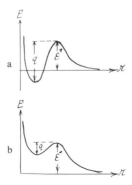

Fig. 6.6. Activation barrier curve in the adsorption on defects of thermal origin: a) exothermic adsorption; b) endothermic adsorption.

$$N_A = \sqrt{N_A^* \beta} = \sqrt{N_A^* \beta_0} \, \exp\left(-\frac{q_A}{2kT}\right). \qquad (6.44)$$

At the same time the adsorption rate in this coverage range is

$$\frac{dN_{CA}}{dt} = \gamma N_A. \qquad (6.45)$$

Substituting (6.44) into (6.45), we obtain the kinetics typical of activated adsorption (with an activation energy of $q_A/2$), although there may be no activation barrier before the surface. This is the result of an increase in the number of activation centers as temperatures increases. While in the usual theories of activated adsorption, which use the concept of an activation barrier near the surface of the crystal, the number of gas particles impinging on the surface increases with temperature and the number of adsorption centers accepting these particles remains constant, in the case at hand the number of accepting centers increases with temperature and the number of impinging particles remains practically constant (see Section 2.2.4).

In the given case adsorption requires preexcitation of the adsorbent, and the extent of this excitation increases as a result of adsorption. If in the process $q_A > q_C$, i.e., the energy required to create an adsorption center is higher than the energy released in the adsorption of a particle at this center (which generally is possible), we will have endothermic adsorption (although each act of adsorption remains exothermic). The possibility of endothermic adsorption was pointed out by de Boer [19].

Within the concept of an activation barrier (see Fig. 6.6, where E is the energy of the system, and r the distance between the particle being adsorbed and the surface) the fact that adsorption is endothermic indicates that the activation barrier

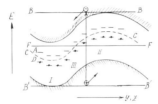

Fig. 6.7. Energy band structure in the adsorption on defects of thermal origin.

height ϵ^* on the adsorption curve is greater than the activation well depth q. Figure 6.6a corresponds to the usual case of exothermic adsorption ($q > \epsilon^*$), while Fig. 6.6b corresponds to endothermic adsorption ($q < \epsilon^*$). We see that thermal disorder (adsorption on centers of thermal origin) may be another cause of endothermic adsorption and non-Langmuir behavior.

6.3.3. Adsorption on the Surface of a Disordered Semiconductor

Let us take a semiconductor where, in the bulk and on the surface, defects of a certain type are distributed irregularly, or form clusters, i.e., a semiconductor with regions of increased and reduced defect concentrations. Such a cluster of microdefects may be considered as a macrodefect. If the microdefects carry an electric charge, the cluster is a region occupied by a space charge (or a surface charge if we are speaking of surface defects) whose electric field has a long range and is superimposed on the electric fields of other clusters. As a result the semiconductor appears to be "immersed" in a nonuniform electric field, whose nature is largely chaotic and reflects the specific features of each given sample. Such semiconductors are called disordered and possess a number of specific features, some of them surface features.

The presence of internal electric fields in a semiconductor is reflected in its energy spectrum; namely, the energy bands prove to be bent and have depressions and mounds (see Fig. 6.7). Depressions appear where positively charged defects are present. These depressions collect electrons, which to a certain extent compensate for the positive charge of the defects. The mounds indicate the places where negatively charged defects are present. They collect holes, which screen the charge of the defects.

The complex pattern of energy bands leads to special effects. If an external potential difference is applied, the trajectory of a carrier (electron or hole) ceases to be a straight line connecting the electrodes but, instead, becomes a complex curve bending around the potential barriers. In the process the electron does not move near the bottom of the conduction band (and the hole does not move near the top of the valence band), as it would in a homogeneous semiconductor, but along a certain "level of flow" corresponding to the lowest value of the energy at which the curved trajectory from electrode to electrode is continuous. Figure 6.7, which corresponds to a one-dimensional model of the crystal, shows the "levels of flow" of an electron and a hole (straight lines BB and $B'B'$, respectively). The "level of flow" for an electron is situated at different distances from the bottom of the conduction band (above it) in the crystal, while the "level of flow" for a hole is situated at different distances from the top of the valence band (below it). This means that the effective carrier mass changes from point to point within the crystal.

The bending of the energy bands in the bulk and at the surface of a semiconductor results in the distance from the Fermi level to the bottom of the conduction band becoming dependent on the coordinates,

$$\epsilon = \epsilon(x, y, z).$$

We will assume that the crystal occupies the half space $x \geq 0$, with the x axis directed normal to the surface of the crystal inward and the y and z axes lying in the surface plane. We will then have

$$\epsilon_s = \epsilon_s(y, z) = \epsilon(0, y, z),$$

which means that every surface property determined by the position of the Fermi level proves to be a function of y and z, i.e., changes from point to point on the surface. Using the terminology of adsorption theory we can say that such a surface is inhomogeneous with respect to this property, while if we use the terminology of semiconductor theory, we can say that such a surface, in an inhomogeneous electric field created by defect clusters, is disordered.

An example of a property that depends on ϵ_s is the adsorptivity of a surface, characterized by the number of gas molecules retained per unit area of the surface at given temperature and pressure at equilibrium between the surface and the gaseous phase. With a disordered surface the adsorptivity proves to be different at different points on the surface.

Another example is the mechanism of chemisorption bonding, which ties a chemisorbed particle to the semiconductor surface. We again turn to Fig. 6.7. Here the levels A and D are the acceptor and donor levels of chemisorbed particles of a given kind, the level CC lies in the middle between A and D, and the local intrinsic levels causing the energy bands to bend are not shown. Let us assume that in the region I the level CC lies considerably below the Fermi level FF, in the region II considerably above it, and in region III near it. In this case practically all the particles in region I are bound to the surface by a "strong" acceptor bond, in region II by a "strong" donor bond, and in region III by a "weak" bond (see Section 3.1.3 and Fig. 3.4). For instance, if we are dealing with CO_2 molecules, then in these three regions of the surface all three forms of chemisorption of CO_2 molecules occur (Figs. 2.11c, 2.11b, and 2.11a). Thus, we have regions of a surface on which the same chemisorbed particles manifest different properties. Regions I and II in Fig. 6.7 can be called the acceptor and donor regions, respectively.

The third example is the behavior of the catalytic activity of a surface. Regions with different values of ϵ_s differ in their catalytic activity. For instance, in an acceptor reaction region I in Fig. 6.7 is more active than region II, while in a donor reaction region II is more active than region I. This follows from the very concepts of acceptor and donor reactions.

In all these examples the surface was inhomogeneous in its properties. The inhomogeneity is due to a single cause, the uneven distribution of defects. But as adsorption proceeds, this inhomogeneity evens out to some extent. Indeed, the appearance of acceptor forms of bonding in the "acceptor" region (region I in Fig. 6.7) lowers the Fermi level, while the appearance of donor forms of bonding in the "donor" region (region II in Fig. 6.7) raises the Fermi level. As a result of these two factors the energy bands are to some extent straightened out.

The inhomogeneity is somewhat evened out when the temperature is raised due to the resulting migration of defects and equalization of their concentrations.

Note that when the surface of a crystal is illuminated by light of a frequency lying in the intrinsic absorption range, an electron and a hole are created inside the crystal, as shown by slanted arrows in Fig. 6.7, which in the case of a disordered surface immediately part. The depressions in the energy bands are enriched by electrons and the mounds by holes, which also leads to a straightening of the energy bands. In this way illumination of a semiconductor evens out the inhomogeneity. We also see that Mott's excitons created in a disordered semiconductor may be unstable, i.e., tend to disintegrate into a free electron and a free hole.

6.4. ADSORPTION ON DISPERSED SEMICONDUCTORS

6.4.1. Adsorptive Properties of a Dispersed Semiconductor

Here we will discuss several so-called dimensional effects in adsorption, i.e., effects caused by the geometric dimensions of the adsorbent.

An effect of this type is adsorption on a dimensionally quantized film. In a semiconductor whose dimensions L are on the order of or less than the electron de Broglie wavelength, $\lambda \approx 10^{-6}$ cm, the carriers are locked within a region of the size of the carrier's wavelength. For this reason their motion is quantized, whence the name dimensionally quantized [20, 21]. A film whose thickness L is of the order of λ combines a quasimolecular size along the normal to the surface (the x axis) with macroscopic size along the y and z axes. Adsorption on such a film results in a sort of a sandwich, consisting of the adsorbate, the two-dimensional gas of electrons or holes, and again the adsorbate. The longitudinal conductivity of this "sandwich" depends on the conditions at its boundaries, i.e., on the state of the adsorbed molecules [22]. The same type of gas may exist on the surface of a massive crystal, e.g., in the inversion layer.

The distribution of adsorbed molecules over the various states characterizes the adsorbate as a whole. Owing to the interaction of the adsorbate with the electrons in the film, the mobility of the electrons and, hence, the longitudinal conductivity depend on the state of the adsorbate. If we are able to change the states of the adsorbed molecules or the occupancy of these states (e.g., by illuminating the sample) and measure the longitudinal conductivity, then we have a tool for studying the energy spectrum of the adsorbed molecules. This problem has been investigated in detail by Peshev [22–25].

In [22, 23] Peshev considers a model in which a current carrier in a dimensionally quantized film with an adsorbate consists of a conduction electron (whose coordinates are x, y, and z) and the molecules that are adsorbed on the section of the surface with coordinates (y, z) of area λ^2. In the lateral direction the electron is bound to the adsorbed molecules via an interaction that depends only on x, while along the y and z axes it is in free motion and changes its adsorption "frame." We assume that the longitudinal and lateral motions of a carrier are separated, just as in a film without an adsorbate [26]. The difference between the two is that the lateral part now contains adsorbed molecules and, therefore, its state determines the carrier mobility. Peshev called such an entity an adsorbon.

Another example of a dimensional effect in adsorption is the adsorptivity of a thin semiconductor film covering a metal (see Section 4.6).

Still another example of a dimensional effect is the dependence of specific (i.e., per unit surface area) adsorptivity on the degree of dispersion (continuation) of the adsorbent. This problem was studied by Kogan [26] and in greater detail by Peshev [27, 28]. The effect starts to manifest itself only when the number of carrier (electrons or holes) localized on the surface becomes of the order of, or greater than, the total number of corresponding carriers contained in the bulk of the crystal (in the energy bands and on local levels). This happens when the separate crystals are small, i.c., $B/S \leq l$, where S is the surface area, B the volume of the crystal, and l, the screening length (usually $l \approx 10^{-4}$–10^{-6} cm). In this case the position of the Fermi level at the crystal surface, and hence the adsorptivity, prove to depend on B/S.

We will explain the mechanism of this effect by employing a model of a one-dimensional dispersed semiconductor, viz., a plane parallel semiconductor slab of thickness $2L$ both sides of which contain chemisorbed particles.

Of course, such a model is far from reality, but we can think of it as reflecting all the adsorption and catalytic features specific of a dispersed semiconductor. The band structure of such a semiconductor is shown in Fig. 6.8 (the surfaces are charged negatively). The horizontal axis in Fig. 6.8 is assumed to coincide with the Fermi level. Let us first assume that $L \gg l$ (Fig. 6.8a). In this case the bulk of the semiconductor is electrically neutral and its energy bands (in the bulk) are horizontal, as shown in Fig. 6.8a. This condition of electroneutrality determines the position of the Fermi ϵ_0 in the bulk and, thus, proves to be insensitive to the presence of a surface. The appearance of chemisorbed particles on the surface and the resulting change in the surface charge does not influence ϵ_0 in this case.

Next let us turn to a thin plate, where $L \leq l$ (Fig. 6.8b). Now the crystal's center is not electrically neutral and the bands in it, for the same surface charge, may be less bent than in the previous case. The Fermi levels in the bulk and at the surface prove to be shifted with respect to their positions in the previous case. It is easy to show that as the plate gets thinner, the potential difference between the

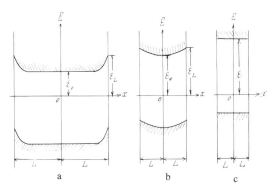

Fig. 6.8. Energy band structure for a dispersed semiconductor: a) $L \gg l$; b) $L \le l$; c) $L \ll l$.

surface and the crystal's center becomes smaller and the bands gradually straighten out.

In the case of a very thin plate, where $L \ll l$ (Fig. 6.8c), the bands may be considered to be practically straight and the entire bulk of the semiconductor charged uniformly [29]. We will study this limiting case and will assume that $\epsilon_0 = \epsilon_L = \epsilon$. The position of the Fermi level ϵ in this case can easily be determined from the condition of the electroneutrality of the crystal as a whole; this condition has the form

$$f(\epsilon) = \sigma(\epsilon) + \rho(\epsilon) L = 0, \tag{6.46}$$

where, as before, σ and ρ are the surface and bulk charge densities, and f is the total charge of the plate.

Equation (6.46) yields

$$\frac{d\epsilon}{dL} = -\frac{df/dL}{df/d\epsilon} = -\frac{\rho}{(d\sigma/d\epsilon) + (d\rho/d\epsilon)L}. \tag{6.47}$$

Note that (see Section 4.6)

$$\frac{d\sigma}{d\epsilon} > 0, \qquad \frac{d\rho}{d\epsilon} > 0$$

always and that, according to (6.46), ρ is positive if σ is negative, and vice versa. This together with (6.47) yields

$$\begin{cases} \dfrac{d\epsilon}{dL} < 0 & \text{for} \quad \sigma < 0, \\[2mm] \dfrac{d\epsilon}{dL} > 0 & \text{for} \quad \sigma > 0. \end{cases}$$

We see that as the crystal gets smaller, i.e., L decreases, the Fermi level moves downward if the surface is negatively charged ($\sigma < 0$) and upward if the surface is positively charged ($\sigma > 0$).

From this we can find how the specific (per unit surface area) adsorptivity of a sample changes when the sample is broken up. If the charge of the surface is of an intrinsic origin and retains its sign when the sample is broken up, the adsorption properties with respect to an acceptor gas and to a donor gas will vary in opposite directions when the sample is broken up: adsorption of the acceptor gas will decrease, while that of the donor gas will increase. If the surface is charged positively, the pattern is reversed.

If the sign of the surface charge is determined not by the history of the surface but by the nature of the adsorbed gas, the adsorptivity will fall when the sample is broken up both for the acceptor gas and for the donor gas.

This result, which was obtained on the assumption that $L \ll l$, remains valid, as shown in [26, 28], when $L < l$. This mechanism may be responsible for the variation of the adsorptivity of semiconductors with the degree of their dispersion, an effect repeatedly cited in the literature (e.g., see [30, 31]). Obviously, this effect will manifest itself sooner (i.e., the smaller the degree of dispersion) the greater the screening length l in the semiconductor and can be expected to occur at $S/B > 10^6$ cm^{-1}.

6.4.2. The Compensation Effect on Dispersed Semiconductors*

For a catalytic reaction, which is sensitive to the position of the Fermi level at the surface (ϵ_L in Fig. 6.8), variation of just the crystal size L leads to changes in the preexponential factor K_0 and the activation energy E in Eq. (5.48) for the reaction rate constant (Arrhenius's equation). (The reader must distinguish between the energy E in Fig. 6.8 and the activation energy E in the formulas of Chapter 5 and in the following formulas in this section.) This is a condensation effect, which occurs because electrons, holes, or surface defects take part in reactions on semiconductors [32]. (We have mentioned this effect in Section 5.5.5.) The equilibrium values of the electron concentration n_s and hole concentration p_s at the surface depend on ϵ_L exponentially [see Eq. (3.21) and Fig. 6.8]. The same is true for the concentration ratio of the charged and neutral forms of any type of surface defects [see Eq. (1.55)]. If a reaction, proceeding slowly, does not violate these relationships strongly, the compensation effect follows directly from the form of the dependence of ϵ_L on L.

The specific rate constant [see Eq. (5.48)] will be proportional to a factor of the form $\exp(\epsilon_L/kT)$ or $\exp(-\epsilon_L/kT)$ depending on what current carriers or surface defects act as reagents:

$$K = K_0 \exp\left(-\frac{E}{kT}\right) \sim \exp\left(\frac{\epsilon_L}{kT}\right) \quad \text{or} \quad \exp\left(-\frac{\epsilon_L}{kT}\right), \tag{6.48}$$

---

*This section was written by O. Peshev.

where

$$\epsilon_L = \epsilon_0(L) + V_s(L), \qquad \epsilon_0(L) = \epsilon_0(\infty) + [\epsilon_0(L) - \epsilon_0(\infty)]. \qquad (6.49)$$

As (6.49) and Fig. 6.8 show, ϵ_L consists of three terms: (1) the position $\epsilon_0(\infty)$ of the Fermi level in the bulk of a large crystal, i.e., in the middle of a thick ($L \gg l$; Fig. 6.8a) film; (2) the shift $\epsilon_0(L) - \epsilon_0(\infty)$ of the Fermi level at the middle of a thin ($L < l$) film in relation to the position of the Fermi level in the bulk of a large crystal; and (3) the band bending V_s (Fig. 6.8b; in Fig. 6.8c $V_s = 0$).

   If while the semiconductor is dispersed the impurity content remains constant, the term $\epsilon_0(\infty)$ will not depend on L:

$$\frac{d\epsilon_0(\infty)}{dL} = 0. \qquad (6.50)$$

   The following considerations can be employed to find the shift $\epsilon_0(L) - \epsilon_0(\infty)$ as a function of L. Let us ignore the bending of bands (see Fig. 6.8c). In this case the charge distribution in the bulk of the film is homogeneous and has a density ρ_L, while the electroneutrality condition for the crystal as a whole assumes the form

$$\sigma(L) + L\rho_L = 0. \qquad (6.51)$$

We will assume that the distributions of the electrons and holes in the bands and on local levels are Boltzmann-like, so that

$$\rho_L = \rho(\infty)\left\{ \exp\left[-\frac{\epsilon_0(L) - \epsilon_0(\infty)}{kT} \right] - \exp\left[-\frac{\epsilon_0(L) - \epsilon_0(\infty)}{kT} \right] \right\}, \qquad (6.52)$$

where $\rho(\infty)$ is the absolute value of the (equal) densities of positive and negative charges in the bulk of a large crystal. If condition (6.50) is met, we also find that

$$\frac{d\rho(\infty)}{dL} = 0. \qquad (6.53)$$

   If $\epsilon_0(L) - \epsilon_0(\infty)$ is not small in comparison with kT, the only term left on the right-hand side of (6.52) is the first if $\sigma < 0$, i.e., $\epsilon_0(L) > \epsilon_0(\infty)$, or the second if $\sigma > 0$, i.e., $\epsilon_0(L) < \epsilon_0(\infty)$. On the basis of (6.51) and (6.52) this yields

$$\epsilon_0(L) - \epsilon_0(\infty) = \mp kT \ln\left[\frac{\pm\sigma(L)}{L\rho(\infty)} \right], \qquad (6.54)$$

where the upper sign corresponds to σ being positive and the lower to σ being negative.

   Finally, in a thin film the bending of bands, V_s, divided by L yields, in order of magnitude, the gradient dV/dx of the electron potential energy [see (4.5)], where, according to (4.1) and (4.9), in the surface plane we have

$$\sigma(L) = -\frac{\chi}{4\pi e}\left(\frac{dV}{dx}\right)_s$$ (6.55)

and, hence, $V_s/L \approx (dV/dx)_s = -4\pi e\sigma(L)/\chi$, i.e.,

$$V_s = -\frac{4\pi e\sigma(L)}{\chi}L.$$ (6.56)

Let us for the present assume the rough approximation that the surface charge density also remains constant as L gets smaller (these restrictions will subsequently be lifted):

$$\frac{d\sigma}{dL} = 0.$$ (6.57)

Then ϵ_L in (6.48) consists of a constant term $[\epsilon_0(\infty)]$, a logarithmic term $[\epsilon_0(L) - \epsilon_0(\infty)]$, and the term V_s, which is linear in L. In view of this, $\exp\{[\epsilon_0(L) - \epsilon_0(\infty)]/kT\}$ becomes part of the preexponential factor K_0, while the band bending V_s remains a term in the activation energy E. If we compare (6.54) and (6.56), we can see that V_s and $\epsilon_0(L) - \epsilon_0(\infty)$ have the same sign and their absolute values vary in opposite fashion when L gets smaller. In other words, the term in the activation energy that depends on L, namely V_s, and the part of the preexponential factor that depends on L, namely $\exp\{[\epsilon_0(L) - \epsilon_0(\infty)]/kT\}$, increase or decrease simultaneously as the film thickness varies.

This constitutes the compensation effect for semiconductors. It manifests itself when $L < l$ and has a clear-cut physical origin: the shift $\epsilon_0(L) - \epsilon_0(\infty)$ is the measure of accumulation of electrons or holes by the thin film, while V_s characterizes the barrier that the electrons or holes must overcome in order to take part in a catalytic reaction (directly or through surface defects). For this reason the number of current carriers increases (decreases) as L gets smaller, but at the same time they find it more difficult (easier) to reach the surface.

This effect is not connected with the approximation (6.57), which was introduced only to simplify calculations [27]. By integrating the Poisson equation (4.5), where $\rho(\epsilon)$ is given in (6.52), we can find the derivative dV/dx and substitute it into (6.55). Differentiating (6.55) with respect to L leads to the following expression [33]:

$$\frac{d[\epsilon_0(L) - \epsilon_0(\infty)]}{dL} = -\gamma(L)\frac{dV_s}{dL}, \quad \text{where } \gamma(L) > 1.$$ (6.58)

The factor $\gamma(L)$ depends on $\epsilon_0(L) - \epsilon_0(\infty)$, V_s, σ, and $d\sigma/d\epsilon_L$. Its value is closer to unity at $d\sigma/d\epsilon_L \neq 0$ than at $d\sigma/d\epsilon_L = 0$. Hence, the fact that $\epsilon_0(L) - \epsilon_0(\infty)$ and V_s vary in opposite fashion as L gets smaller, which manifests itself as a compensation effect, follows from the thermal equilibrium of the current carriers in a small $(L < l)$ crystal with a charged surface.

Now we will lift the restriction (6.50); i.e., we assume that the impurity content in the semiconductor may change as L gets smaller. But how do dispersion and alloying act simultaneously? The answer is given below (see also [34]). The Debye length l is a natural scale of length in the studies of dispersed semiconductors, since all distances r can be expressed in the corresponding equations in the dimensionless form r/l. At the same time the Debye length can be taken as the measure of impurity content in the semiconductor, since the impurity content fixes the values of $\rho(\infty)$, which in turn determines the parameters $\epsilon_0(\infty)$ and l. Increasing (decreasing) l by introducing impurities into a thin ($L < l$) film makes it, so to say, thinner (thicker). In other words, by influencing the length scale, alloying changes the effective size of a small crystal. Thus, we have arrived at the rule that the introduction of impurities into a semiconductor with a constant degree of dispersion is equivalent to a change in the degree of dispersion with the impurity content kept constant.

The compensation effect in a dispersed semiconductor, theoretically predicted by Peshev, was discovered by Herrman, Vergnon, and Teichner in the acceptor reaction of oxidation of CO on nonporous monodisperse particles of n-TiO_2 (anatase) from 100 to 2000 Å in diameter, both pure [35] and doped with Nb or Ga [36]. Comparison of the theoretical and experimental data shows [34, 37, 38] that there is good agreement between the two. Moreover, when the reaction rate is independent of the electron concentration, which is the case with samples trained in advance in a CO atmosphere, the effect is not present, which serves as additional proof of its electronic origin.

6.5. CONTROLLING THE STOICHIOMETRY OF CRYSTALS*

6.5.1. Theoretical Aspects of the Problem

Electronic processes on the surface of a crystal determine not only the catalytic and adsorptive properties of the surface. These processes, as shown by Pikus and his collaborators, also play the main role in the mechanism of thermal evaporation and the related change of the stoichiometry of semiconductor crystals of binary compounds of the II–VI type with ionic bonding when the crystals are heated to a high temperature in a vacuum and there is no equilibrium between the solid phase and the vapor.

According to [39, 40], we must distinguish three stages in the evaporation of an atom from the surface of a ionic crystal. At first a surface ion of the lattice leaves a normal site and thus creates a surface defect to which there corresponds a local energy level in the semiconductor's forbidden band. In the second stage there is electron exchange between the defect and the conduction band, which results in the appearance of a neutral atom on the surface. This atom is weakly bound to the surface (e.g., by van der Waals' forces) and desorbs (evaporates).

---

*This section was written by G. Ya. Pikus.

Thus, the surface concentration of the neutral atoms and the evaporation rate, which depends on the surface concentration, are determined not only by the rate at which surface defects are created but by the equilibrium between charged and neutral defects, which in essence are analogs of chemisorbed particles bound to the surface by "strong" and "weak" bonds, respectively.

What has been said implies that we can control the evaporation rate of the components of a crystal as well as its adsorptivity by changing the position of the Fermi level at the surface. Indeed, since the neutralization of an anion is achieved by tearing away an excess electron and transferring it from the defect level into the conduction band, we find that according to the Fermi statistics the elevation of the Fermi level at the surface (or the increase in concentration of free electrons, n_s), irrespective of the cause, leads to a decrease in the evaporation rate of the metalloid.

At the same time the concentration of neutral atoms and the rate of metal evaporation must increase with n_s, in contrast to the previous case, since the neutralization of a cation occurs via the addition of a missing electron. Thus, neutralization of surface defects is the main stage in the mechanism of ionic crystal evaporation. This inevitably implies that evaporation and the resulting change of the stoichiometry of the crystal, caused by appearance of vacancies, must be closely linked through the electronic system of the crystal.

Let us discuss this question in greater detail. First we note that since the evaporation of atoms is a surface effect, the violation of the stoichiometry of crystals starts from the surface and propagates into the bulk as a result of vacancy diffusion. Another aspect of this process is that anion and cation vacancies created in evaporation are electrically active defects of the donor and acceptor types, respectively.

Since the evaporation rates of a metal and a metalloid at the initial stage are generally different, in the heating of a crystal the position of the Fermi level (n_s) at the crystal's surface must change because of both the variation of the concentration ratio of donors and acceptors in the bulk and the violations of the surface's electroneutrality, i.e., the appearance of a surface charge and the resulting electric field and bending of bands at the surface. This, in turn, must lead to a change in the evaporation rates, opposite for atoms of the metal and the metalloid. If, for instance, the initial evaporation rate for a metalloid is greater than that for a metal, then the Fermi level must move upward and n_s must increase, because of the relatively greater increase in the concentration of anion vacancies (donors) and the appearance of a positive surface charge owing to enrichment of the surface with the electropositive component. Correspondingly, the evaporation rate for the metalloid must decrease with the passage of time and the evaporation rate for the metal must increase. The variation of the evaporation rates, the position of the Fermi level (n_s), the surface charge, and the electric field will continue up to a value of n_s at which the evaporation rates for the metal and the metalloid become equal; i.e., the ratio of cation and anion vacancy concentrations, which are different in the bulk and at the surface, ceases to change. Although there is no equilibrium between the solid phase and the vapor, a characteristic dynamic-equilibrium state of the crystal with violated stoichiometry is achieved when n_s

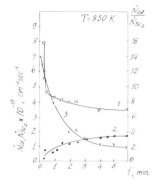

Fig. 6.9. Rates of evaporation of cadmium (curve 1) and selenium (curve 2) and their ratio (curve 3) as functions of time.

reaches the above–mentioned value. In such an equilibrium state the electronic system of the crystal is in dynamic equilibrium with the flow of evaporating atoms, on the one hand, and with the defects (vacancies) created in the process, on the other, while the vacancy concentration gradient occurring in the surface layer of the crystal is balanced by the electric field localized in the same layer [40]. Note that this state must strongly depend on temperature and vary reversibly with it.

6.5.2. Experimental Results

Dynamic-equilibrium states were observed experimentally in vacuum heating of the following crystals of the II–VI group with ionic bonds: BaO and SrO [41], and CdSe and CdS [42]. The dynamics of formation of such a state for a single crystal at $T = 850$ K can clearly be seen in Fig. 6.9 [42], which shows the time dependence of the rates of evaporation of cadmium, $\dot{N}_{Cd}$ [39], and selenium, $\dot{N}_{Se_2}$ [40], and their ratio, $\dot{N}_{Cd}/\dot{N}_{Se_2}$ [43].* We can see by Fig. 6.9 that as the evaporation rate of cadmium decreases, the selenium flux increases, and finally the ratio of these two quantities attains a stable value equal to two. Since the metalloid here evaporates in the form of diatomic molecules, Se_2, the fluxes of cadmium and selenium become equal when the ratio $\dot{N}_{Cd}/\dot{N}_{Se_2}$ stabilizes. Simultaneously with the stabilization of $\dot{N}_{Cd}/\dot{N}_{Se_2}$ the absolute values of $\dot{N}_{Cd}$ and $\dot{N}_{Se_2}$ are stabilized, as was to be expected. Starting from this moment, we can consider the evaporation of the crystal as layer-by-layer destruction without change of composition.

The main consequence of the above facts is that the evaporation processes and, hence, violation of stoichiometry of an ionic crystal can be controlled by changing, in the course of vacuum heating, the free electron concentration n_s at the surface of the crystal via external sources such as light, electric field, and electrically active impurities.

---

*Here we use the common notation in which a dot over a symbol means the time derivative of the corresponding quantity.

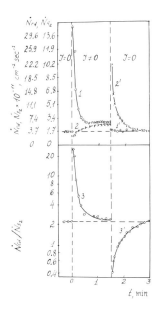

Fig. 6.10. Rates of evaporation of cadmium (curves 1 and 1′) and sulfur (curves 2 and 2′) and their ratio (curves 3 and 3′) at 800 K with "white" light of flux intensity $J \approx 100$ mW/cm² on (primed curves) and off (unprimed curves), as functions of time.

Figure 6.10 [43] shows the time dependence of the evaporation rate for cadmium, $\dot{N}_{Cd}$ (curves 1 and 1′), and sulfur, $\dot{N}_{S_2}$ (curves 2 and 2′), and their ratio $\dot{N}_{Cd}/\dot{N}_{S_2}$ (curves 3 and 3′) for the CdS crystal in a dynamic equilibrium state in the dark at $T = 800$ K if "white" light with a flux intensity $J \approx 100$ mW/cm² is switched on and off. Since after annealing the CdS crystal becomes photosensitive, the free electron concentration at the surface suddenly changes when the light is switched on. Accordingly, when the light is switched on, the evaporation rate of the metal, $\dot{N}_{Cd}$, suddenly increases and that of the metalloid, $\dot{N}_{S_2}$, sharply decreases, owing to the change in the equilibrium between charged and neutral defects, a change in which the concentration of neutral defects in the case of a metal and of charged defects in the case of a metalloid increases.

The experiment described clearly demonstrates the decisive role played by the neutralization of surface defects in the evaporation mechanism, on the one hand, and the connection of the neutralization process with the free electrons (electron transitions between the defect levels and the conduction band), on the other. This connection is also clearly demonstrated by the fact that the initial jump in $\dot{N}_{Cd}$ when the light is switched on is greatest when the energy of the quanta is equal to the width of the crystal's forbidden gap at the temperature of measurements [43], while the magnitude of this jump is proportional to $(\Delta n_s)^2$. (Here Δn_s is the addition to the free electron concentration at the surface due to light; see [44].) The latter indicates that evaporation of a Cd atom follows the neutralization of a doubly charged Cd⁺⁺ ion via attachment of two free electrons.

The considerable change in the ratio of the metal and metalloid evaporation rates ($\dot{N}_{Cd}/\dot{N}_{S_2} \gg 2$; Fig. 6.10 [41]) brought on by illumination must lead to considerable changes in the surface and bulk compositions compared to those obtained after heating in darkness, and the changes result in an increase in the

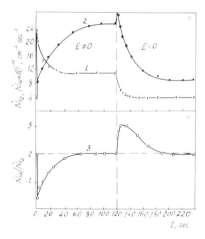

Fig. 6.11. Rates of evaporation of sulfur (curve 1) and cadmium (curve 2) and their ratio (curve 3) as functions of time in a depleting field (and without that field).

metal vacancy concentration in the crystal. At the same time, the predominant evaporation of the metal must, in the final analysis, diminish n_s at the crystal surface by increasing the concentration of cation vacancies (acceptors) and the negative surface charge (due to enrichment of the surface with the electronegative component). Correspondingly, after the initial jump in $\dot{N}_{Cd}$ this quantity decreases while $\dot{N}_{S_2}$ grows. This process continues up to a new dynamic-equilibrium state of the crystal ($\dot{N}_{Cd}/\dot{N}_{S_2} = 2$), with the departure from stoichiometry at the surface and in the bulk differing from the initial values. The value of n_s necessary for the state to occur is sustained by two processes of excitation of electrons into the conduction band (both acting at the same time): thermal and optical. When the light is switched off, the departure from stoichiometry becomes nonequilibrium owing to the rapid drop in n_s. Correspondingly, the metalloid evaporation rate suddenly increases while the metal evaporation rate drops (Fig. 6.10, curves 1 and 2), the excess of metal created during illumination decreases, and after some time the crystal returns to the initial dynamic-equilibrium state.

When n_s decreases due to a depleting electric field normal to the surface, the dynamics of variation of the rate of evaporation and the crystal composition is similar in nature but opposite in direction. As Fig. 6.11 demonstrates [45]—the introduction and switching off of a depleting field $E \approx 4 \times 10^3$ V/cm—the metalloid evaporation rate suddenly increases as n_s drops. (The figure shows the time variation of the evaporation rates for sulfur, $\dot{N}_{S_2}$ [39], and cadmium, $\dot{N}_{Cd}$ [40], and the ratio $\dot{N}_{Cd}/\dot{N}_{S_2}$ [41] for a crystal of CdS brought into dynamic equilibrium at $T = 605$ K.) The resulting increase in the concentration of anion vacancies (donors) and the positive surface charge leads to a decrease in $\dot{N}_{S_2}$ and an increase in $\dot{N}_{Cd}$ by stimulating an increase in n_s. This continues up to a new dynamic-equilibrium state of the crystal ($\dot{N}_{Cd}/\dot{N}_{S_2} - 2$) with a concentration of anion vacancies higher than the initial and a greater positive surface charge. When the field is switched off, the evaporation rate of the metal diminishes more slowly than that of the metalloid, and thanks to this the excess of metalloid vacancies (metal atoms) that appeared earlier decreases and the crystal returns to the initial dynamic-equilibrium state.

Similar changes in the kinetics of evaporation of the II–IV group crystals have also been observed when n_s changes owing to thermal emission [46] or to introduction of electrically active impurities [47]. Thus, via vacuum annealing combined with external excitation of the electron system we can change the departure from stoichiometry at the surface and in the bulk of an ionic crystal within considerable limits and in the necessary direction by rapidly "freezing" the states attained at different annealing stages.

As shown by Pikus and Chaika [40], when the diffusion rate of the vacancies is much higher than the rate of evaporation of atoms from the surface, the dynamic-equilibrium state of a binary ionic crystal attained in the process of vacuum annealing can be considered as quasisteady. This implies that we can use the methods of statistical thermodynamics to calculate this state, provided we know the energies of vacancy formation and thermal ionization of surface defects [39, 40]. Calculations performed for crystals of BaO [48, 49] and CdS and CdSe [47] agree satisfactorily with the experimental data on the peculiarities of evaporation and changes in composition and the electrophysical properties of these crystals during heating in a vacuum. This illustrates the possibility of predicting the violations of stoichiometry and the resulting changes in the electronic properties of the bulk and surface that occur during vacuum heating of ionic crystals.

The possibility of effectively controlling the violations of stoichiometry of the surface and bulk of a crystal by heating it in a vacuum in conjunction with an external excitation of its electron system actually means that we can control the adsorptive and catalytic properties of the crystal's surface.

Chapter 7

THE EFFECT OF ILLUMINATION ON THE ADSORPTIVE AND CATALYTIC PROPERTIES OF A SEMICONDUCTOR

7.1. THE PHOTOADSORPTION EFFECT

7.1.1. Positive and Negative Photoadsorption Effects

It is now experimentally well established that illumination influences the adsorptive properties of a semiconductor surface. This phenomenon will be referred to as the photoadsorption effect. Illumination may influence both the adsorption equilibrium and the adsorption kinetics. Here we must distinguish between positive and negative photoadsorption effects.

Let us start with adsorption equilibrium. If the temperature and pressure are fixed, the semiconductor surface contains a fixed number of adsorbed particles, which characterizes the adsorptivitiy of the surface and may be changed by illumination. In some cases the adsorptivity is increased by illumination, i.e., external light leads to additional adsorption (photoadsorption, or the positive photoadsorption effect), while in other cases the adsorptivity is reduced, i.e., some adsorbed particles leave the surface when subjected to external light (photode-sorption, or the negative photoadsorption effect). Sometimes the adsorptivity is not influenced by illumination at all.

Variations in adsorptivity are commonly registered by pressure variations in the adsorption volume. When illumination is followed by a drop in pressure, we are faced with photoadsorption, while an increase in pressure implies photo-desorption. If the pressure remains unchanged, we have photoadsorption–inactive light absorption.

Let us now consider adsorption kinetics.* Let us assume that a semiconductor is brought into contact with a gas, so that adsorption is initiated. Illumination often influences the adsorption rate, sometimes increasing it (the positive effect) or decreasing it (the negative effect). Moreover, the very time dependence of the adsorption rate (the kinetic law) may be changed by illumination. The adsorption activation energy may be influenced by illumination, too. Sometimes it becomes negative, which means that, in the presence of light, heating (in this case) hinders the adsorption process instead of accelerating it.

In some cases adsorption proceeds for some time after the light is turned off as though there was still illumination, which means that we are dealing with an aftereffect. Sometimes the adsorptive properties of the surface in the dark are modified if the surface was previously illuminated for a certain period of time, which means we are dealing with a memory effect.

Photoadsorption has been thoroughly studied. Various authors have investigated it using different adsorbents and adsorbates. The sign and magnitude of the effect depend not only on the experimental conditions, such as pressure, temperature, and light frequency, but also on the prehistory of the sample. The relevant experimental data and theory of the phenomenon are given below.

Note that often the experimentally observed effect of illumination on the adsorptive properties of a semiconductor is only an apparent effect. For instance, photoadsorption in some cases is just photodesorption in disguise. An example is oxygen "photoadsorption" on SiO_2, first observed by Solonitsyn [1, 2]. In this case the light seems to break the Si–OH bond, which leads to desorption of the OH groups covering the surface. As a result new free valences appear on the surface, and these valences act as adsorption centers near which additional oxygen molecules are localized.

Photodesorption may also be only an apparent effect and, in fact, have a trivial origin. It may be due to heating of the adsorbent by light absorption. Illumination here is only an indirect factor. As shown by Kotel'nikov [3], this was probably the case in the experiments in oxygen "photodesorption" from NiO described by Haber and Stone [4].

According to the experimental data, photoadsorption can be observed only when the light is absorbed by the semiconductor. Moreover, the light must be (and this should be especially stressed) photoelectrically active, i.e., an internal photoeffect has to take place in the semiconductor. In other words, the semiconductor must be enriched with free electrons or free holes or both.

7.1.2. The Photoadsorption Effect on Ideal and Real Surfaces

As is known, there are two mechanisms of light absorption, the electronic and the excitonic. In "electronic absorption" electrons undergo transitions from the

---

*Here, as everywhere in the book, adsorption is always to be understood as chemisorption.

valence band into the conduction band (intrinsic absorption) or from a local (donor) level into the conduction band or, finally, from the valence band into a local (acceptor) level (extrinsic absorption). All these processes lead to photoelectrically active light absorption. In intrinsic absorption we are dealing with the creation of charge carriers of both signs (electrons and holes) simultaneously, while in extrinsic absorption carriers of only one sign are produced (electrons or holes).

If the excitonic absorption mechanism is involved, the immediate result of phonon absorption is formation of an exciton. This process is photoelectrically inactive and does not lead to a photoadsorption effect. However, while wandering through the crystal and colliding with the defects of the crystal lattice, the free exciton may be annihilated, ionizing the defect and thus producing a free carrier. Hence, the excitonic light absorption mechanism can in the final analysis also lead to an internal photoelectric effect and thereby result in a photoadsorption effect as a secondary effect.

On a real surface the chemisorption of gas particles can occur both on the atoms (ions) of the lattice proper and on the microdefects of various types. In this chapter we will consider only two models of the surface, models that in a certain sense are limiting cases. The theoretician is often forced to operate with approximate models, and the choice of a model depends on the body of experimental facts that the given model encompasses and on how good the model describes these facts.

We start with the case where the adsorption centers are the atoms (ions) of the lattice proper, ignoring the adsorption at defects. This can be done when the concentration of the atoms or ions is considerably higher then the concentration of defects. In other words, we will consider adsorption on sections of the ideal surface and take the lattice sites as adsorption centers.

Later we will see that illumination (when the surface coverage is not too large) changes the concentration of the adsorbed particles that are in a charged state, while the concentration of the neutral particles remains unchanged. This process changes the overall adsorptivity of the surface. The magnitude and sign of the photoadsorption effect prove to depend on the position of the Fermi level of the crystal without illumination. In such an idealized model the defects of the surface, although not acting as adsorption centers, do affect the adsorptive properties and, particularly, the photoadsorptive properties of the surface since the position of the Fermi level depends on the nature and concentration of these defects. We will study such a model in Sections 7.2 and 7.3. The model proves to be sufficient for understanding the basic laws governing the photoadsorption effect (the dependence of the magnitude and sign of the effect on the experimental conditions and the prehistory of the sample).

Sections 7.4, 7.5, and 7.6 are devoted to the opposite idealized model, in a sense. We will assume that the defects of the surface are adsorption centers and ignore the interaction between the defects and the adsorption of atoms (or ions) of the lattice proper. This can be done if the binding energy (the adsorption heat) on a defect is much larger than on an atom (or ion) of the lattice.

Illumination, as we will see, changes the occupancy of the defects by electrons and holes and thus changes the concentration of the adsorption centers of each given type and the adsorptivity of the surface as a whole. The model therefore explains the regularities as the previous model, but it also describes some peculiarities of the adsorption kinetics, namely, the aftereffects and memory effects.

7.1.3. Review of Basic Experimental Data

We begin by giving a short summary of the observed facts.

(1) Much experimental work has been done to study the effect of surface treatment on the magnitude and sign of the photoadsorption effect under adsorption equilibrium conditions.

Romero-Rossi and Stone [5, 6] studied oxygen adsorption on SnO_2. A positive effect (photoadsorption) was observed at room temperature and low oxygen pressure. On raising the pressure the effect diminished, becoming negative (photodesorption) at high enough pressures. The same authors observed the opposite effect at 400°C with the same system; i.e., photodesorption was observed at low pressures but was replaced by photoadsorption as the pressure was increased.

The same result was obtained by Kwan [7], who studied O_2 adsorption on TiO_2 at 500°C: photodesorption occurred at low oxygen pressures and photoadsorption at higher pressures.

Stone [6, 8] also studied O_2 adsorption on TiO_2. Photoadsorption was observed. It became noticeably weaker when water was removed from the surface and was partly reestablished when the sample was heated in an atmosphere of water vapor.

Lisachenko and Vilesov [36] noted a sharp increase in oxygen photoadsorption on MgO with hydroxylation of the surface.

Bickley and Jayanty [39] investigated the role of adsorbed water in photoadsorption of oxygen on TiO_2 and found that the photoadsorption of O_2 increased with the number of H_2O molecules on the surface.

(2) Many experimental papers report the influence of various dopants introduced into crystals on the magnitude and sign of the effect when there is equilibrium between the surface and the gaseous phase. For instance, Kwan [7] used the classical and much studied system $ZnO + O_2$. A sample of ZnO containing Al (donor) as impurity yielded the negative effect, while the effect was positive with Li (acceptor) as impurity.

Romero-Rossi and Stone [5, 6] found the positive effect enhanced when the sample was doped by Li (acceptor) and reduced when Ga (donor) was the dopant (for the same system, $ZnO + O_2$).

Many researchers have noted the dependence of the magnitude and sign on the nature and degree of the deviation from stoichiometry in samples. According to Fujita and Kwan [9], photodesorption of oxygen takes place on ZnO containing some nonstoichiometric zinc excess (reduced samples), while photoadsorption takes place on ZnO samples containing zinc in stoichiometric deficit (oxidized samples).

The results of Barry [10], who studied the influence of preliminary treatment of ZnO samples on the sign of the photoadsorption effect with respect to oxygen, agree with other data. A sample was annealed in an oxygen atmosphere at high temperature and cooled to room temperature, at which adsorption was studied. Photodesorption was observed on untreated samples, while photoadsorption was observed on samples treated in the above-mentioned manner, i.e., saturated with oxygen.

The same result was obtained by Terenin and Solonitsyn [11], i.e., reduced samples of ZnO showed a negative photoadsorption with respect to oxygen, while oxidized samples of ZnO showed a positive photoadsorption effect. The same regularity was observed in the case of oxygen adsorption on TiO_2. According to Kennedy, Ritchil, and MacKenzie [12] and Kazanskii *et al.* [13], photoadsorption is replaced by photodesorption on degassing (reducing) the TiO_2 sample (as in the case with ZnO).

However, quite a different result was obtained by Romero–Rossi and Stone [5], who found that oxygen photodesorption is characteristic of ZnO samples containing lower concentrations of nonstoichiometric zinc excess, while photoadsorption is characteristic of ZnO samples with higher concentrations of such zinc. In agreement with these results are the data of Haber and Kowalska [14], who used the same system, O_2 on ZnO. The researchers found that the positive effect (photoadsorption) is replaced by the negative effect (photodesorption) after the sample is oxidized.

(3) Bykova, Komolov, and Lazneva [37, 38] studied the effect of an external electric field on oxygen photodesorption from CdS. The field was normal to the adsorbing surface. A sharp increase in oxygen photodesorption was observed for such a direction of the field at which the surface layer of the semiconductor was enriched with holes, while for the reverse direction photodesorption was somewhat weaker.

(4) There is a certain general regularity observed by many investigators working with various systems. As a rule, photoadsorption is irreversible, which means that the molecules adsorbed by the surface during illumination remain on it for a sufficient length of time after the light is switched off. However, these molecules can be removed by heating the sample. Such an effect has been observed, for example, in oxygen photoadsorption on TiO_2 [6, 8] and on ZnO [9], and in many other cases. On the other hand, photodesorption usually proves to be reversible. However, Constantinescu and Segal [94] discovered the irreversibility of oxygen photodesorption from ZnO at room temperature.

(5) Studying the effect of ultraviolet light on oxygen adsorption by ZnO, Steinbach and Harborth [35] observed the production on the surface of atomic oxygen, which was desorbed and registered by a mass spectrometer. The researchers believe that the production of atomic oxygen is due to bond rupture in the ZnO lattice, occurring as the result of absorption of photons.

(6) Studying the same system ($NO + O_2$ and ultraviolet light), Solonitsyn [15] observed the following phenomena:

(a) If the semiconductor sample is illuminated and then the light turned off and adsorption observed some time later, the adsorptivity in the dark proves to be somewhat enhanced, and the longer the illumination the more noticeable is the effect. The adsorptivity reaches saturation if the illumination period is long enough.

(b) The photoadsorptivity (i.e., adsorptivity in the presence of light), on the other hand, of such a semiconductor (which has undergone preliminary illumination) proves to be reduced, and the longer the previous illumination the greater the reduction, until the photoadsorptivity disappears completely.

After the light is turned off, the adsorptivity in the dark remains enhanced for a rather long time, while the photoadsorptivity remains reduced. Gradually these quantities relax to the values which they had prior to illumination.

(7) Coekelbergs *et al.* [16], who studied the adsorption of O_2 and CO on Al_2O_3, found that if during photoadsorption the light is switched off, in some cases adsorption continues for some time "by inertia" (an aftereffect). The magnitude of the aftereffect, i.e., the amount of the substance additionally adsorbed after the light is turned off, is smaller the higher the temperature. At high enough temperatures the aftereffect disappears completely.

(8) One more regularity should be mentioned (see [16]). Consider the photoadsorption kinetic curve $N(t)$, with t the time measured from the moment the light was switched on and N the surface density of adsorbed particles. This curve often contains a region corresponding to an "induction period" during which the adsorption rate does not decrease (contrary to adsorption in the dark) but increases with time, i.e., dN/dt is positive in this region.

The aim of the present chapter is to give a general theory of photoadsorption phenomena and explain from a unified point of view the variety of experimental findings, which often seem to contradict each other.

7.2. THE PHOTOADSORPTION EFFECT AT AN IDEAL SURFACE

7.2.1. The Effect of Illumination on the Amount of Various Forms of Chemisorption

The relative amount of the electrically neutral, negative, and positive forms of chemisorption in the presence of light will be denoted by η^0, η^-, and η^+, respectively. The same quantities without illumination will be denoted by η_0^0, η_0^-, and η_0^+. In other words, we wish to see how the relative coverages by the various forms of chemisorption react to light.

Under electronic equilibrium for the acceptor and donor levels E_A and E_D, which depict the particle of the given type, we have, respectively,

$$\alpha_1^- N^0 - \alpha_3^- p_s N^- = \alpha_2^- N^- - \alpha_4^- n_s N^0, \qquad (7.1a)$$

$$\alpha_1^+ N^0 - \alpha_3^+ n_s N^+ = \alpha_2^+ N^+ - \alpha_4^+ p_s N^0, \qquad (7.1b)$$

where n_s and p_s are concentrations of free electrons and holes at the surface plane in the presence of light, and α_i^- and α_k^+ (with $k = 1, 2, 3, 4$) are factors whose interrelationship we will discuss below.

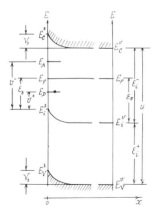

Fig. 7.1. Energy band structure and band bending at the surface of a semiconductor under illumination.

The first term on the left-hand side of Eq. (7.1a) gives the number of electron transitions from the valence band onto the level E_A per unit time (i.e., transition rate) per unit surface area (see Fig. 7.1) and the second describes the reverse transitions. The first term on the right-hand side of Eq. (7.1a) expresses the transitions from level E_A into the conduction band, while the second corresponds to transitions in the reverse direction. Likewise, Eq. (7.1b) describes transitions between the E_D level and conduction band [the left-hand side of Eq. (7.1b)] and between the E_D level and valence band (the right-hand side of the same equation).

In Fig. 7.1 E_C^s and E_V^s are the bottom of the conduction band and the top of the valence band in the surface plane, E_C^v and E_V^v are the same quantities in the bulk of the crystal, E_F is the Fermi level without light, and E_i^s and E_i^v the Fermi levels in an intrinsic semiconductor at the surface and in the bulk of the crystal, respectively.

Using Eqs. (7.1a) and (7.1b), we obtain

$$\frac{N^-}{N^0} = \frac{\eta^-}{\eta^0} = \frac{\alpha_1^- + \alpha_4^- n_s}{\alpha_2^- + \alpha_3^- p_s}, \tag{7.2a}$$

$$\frac{N^+}{N^0} = \frac{\eta^+}{\eta^0} = \frac{\alpha_1^+ + \alpha_4^+ p_s}{\alpha_2^+ + \alpha_3^+ n_s}. \tag{7.2b}$$

The factors $\alpha_1^-, \alpha_2^-, \alpha_3^-$, and α_4^-, as well as $\alpha_1^+, \alpha_2^+, \alpha_3^+$, and α_4^+ are interrelated by the principle of detailed balance applied to the conditions prior to illumination:

$$\alpha_1^- N_0^0 + \alpha_3^- p_{s0} N_0^- = \alpha_2^- N_0^- - \alpha_4^- n_{s0} N_0^0 = 0, \tag{7.3a}$$

$$\alpha_1^+ N_0^0 - \alpha_3^+ n_{s0} N_0^+ = \alpha_2^+ N_0^+ - \alpha_4^+ p_{s0} N_0^0 = 0, \tag{7.3b}$$

where N_0^0, N_0^-, and N_0^+ are the surface concentrations of the neutral, negatively charged, and positively charged chemisorbed particles prior to illumination, and n_{s0} and p_{s0} are the free carrier concentrations prior to illumination. Combining (7.3a) and (7.2a), we obtain

$$\alpha_3^- = \alpha_1^- \frac{\eta_0^0}{\eta_0^-} \frac{1}{p_{s0}}, \text{ where } \alpha_1^- = \beta_1^- \exp\left(-\frac{E_A - E_V}{kT}\right),$$

(7.4a)

$$\alpha_4^- = \alpha_2^- \frac{\eta_0^-}{\eta_0^0} \frac{1}{n_{s0}}, \text{ where } \alpha_2^- = \beta_2^- \exp\left(-\frac{E_C - E_A}{kT}\right).$$

Likewise, it follows from (7.3b) that

$$\alpha_3^+ = \alpha_1^+ \frac{\eta_0^0}{\eta_0^+} \frac{1}{n_{s0}}, \text{ where } \alpha_1^+ = \beta_1^+ \exp\left(-\frac{E_C - E_D}{kT}\right),$$

$$\alpha_4^+ = \alpha_2^+ \frac{\eta_0^+}{\eta_0^0} \frac{1}{p_{s0}}, \text{ where } \alpha_2^+ = \beta_2^+ \exp\left(-\frac{E_D - E_V}{kT}\right),$$

(7.4b)

where we may assume that, by order of magnitude,

$$\beta_1^- = \beta_2^- = \beta^-, \qquad \beta_1^+ = \beta_2^+ = \beta^+.$$

(7.5)

Substituting (7.4a) into (7.2a) and (7.4b) into (7.2b) and introducing the notations

$$\alpha^- = \frac{\alpha_1^-}{\alpha_2^-} \frac{\eta_0^0}{\eta_0^-} = \exp\left[-\frac{(E_F - E_V) - (E_C - E_A)}{kT}\right],$$

(7.6a)

$$\alpha^+ = \frac{\alpha_1^+}{\alpha_2^+} \frac{\eta_0^0}{\eta_0^+} = \exp\left[-\frac{(E_F - E_V) - (E_C - E_D)}{kT}\right],$$

(7.6b)

we obtain

$$\frac{\eta^-}{\eta^0} = \frac{\eta_0^-}{\eta_0^0} \mu^-,$$

(7.7a)

$$\frac{\eta^+}{\eta^0} = \frac{\eta_0^+}{\eta_0^0} \mu^+,$$

(7.7b)

where

$$\mu^- = \frac{1 + \alpha^- + (\Delta n_s / n_{s0})}{1 + \alpha^- + \alpha^-(\Delta p_s / p_{s0})},$$

(7.8a)

$$\mu^+ = \frac{1 + \alpha^+ + (\Delta p_s / p_{s0})}{1 + \alpha^+ + \alpha^+(\Delta n_s / n_{s0})}.$$

(7.8b)

and

$$\Delta n_s = n_s - n_{s0}, \qquad \Delta p_s = p_s - p_{s0}.$$

(7.9)

Obviously, Δn_s and Δp_s are the light-induced contributions to the corresponding concentrations.

Using (7.7a), (7.7b), and the fact that $\eta^+ + \eta^- + \eta^0 = 1$, we finally obtain

$$\frac{\eta^0}{\eta_0^0} = [1 + \eta_0^- (\mu^- - 1) + \eta_0^+ (\mu^+ - 1)]^{-1},$$

(7.10)

$$\frac{\eta^-}{\eta_0^-} = \frac{\eta^0}{\eta_0^0} \mu^-, \qquad \frac{\eta^+}{\eta_0^+} = \frac{\eta^0}{\eta_0^0} \mu^+.$$

Equations (7.10) describe the variation of the surface contents of various forms of chemisorption under illumination. This variation, as we have seen, is due to the appearance of nonequilibrium free carriers generated by the light. Indeed, at $\Delta n_s = \Delta p_s = 0$ we would have, according to (7.8a) and (7.8b), $\mu^- = \mu^+ = 1$. Then, in view of (7.10), we would have

$$\eta^0 = \eta_0^0, \qquad \eta^- = \eta_0^-, \qquad \eta^+ = \eta_0^+,$$

i.e., the contents of the neutral, negatively charged, and positively charged forms of chemisorption would be the same as in the dark.

7.2.2. Allowing for the Annihilation of Excitons at Chemisorbed Particles

In deriving (7.10) we could ignore the origin of the nonequilibrium carriers, i.e., the light absorption mechanism in the crystal, which might be either electronic (when absorption of a quantum is accompanied by electron transfer from a lower level onto a higher level) or excitonic (when absorption of a quantum is accompanied by creation of an exciton, which then annihilates at a lattice defect, causing ionization of the defect and thus creates a free charge carrier).

One aspect of the excitonic mechanism of light absorption is worth mentioning, however. The point is that an exciton may annihilate not only on a structural defect of the lattice but, generally speaking, at a chemisorbed particle as well, changing its charge state and, hence, changing the relative content of various forms of chemisorption at the surface (the free carriers take no direct part in the process. Allowing for this effect requires special considerations (see [19]) and leads, as we will see, to the same formulas (7.7a), (7.7b), and (7.10), in which, however, the parameters μ^- and μ^+ differ from those given by (7.8a) and (7.8b).

Indeed, using the nomenclature of the electronic theory of chemisorption [17, 18], i.e., denoting by eL, pL, and epL a free electron, hole, and exciton, respectively, and by CL, CeL, CpL a chemisorbed particle in the neutral, negatively charged, and positively charged states, we can write the process of exciton annihilation at a chemisorbed particle thus:

$$CL + epL \rightarrow CeL + pL,$$
$$CeL + epL \rightarrow CL + eL,$$

or

$$CL + epL \rightarrow CpL + eL,$$
$$CpL + epL \rightarrow CL + pL.$$

Correspondingly, the electronic equilibrium conditions (7.1a) and (7.1b) on the levels E_A and E_D (see Fig. 7.1) are replaced by

$$(\alpha_1^- N^0 - \alpha_3^- p_s N^-) + \gamma_1^- e_s N^0 = (\alpha_2^- N^- - \alpha_4^- n_s N^0) + \gamma_2^- e_s N^-, \qquad (7.11a)$$

$$(\alpha_1^+ N^0 - \alpha_3^+ n_s N^+) + \gamma_1^+ e_s N^0 = (\alpha_2^+ N^+ - \alpha_4^+ p_s N^0) + \gamma_2^+ e_s N^+, \qquad (7.11b)$$

where e_s is the free exciton concentration in the surface plane (the factors γ_1^-, γ_2^-, γ_1^+, and γ_2^+ are of no interest at the moment).

Starting from Eqs. (7.11a) and (7.11b), we can easily obtain [just as we did by employing Eqs. (7.1a) and (7.1b)] Eqs. (7.7a) and (7.7b) and, hence, Eqs. (7.10). However, the parameters μ^- and μ^+ are now given by

$$\mu^- = \frac{1 + \alpha^- + (\Delta n_s/n_{s0}) + \delta_1^- e_s}{1 + \alpha^- + \alpha^-(\Delta p_s/p_{s0}) + \delta_2^- e_s}, \qquad (7.12a)$$

$$\mu^+ = \frac{1 + \alpha^+ + (\Delta p_s/p_{s0}) + \delta_1^+ e_s}{1 + \alpha^+ + \alpha^+(\Delta n_s/n_{s0}) + \delta_2^+ e_s}, \qquad (7.12b)$$

where

$$\delta_1^- = \alpha^-(\gamma_1^-/\alpha_1^-), \qquad \delta_2^- = \gamma_2^-/\alpha_2^-,$$
$$\delta_1^+ = \alpha^+(\gamma_1^+/\alpha_2^+), \qquad \delta_2^+ = \gamma_2^+/\alpha_2^+. \qquad (7.13)$$

In the absence of excitons in the crystal ($e_s = 0$) or neglecting their annihilation at chemisorbed particles ($\gamma_1^- = \gamma_2^- = 0$ and $\gamma_1^+ = \gamma_2^+ = 0$), Eqs. (7.11a) and (7.11b) reduce to Eqs. (7.1a) and (7.1b), respectively, and Eqs. (7.12a) and (7.12b), according to (7.13), reduce to Eqs. (7.8a) and (7.8b).

7.2.3. The Mechanism of the Influence of Illumination on the Adsorptivity of a Surface

We will limit the discussion to the case where adsorption equilibrium has set in. To describe the photoadsorption effect, we introduce the quantity Φ, which is the light-induced fractional change in the surface adsorptivity:

$$\Phi = \frac{N - N_0}{N_0}, \qquad (7.14)$$

where, as usual, N_0 and N are the surface concentrations of the chemisorbed particles of a given type taken in the dark and under illumination (all other conditions being fixed).

As noted earlier (see Section 7.1), if the adsorptivity of a surface is increased by illumination ($N > N_0$), the photoadsorption effect is positive ($\Phi > 0$); if the opposite case is present ($N < N_0$), we have a negative photoadsorption effect ($\Phi < 0$); finally, if $N = N_0$, light absorption is photoadsorption-inactive.

To calculate the magnitude of the photoadsorption effect Φ we must determine N and N_0. If we assume that adsorption is not accompanied by desorption, we have (see Section 3.2.1)

$$aP(N^* - N) = b^0 N^0 \exp\left(-\frac{q^0}{kT}\right) + b^- N^- \exp\left(-\frac{q^-}{kT}\right) + b^+ N^+ \exp\left(-\frac{q^+}{kT}\right), \quad (7.15)$$

where P is the pressure and N^* the maximal number of particles that can be adsorbed on a unit surface area (in other words, N^* is the surface concentration of the adsorption centers, whose role in our model is taken by lattice atoms proper). The left-hand side of Eq. (7.15) gives the number of particles adsorbed every second on a unit surface area, while the first, second, and third terms on the right-hand side give the number of particles desorbed every second from a unit surface area from, respectively, the neutral, negatively charged, and positively charged states. Here q^0, q^-, and q^+ are the binding energies of the corresponding states, with

$$q^- = q^0 + (E_C^s - E_A), \qquad q^+ = q^0 + (E_D - E_V^s), \quad (7.16)$$

where (see Fig. 7.1) $E_C^s - E_A$ and $E_D - E_V^s$ are the energies of the free electron and free hole affinities of the chemisorbed particle, respectively. The coefficients a, b^0, b^-, and b^+ in (7.15) are of no interest to us at present; we only note that, as far as the order of magnitude is concerned, we may assume that

$$b^0 = b^- = b^+. \quad (7.17)$$

On the basis of (7.16) we can write the equation of equilibrium (7.15) thus:

$$aP^*(N^* - N) = \left\{ b^0 \eta^0 + b^- \eta^- \exp\left(-\frac{E_C^s - E_A}{kT}\right) + b^+ \eta^+ \exp\left(-\frac{E_D - E_V^s}{kT}\right) \right\} N \exp\left(-\frac{q^0}{kT}\right),$$

where

$$N = \frac{N^*}{1 + b/p} \quad (7.18)$$

and, similarly,

$$N_0 = \frac{N^*}{1 + b_0/P} , \quad (7.19)$$

where we have employed the notation

$$b_0 = \frac{1}{a}\left[b^0\eta_0^0 + b^-\eta_0^- \exp\left(-\frac{E_C^s - E_A}{kT} \right) + b^+\eta_0^+ \exp\left(-\frac{E_D - E_V^s}{kT} \right) \right] \exp\left(-\frac{q^0}{kT} \right),$$

$$(7.20)$$

$$b = \frac{1}{a}\left[b^0\eta^0 + b^-\eta^- \exp\left(-\frac{E_C^s - E_A}{kT} \right) + b^+ \eta^+ \exp\left(-\frac{E_D - E_V^s}{kT} \right) \right] \exp\left(-\frac{q^0}{kT} \right).$$

Assuming that (7.17) is true, we can, on the basis of (3.5), (7.7a), and (7.7b), rewrite (7.20) thus:

$$b_0 = \left[1 + \exp\left(-\frac{E_C^s - E_F}{kT} \right) + \exp\left(-\frac{E_F - E_V^s}{kT} \right) \right] \frac{b^0}{a}\, \eta_0^0 \exp\left(-\frac{q^0}{kT} \right),$$

$$(7.21)$$

$$b = \left[1 + \mu^- \exp\left(-\frac{E_C^s - E_F}{kT} \right) + \mu^+ \exp\left(-\frac{E_F - E_V^s}{kT} \right) \right] \frac{b^0}{a}\, \eta^0 \exp\left(-\frac{q^0}{kT} \right).$$

Equations (7.18) and (7.19) express the adsorptivity N_0 in terms of the parameters η_0^0, η_0^-, and η_0^+ or N in terms of η^0, η^-, and η^+. If we assume that $\eta_0^- = \eta_0^+ = 0$ and $\eta_0^0 = 1$, i.e., in the dark all the particles are in the neutral state, we return to the classical case and Eq. (7.19) is reduced to the equation of the Langmuir isotherm. According to Eqs. (7.7a), (7.7b), and (7.10), $\eta^- = \eta^+ = 0$ and $\eta^0 = 1$ in this case and, hence, in view of (7.20), $b = b^0$, i.e., $N = N^0$, which implies that the photoadsorption effect vanishes.

Let the electron and hole gases at the semiconductor surface be nondegenerate. Then, by definition,

$$\exp\left(-\frac{E_C^s - E_F}{kT} \right) \ll 1,$$

$$(7.22)$$

$$\exp\left(-\frac{E_F - E_V^s}{kT} \right) \ll 1.$$

Moreover, we assume that

$$\mu^- \ll \exp\left(-\frac{E_C^s - E_F}{kT} \right), \qquad \mu^+ \leqslant \exp\left(\frac{E_F - E_V^s}{kT} \right). \qquad (7.23)$$

the meaning of which will be made clear in what follows. Note that, in view of (7.22), conditions (7.23) are certainly met if $\mu^- \leq 1$ and $\mu^+ \leq 1$. On the basis of (7.22) and (7.23) we can write (7.21) as follows:

$$b_0 = \frac{b^0}{a}\, \eta_0^0 \exp\left(-\frac{q^0}{kT} \right),$$

$$(7.24)$$

$$b = \frac{b^0}{a}\, \eta^0 \exp\left(-\frac{q^0}{kT} \right).$$

If we consider only the low-pressure case (the Henry region), instead of (7.18) and (7.19) we have

$$N = \frac{N^*}{b} P, \qquad N_0 = \frac{N^*}{b_0} P. \tag{7.25}$$

This implies, among other things, that

$$\eta^0 N = \eta_0^0 N_0 \qquad \text{or} \qquad N^0 = N_0^0,$$

[on the basis of (7.24)], i.e., the surface content of the neutral form of chemisorption does not change under illumination.

7.2.4. The Magnitude of the Photoadsorption Effect

Substituting (7.24) into (7.25) and the result into (7.14), we obtain

$$\Phi = (\eta_0^0 / \eta^0) - 1 \tag{7.26}$$

or, in view of (7.10),

$$\Phi = \eta_0^-(\mu^- - 1) + \eta_0^+(\mu^+ - 1). \tag{7.27}$$

Note that (7.27) is valid only for positive values of Φ that are not too large. Indeed, on the basis of (7.27) we can rewrite condition (7.23) as follows:

$$\Phi \ll \Phi^*,$$

where

$$\Phi^* = \eta_0^- \left[\exp\left(\frac{E_C^s - E_F}{kT} \right) - 1 \right] + \eta_0^+ \left[\exp\left(\frac{E_F - E_V^s}{kT} \right) - 1 \right],$$

or, if (7.22) and (3.5) are taken into account,

$$\Phi^* = \eta_0^0 \left[\exp \frac{E_C^s - E_A}{kT} + \exp \frac{E_D - E_V^s}{kT} \right]. \tag{7.28}$$

Let us first consider the case of acceptor particles. Here $\eta_0^+ = 0$. Let $E_F - E_A$ satisfy the condition

$$\exp\left(-\frac{E_F - E_A}{kT} \right) \ll 1.$$

In view of (3.6b) we can put $\eta_0^- = 1$ and, according to (7.27),

$$\Phi^* = \exp{\frac{E_C^s - E_F}{kT}},$$

i.e., in view of (7.22), $\Phi \gg 1$. In the case at hand Eq. (7.27) for Φ takes the simple form

$$\Phi = \mu^- - 1. \tag{7.29a}$$

In the case of donor particles we have $\eta_0^- = 0$. Let

$$\exp\left(-\frac{E_D - E_F}{kT}\right) \ll 1.$$

Then we can assume [see (3.6a)] that $\eta_0^+ = 1$. From (7.28) we have

$$\Phi^* = \exp{\frac{E_F - E_V^s}{kT}}$$

and, consequently [see (7.22)], $\Phi^* \gg 1$. A simple expression follows for Φ, viz.,

$$\Phi = \mu^+ - 1. \tag{7.29b}$$

Thus, to obtain the magnitude of the photoadsorption effect from (7.29a) and (7.29b), we have to calculate the parameters μ^- and μ^+.

7.3. THE SIGN AND ABSOLUTE VALUE OF THE PHOTOADSORPTION EFFECT AT AN IDEAL SURFACE

7.3.2. Statement of the Problem

The criteria for the sign of the photoadsorption effect follow directly from (7.29a) and (7.29b) on the basis of (7.8a) and (7.8b). When the chemisorbed particles are of an acceptor nature, we obtain

$$\Phi \geqslant 0, \quad \text{if} \quad (\Delta n_s/n_{s0}) \geqslant \alpha^- (\Delta p_s/p_{s0}),$$

$$\Phi \leqslant 0, \quad \text{if} \quad (\Delta n_s/n_{s0}) \leqslant \alpha^- (\Delta p_s/p_{s0}), \tag{7.30a}$$

while in the case of donor particles we have

$$\Phi \geqslant 0, \quad \text{if} \quad (\Delta p_s/p_{s0}) \geqslant \alpha^+ (\Delta n_s/n_{s0}),$$

$$\Phi \leqslant 0, \quad \text{if} \quad (\Delta p_s/p_{s0}) \leqslant \alpha^+ (\Delta n_s/n_{s0}), \tag{7.30b}$$

where, according to (7.6a), (7.6b), and (7.5),

$$\alpha^- = \exp\left[-\frac{(E_F - E_V^s) - (E_C^s - E_A)}{kT} \right],$$

(7.31)

$$\alpha^+ = \exp\left[\frac{(E_F - E_V^s) - (E_C^s - E_D)}{kT} \right]$$

and, assuming that the electron and hole gases are nondegenerate in the surface plane (see Section 1.6.2),

$$n_{s0} = n_i \exp\left(\frac{E_F - E_i^s}{kT} \right), \qquad p_{s0} = n_i \exp\left(\frac{E_i^s - E_F}{kT} \right).$$

(7.32)

Here n_i is the electron (hole) concentration in an intrinsic semiconductor.

The problem now is to make the criteria (7.30a) and (7.30b) explicit, i.e., to express them in terms of the parameters characterizing the experimental conditions and the sample's history. To this end we must calculate the light-induced contributions Δn_s and Δp_s.

Let the semiconductor sample occupy the half space $x \geq 0$, while the half space $x < 0$ corresponds to the gaseous phase. The semiconductor surface $x = 0$ is assumed to be illuminated by light, causing the valence electrons to go over to the conduction band (intrinsic absorption). Suppose that $n_0(x)$ and $p_0(x)$ are the concentrations of the free electrons and holes in the plane x (with $x \geq 0$) in the dark, while $\Delta n(x)$ and $\Delta p(x)$ are the respective light-induced variations. Clearly,

$$n_0(0) = n_{s0}, \qquad p_0(0) = p_{s0},$$

with

$$\Delta n(0) = \Delta n_s, \qquad \Delta n(\infty) = 0,$$
$$\Delta p(0) = \Delta p_s, \qquad \Delta p(\infty) = 0.$$

Let $F_0(x)$ and $V_0(x)$ be the electric field strength and electron potential energy in the plane x in the dark. Obviously, $eF_0 = dV_0/dx$, with e the magnitude of the electron charge. Let $\Delta F(x)$ and $\Delta V(x)$ be the respective light-induced variations. We adopt the notation

$$F_0(0) = F_{s0}, \qquad \Delta F(0) = \Delta F_s,$$
$$V_0(0) = V_{s0}, \qquad \Delta V(0) = \Delta V_s,$$

with

$$F_0(\infty) = 0, \qquad \Delta F(\infty) = 0,$$
$$V_0(\infty) = 0, \qquad \Delta V(\infty) = 0,$$

which means that in the bulk of the crystal sample the bands are flat (no electric field present). Note that (see Fig. 7.1)

$$V_{s0} = E_C^s - E_C^v = E_V^s - E_V^v \tag{7.33}$$

To calculate $\Delta n(x)$ and $\Delta p(x)$, we use the Poisson equation and the continuity equations for the electrons and holes [20]. In the present case the solution of this set of equations will be obtained by assuming the excitation to be weak, so that the light-induced variations of the carrier concentrations can be assumed to be small compared to the concentration of the majority carriers (i.e., at any $x \geq 0$):

$$
\begin{aligned}
\Delta n(x), \quad \Delta p(x) &\ll n_0(x), \quad \text{if} \quad p_0(x) \leq n_0(x), \\
\Delta n(x), \quad \Delta p(x) &\ll p_0(x), \quad \text{if} \quad n_0(x) \leq p_0(x).
\end{aligned}
\tag{7.34}
$$

Moreover, we assume that the impurities in the bulk are completely ionized and that

$$|\Delta V(x)| \ll kT,$$

for all $x \geq 0$.

The last condition means that the behavior of the potential in the lattice and, hence, the charge on the surface are practically unchanged by illumination. This is true if the magnitude of the photoadsorption effect is small or if the surface charge is mostly of an intrinsic origin.

On the above assumptions the Poisson equation has the form (see [20])

$$\frac{d\Delta F}{dx} = \frac{4\pi e}{\chi}(\Delta p - \Delta n), \tag{7.35}$$

while the continuity equations are (see any textbook on semiconductor physics)

$$\frac{dj_n}{dx} = \frac{dj_p}{dx} = g - \frac{\Delta n}{\tau_n} - \frac{\Delta p}{\tau_p}, \tag{7.36}$$

where χ is the dielectric constant, $j_n(x)$ and $j_p(x)$ the electron and hole fluxes, respectively $(j_n = j_p)$, τ_n and τ_p the electron and hole lifetimes, and g the pair generation rate, i.e., the number of light-generated electron–hole pairs per unit time per unit volume. We have (see [20])

$$j_p = D_p\left(\frac{d\Delta n}{dx} - \Delta p\,\frac{eF_0}{kT} - p_0\,\frac{e\Delta F}{kT}\right), \tag{7.37}$$

$$g = \eta\kappa I_s \exp(-\kappa x), \tag{7.38}$$

where D_n and D_p are the electron and hole diffusion coefficients, I_s the photon flux at the surface plane $x = 0$, κ the light absorption coefficient, and η the quantum yield ($\eta = 1$ in what follows); $j_s = j_n(0) = j_p(0)$ is proportional to the surface recombination rate.

7.3.2. Solution for a Simplified Potential Function

Solution of the simultaneous equations (7.35) and (7.36) generally involves mathematical difficulties. We will therefore employ an explicit approximation for the potential, viz.,

$$F_0(x) = \begin{cases} F_s & \text{for} \quad 0 \leqslant x < x_0, \\ 0 & \text{for} \quad x_0 < x < \infty, \end{cases}$$

where

$$x_0 = -\frac{V_s}{eF_s}.$$

Such an approximation is often used in solving continuity equations (e.g., see [21]). The solutions are obtained separately for $0 \leq x < x_0$ and $x_0 < x < \infty$ and are then matched in the plane $x = x_0$ (note that x_0 drops out of the final result). We omit the intermediate calculations (to be found in [20]). Putting $x = 0$ in the final expression for $\Delta n(x)$ and $\Delta p(x)$, we obtain

$$\Delta n_s = \left(\sqrt{\frac{\tau_i}{D_i}} + \frac{kT}{eF_s} \frac{1}{D_n} \right) (I_s + j_s) \exp\left(-\frac{V_s}{kT} \right) - \frac{kT}{eF_s} \frac{1}{D_n} \left(\frac{I_s}{1 - eF_s/kT\kappa} + j_s \right),$$

(7.39)

$$\Delta p_s = \left(\sqrt{\frac{\tau_i}{D_i}} - \frac{kT}{eF_s} \frac{1}{D_p} \right) (I_s + j_s) \exp\left(\frac{V_s}{kT} \right) + \frac{kT}{eF_s} \frac{1}{D_p} \left(\frac{I_s}{1 + eF_s/kT\kappa} + j_s \right),$$

where the subscript $i = p$ in the case of an n-type semiconductor and $i = n$ if the semiconductor is p-type. It can be shown that, in most cases of practical interest,

$$\kappa \gg \frac{e | F_s |}{kT} \gg \frac{1}{\sqrt{\tau_i D_i}} \exp\frac{| V_s |}{kT}.$$

(7.40)

so that (7.39) takes the simpler form

$$\Delta n_s = (I_s + j_s) \sqrt{\frac{\tau_i}{D_i}} \exp\left(-\frac{V_s}{kT} \right), \quad \Delta p_s = (I_s + j_s) \sqrt{\frac{\tau_i}{D_i}} \exp\left(\frac{V_s}{kT} \right).$$

(7.41)

Note that in most cases, as shown in [22],

$$j_s \ll I_s.$$

(7.42)

We will adopt a system of notations that will subsequently be useful. Let us reckon the energies from the Fermi level in an intrinsic semiconductor, which will therefore be the reference point on the energy scale. We put (see Fig. 7.1)

$$v^- = E_A - E_i^s, \quad v^+ - F_D - E_i^s,$$

$$\epsilon_i^- = E_C^v - E_i^v, \quad \epsilon_i^+ = E_i^v - E_V^v, \tag{7.43}$$

$$\epsilon_s = E_F - E_i^s, \quad \epsilon_v = E_F - E_i^v.$$

We will denote the forbidden band gap by u, i.e.,

$$u = E_C^s - E_V^s = E_C^v - E_V^v. \tag{7.44}$$

and assume that

$$\varphi^- = -(\epsilon_v + V_s - v^-), \quad \varphi^+ = +(\epsilon_v + V_s - v^+)$$

or, in abbreviated form,

$$\varphi^\mp = \mp(\epsilon_v + V_s - v^\mp), \tag{7.45}$$

where the minus sign must be taken in the case of acceptor particles and the plus sign in the case of donor particles. If ϵ_v is positive, we are dealing with an n-type semiconductor, while if $\epsilon_v < 0$, then a p-type semiconductor is involved.

Combining (7.31), (7.32), and (7.41) with (7.43) and (7.45), we have

$$\alpha^- \frac{\Delta p_s}{p_{s0}} \frac{n_{s0}}{\Delta n_s} = \exp\left(-\frac{\varphi^-}{kT}\right),$$

$$\alpha^+ \frac{\Delta n_s}{n_{s0}} \frac{p_{s0}}{\Delta p_s} = \exp\left(-\frac{\varphi^+}{kT}\right). \tag{7.46}$$

According to (7.30a), (7.30b), and (7.46), the criteria for the sign of the photoadsorption effect take the following form:

Acceptor particles: $\Phi \gtrless$ if $\varphi^- \gtrless 0$,

Donor particles: $\Phi \gtrless$ if $\psi^+ \gtrless 0$. $\tag{7.47}$

Thus, if the adsorbent and adsorbate energy levels (i.e., v^- and v^+) are fixed, the sign of the photoadsorption effect, as seen from (7.47) and (7.45), is determined by the position of the bulk Fermi level in the dark, ϵ_v, and the band bending at the surface, V_s (see Fig. 7.1). Both quantities depend on the temperature and the prehistory of the sample (i.e., on the treatment the sample underwent prior to the experiment); in addition, V_s depends on the surface coverage by the adsorbed particles (i.e., on the pressure in the gaseous phase). Thus, (7.47) and (7.45) describe the way in which the sign of the photoadsorption effect depends on the natures of the adsorbent and adsorbate, on the experimental conditions, and on the prehistory of the sample.

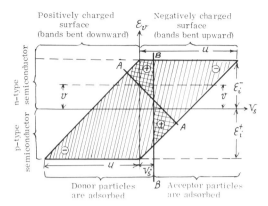

Fig. 7.2. Range of the position of the bulk Fermi level and the band bending at the surface with the free carriers obeying the Boltzmann distribution everywhere in the sample.

Figure 7.2 shows the range of ϵ_v and V_s where the free carriers obey the Boltzmann distribution everywhere in the sample (the case where the Fermi level crosses the bands nowhere). The heavy lines AA and BB divide, according to (7.45), into the regions of positive and negative photoadsorption (marked by the appropriate signs in Fig. 7.2). Along the line AA we have $\Phi = 0$. The segment v shows the position of the surface local level corresponding to the adsorbed particle of a given type in the energy spectrum of the crystal. We have (see Fig. 7.1)

$$v = \begin{cases} v^- & \text{for acceptor particles,} \\ v^+ & \text{for donor particles.} \end{cases} \tag{7.48}$$

The segment V_s^* describes the initial bending of the bands observed in the absence of chemisorbed particles on the surface and, hence, caused by surface states of nonadsorptive origin (obviously, depending on the initial treatment of the surface, we may observe both positive and negative values of V_s^*).

We see that if the Fermi level ϵ_v moved (for V_s constant) or if the bending of the bands, V_s, varies (for ϵ_v constant) or if, finally, both parameters vary simultaneously, the sign of the photoadsorption effect may be changed.

Figure 7.2 also shows that the regions of the positive and negative effects may be broadened or narrowed depending on the previous treatment of the sample (i.e., depending on the value of V_s^*). Indeed, when V_s^* is varied, the vertical line BB is shifted to the left or to the right (remaining parallel to itself).

7.3.3. The Case of the Excitonic Mechanism of Light Absorption

Let us now consider the case of the excitonic mechanism of light absorption, taking into account the possibility of exciton annihilation at the chemisorbed particles. The criteria for the sign of the photoadsorption effect follow directly from (7.29a), (7.29b), (7.12a), and (7.12b). In contrast to (7.30a) and (7.30b), these criteria now have the following form:
(a) acceptor-like chemisorbed particles:

$$\Phi \geqslant 0, \quad \text{if} \quad (\Delta n_s/n_{s0}) + \delta_1^- e_s \geqslant \alpha^- (\Delta p_s/p_{s0}) + \delta_2^- e_s,$$

$$\Phi \leqslant 0, \quad \text{if} \quad (\Delta n_s/n_{s0}) + \delta_1^+ e_s \leqslant \alpha^- (\Delta p_s/p_{s0}) + \delta_2^- e_s, \quad (7.49a)$$

(b) donor-like chemisorbed particles:

$$\Phi \geqslant 0, \quad \text{if} \quad (\Delta p_s/p_{s0}) + \delta_1^+ e_s \geqslant \alpha^+ (\Delta n_s/n_{s0}) + \delta_2^+ e_s,$$

$$\Phi \leqslant 0, \quad \text{if} \quad (\Delta p_s/p_{s0}) + \delta_1^+ e_s \leqslant \alpha^+ (\Delta n_s/n_{s0}) + \delta_2^+ e_s, \quad (7.49b)$$

where e_s is the exciton concentration in the surface plane $(x = 0)$, while α^-, α^+ and n_{s0}, p_{s0} are given by (7.31) and (7.32), respectively, and $\delta_1^\pm$ and $\delta_2^\pm$ are given by (7.13).

To make the conditions (7.49a) and (7.49b) explicit, it is necessary to determine e_s in addition to Δn_s and Δp_s. We will consider the same model of a semiconductor as that used in Section 7.3.1. The exciton concentration in the plane x (with $x \geq 0$) will be denoted by $e(x)$. Obviously,

$$e(0) = e_s. \qquad (7.50)$$

Assuming that excitons at the surface and in the bulk of the sample are annihilated at defects of different types (we will call them annihilation centers) and neglecting spontaneous thermal and radiative exciton annihilation and exciton diffusion (i.e., assuming that the exciton diffusion length is small compared to the inverse light absorption coefficient κ^{-1}), we will have

$$g(x) = e(x) \cdot Z(x), \qquad (7.51)$$

where $g(x)$ is the number of excitons generated in a unit volume per unit time in the plane x, and the quantity on the right-hand side of (7.51) is the number of excitons being annihilated there (per unit volume and per unit time). The quantity $g(x)$ has the form (7.38), while

$$Z(x) = \sum_m \zeta_m Z_m(x),$$

where ζ_m is the exciton annihilation probability at a center of the type m and Z_m is the concentration of such centers. Thus, in view of (7.38), (7.50), and (7.51), we have

$$e_s = \kappa I_s / Z_s, \tag{7.52}$$

with $Z_s = Z(0)$.

The light-induced variations $\Delta n(x)$ and $\Delta p(x)$ may be obtained, as in Section 7.3.1, from solving Poisson's equation and the continuity equations simultaneously. Assuming that practically all the annihilation centers are ionized in the dark, we can show (see [19]) that the above-mentioned equations have the same form as those considered in Section 7.3.1 [see Eqs. (7.35) and (7.36)]. Thus, using the same approximations as in Section 7.3.2, we can again use (7.41) for Δn_s and Δp_s. Substituting (7.41), (7.32), (7.46), and (7.31) into (7.49a) and (7.49b) and employing the notations (7.45), we find that the positive and negative effect criteria take the following forms:

(a) in the case of acceptor-like particles we have $\Phi \gtrless 0$ if

$$\exp\left(\frac{\varphi^-}{kT}\right) \gtrless \frac{\gamma_0^- + \gamma_1^- \exp(-V_s/kT)}{\gamma_0^- + \gamma_1^- \exp(V_s/kT)}, \tag{7.53a}$$

or, in other words, if

$$\epsilon_v \gtrless f^-(V_s), \tag{7.54a}$$

where

$$f^-(V_s) = v^- + kT \ln \frac{\gamma_0^- + \gamma_1^- \exp(V_s/kT)}{\gamma_2^- + \gamma_0^- \exp(V_s/kT)}; \tag{7.55a}$$

(b) in the case of donor-like particles we have $\Phi \gtrless 0$ if

$$\exp\frac{\varphi^+}{kT} \gtrless \frac{\gamma_0^+ + \gamma_2^+ \exp(V_s/kT)}{\gamma_0^+ + \gamma_1^+ \exp(-V_s/kT)}, \tag{7.53b}$$

or, in other words, if

$$\epsilon_v \gtrless f^+(V_s), \tag{7.54b}$$

where

$$f^+(V_s) = v^+ + kT \ln \frac{\gamma_0^+ + \gamma_2^+ \exp(V_s/kT)}{\gamma_1^+ + \gamma_0^+ \exp(V_s/kT)}. \tag{7.55b}$$

Here we have used the notations

$$\gamma_0^- = (\beta^- Z_s / \kappa C_n) \sqrt{\tau/D},$$

$$\gamma_0^+ = (\beta^+ Z_s / \kappa C_p) \sqrt{\tau/D},$$

where τ and D are the lifetime and diffusion length of the minority carriers, the

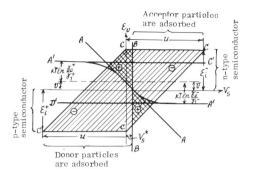

Fig. 7.3. Range of the position of the bulk Fermi level and the band bending at the surface for the excitonic mechanism of light absorption.

meaning of C_n and C_p is the same as in Section 1.6.2, and the meaning of β^-, β^+, $\gamma_1^{\mp}$, and $\gamma_2^{\mp}$ is clear from (7.4a), (7.4b), (7.11a), and (7.11b).

If the excitons undergo only elastic collisions with the adsorbed particles, i.e., if they are annihilated neither at neutral nor at charged particles, then [see Eqs. (7.11a) and (7.11b)] both γ_1 and γ_2 are zero and the conditions (7.53a) and (7.53b) take the form, respectively,

$$\exp\left(\frac{\varphi^-}{kT}\right) \lessgtr 1 \qquad \text{if} \qquad \epsilon_v \lessgtr v^- - V_s,$$

$$\exp\left(\frac{\varphi^+}{kT}\right) \gtrless 1 \qquad \text{if} \qquad \epsilon_v \gtrless v^+ - V_s,$$

which, as expected, are identical to the criteria (7.47) obtained in Section 7.3.2.

7.3.4. A Graphic Representation of the Results

The results obtained have a simple graphic representation similar to that given in Section 7.3.2. Figure 7.3 shows the ϵ_v, V_s plane, where the straight lines $CCCC$ border the region inside which the electron and hole gases are nondegenerate (i.e., the Fermi level never crosses the band edges). Here again, V_s^* is the initial bending of the bands in the absence of adsorbed particles and due, therefore, to surface states of nonadsorption origin. The straight line $V_s = V_s^*$ is denoted in Fig. 7.3 by BB and may be shifted to the left or right, depending on the previous treatment of the sample. The relation

$$\epsilon_v = v^- - V_s \qquad \text{or} \qquad \epsilon_v = v^+ - V_s \qquad (7.56)$$

is shown by the line AA, while the curve

$$\epsilon_v = f^-(V_s) \qquad \text{or} \qquad \epsilon_v = f^+(V_s), \tag{7.57}$$

where $f^-(V_s)$ and $f^+(V_s)$ have the form (7.55a) and (7.55b), respectively, is marked $A'A'$. This curve can easily be constructed if we assume that

$$\gamma_1^-/\gamma_0^- \ll 1 \ll \gamma_0^-/\gamma_2^-,$$

$$\gamma_1^+/\gamma_0^+ \ll 1 \ll \gamma_0^+/\gamma_2^+.$$

The function (7.57) then takes a simple form in the following three regions:

(a) V_s is large and negative (adsorption of donor-like particles). Here

$$\exp(V_s/kT) \ll \gamma_1^+/\gamma_0^+ \ll \gamma_0^+/\gamma_2^+,$$

and it follows from (7.57) and (7.55b) that

$$\epsilon_v = v^+ + kT \ln(\gamma_0^+/\gamma_1^+) - kT(\gamma_0^+/\gamma_1^+)\exp(V_s/kT).$$

Thus, the asymptote to the curve $A'A'$ as $V_s \to \infty$ is the straight line $A'C'$:

$$\epsilon_v = v^+ + kT \ln(\gamma_0^+/\gamma_1^+). \tag{7.58a}$$

(b) V_s is large and positive (adsorption of acceptor-like particles). Here

$$\gamma_2^-/\gamma_0^- \ll \gamma_0^-/\gamma_1^- \ll \exp(V_s/kT),$$

and Eqs. (7.57) and (7.55a) yield

$$\epsilon_v = v^- - kT \ln(\gamma_0^-/\gamma_1^-) + kT(\gamma_0^-/\gamma_1^-)\exp(-V_s/kT),$$

Thus, the asymptote to the curve $A'A'$ as $V_s \to \infty$ is the straight line $D'A'$:

$$\epsilon_v = v^- - kT \ln(\gamma_0^-/\gamma_1^-) \tag{7.58b}$$

(c) Finally, the intermediate range of values of V_s, where

$$(\gamma_1^+/\gamma_0^+) \ll \exp(V_s/kT) \ll (\gamma_0^+/\gamma_2^+),$$

$$(\gamma_2^-/\gamma_0^-) \ll \exp(V_s/kT) \ll (\gamma_0^-/\gamma_1^-).$$

Equations (7.57), (7.55a), and (7.55b) then yield

$$\epsilon_v = v^- - V_s \qquad \text{or} \qquad \epsilon_v = v^+ - V_s,$$

i.e., the curve $A'A'$ coincides with the straight line AA.

The curves $A'A'$ and BB divide the entire plane into regions of positive and negative photoadsorption effects, shaded differently in Fig. 7.3.

According to Eqs. (7.11a) and (7.11b), $\gamma_2^- = \gamma_2^+ = 0$ if the surface annihilation of excitons proceeds only at neutral centers. But if the excitons are annihilated only at charged centers, $\gamma_1^- = \gamma_1^+ = 0$. Note that as $\gamma_1^+ \to 0$, the asymptote $A'C'$, as seen from (7.58a), shifts upward (remaining parallel to itself). Finally, if the excitons do not annihilate on the chemisorbed particles, $\gamma_1^- = \gamma_2^- = \gamma_1^+ = \gamma_2^+ = 0$ and, according to (7.55a) and (7.55b), the curve $A'A'$ becomes a straight line that coincides with AA. We have thus arrived at the same result as in Section 7.3.2, and Fig. 7.3 becomes identical to Fig. 7.2.

We see, therefore, that allowing for exciton annihilation by chemisorbed particles results in the replacement of the straight line AA in Fig. 7.3 by the curve $A'A'$. This does not qualitatively change any results, but produces some broadening of the positive effect region (Fig. 7.2).

7.3.5. The Case of High Excitation

In Sections 7.3.1–7.3.4 we considered the case of low excitation; i.e., the light-induced contributions to the concentrations of the free-charge carriers (electrons and holes) were small compared to the thermal concentration of the majority carriers [see Eq. (7.34)]. In this section we will study the opposite case of high excitation, i.e., the light-induced fractional contributions to the electron and hole concentrations will be relatively large.

If the chemisorbed particles are acceptors, we will assume that

$$\Delta n_s/n_{s0} \gg 1 + \alpha^-, \qquad \Delta p_s/p_{s0} \gg 1 + \frac{1}{\alpha^-}. \qquad (7.59a)$$

In the donor case we put

$$\Delta p_s/p_{s0} \gg 1 + \alpha^+, \qquad \Delta n_s/n_{s0} \gg 1 + \frac{1}{\alpha^+}. \qquad (7.59b)$$

Here the parameters α^- and α^+ have the form (7.31), while the thermal carrier concentrations, n_{s0} and p_{s0}, have the form (7.32). If light absorption proceeds via the excitonic mechanism, we will neglect exciton annihilation at chemisorbed particles.

According to (7.8a), (7.8b), (7.59a), and (7.59b) we have

$$\mu^- = \frac{\Delta n_s/n_{s0}}{\alpha^-(\Delta p_s/p_{s0})}, \qquad (7.60a)$$

$$\mu^+ = \frac{\Delta p_s/p_{s0}}{\alpha^+(\Delta n_s/n_{s0})}. \qquad (7.60b)$$

The problem is to calculate the light-induced contributions Δn_s and Δp_s.

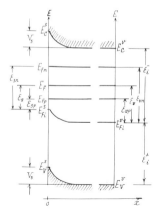

Fig. 7.4. Energy band structure and band bending at the surface of a semiconductor in the case of high excitation.

Let n_s, p_s and n_v, p_v be the free electron and hole concentrations in the presence of light at the surface and in the bulk, respectively. Obviously,

$$n_s = n_i \exp{(\epsilon_{sn}/kT)},$$

$$p_s = n_i \exp{(-\epsilon_{sp}/kT)} \tag{7.61}$$

and

$$n_v = n_i \exp{(\epsilon_v/kT)},$$

$$p_v = n_i \exp{(-\epsilon_v/kT)}, \tag{7.62}$$

where n_i is the electron (hole) concentration in the intrinsic semiconductor, while the meaning of the other symbols follows from Fig. 7.4, where E_F is the Fermi level in the dark (under thermodynamic equilibrium), and E_{Fn} and E_{Fp} are the Fermi quasilevels (under illumination) for electrons and holes, respectively. We assume that within the space-charge region (the region where band bending exceeds kT) the Fermi quasilevels remain constant (in Fig. 7.4 the straight lines E_{Fn} and E_{Fp} are horizontal). Moreover, we assume, as in Sections 7.3.3 and 7.3.4, that the light leaves the potential distribution in the lattice practically unchanged. This yields

$$\epsilon_{sn} = \epsilon_{vn} - V_s,$$

$$\epsilon_{sp} = \epsilon_{vp} - V_s,$$

and, in view of (7.61) and (7.62),

$$n_s = n_v \exp{\left(-\frac{V_s}{kT}\right)},$$

$$p_s = p_v \exp{\left(\frac{V_s}{kT}\right)}.$$

Since

$$n_{s0} = n_{v0} \exp\left(-\frac{V_s}{kT}\right),$$

$$p_{s0} = p_{v0} \exp\left(\frac{V_s}{kT}\right),$$

we will have

$$\Delta n_s = \Delta n_v \exp\left(-\frac{V_s}{kT}\right),$$

(7.63)

$$\Delta p_s = \Delta p_v \exp\left(\frac{V_s}{kT}\right).$$

where, as in (7.9), we have used the notations

$$\Delta n_v = n_v - n_{v0},$$
$$\Delta p_v = p_v - p_{v0}.$$

Let us assume that the bulk of the semiconductor sample (i.e., outside the region of the space charge) stays electrically neutral both in the dark and under illumination. If Δn_v and Δp_v are large compared with the concentration of the acceptor or donor impurity responsible for the conductivity of the semiconductor, the above-mentioned condition of electroneutrality can be written thus:

$$\Delta n_v = \Delta p_v.$$

(7.64)

Substituting (7.31), (7.32), and (7.63) into (7.60a) and (7.60b) and taking into account (7.64) and (7.33), we obtain

$$\mu^- = \exp\left(\frac{\varphi^-}{kT}\right),$$

(7.65a)

$$\mu^+ = \exp\left(\frac{\varphi^+}{KT}\right),$$

(7.65b)

and, hence, according to (7.29a) and (7.29b),

$$\Phi = \exp\left(-\frac{\varphi^-}{kT}\right)$$

(7.66a)

in the case of acceptor chemisorbed particles and

$$\Phi = \exp\left(-\frac{\varphi^+}{kT}\right)$$

(7.66b)

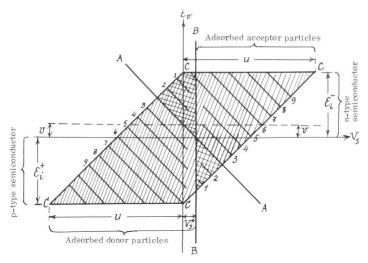

Fig. 7.5. Range of the position of the bulk Fermi level and the band bending at the surface in the case of high excitation.

in the case of donor chemisorbed particles [here we have used the notations (7.45)].

We see that both the sign and magnitude of the photoadsorption effect depend on the position of the Fermi level in the bulk of the crystal and the surface potential (i.e., the surface bending of the bands) prior to illumination.

Figure 7.5 shows the dependence of ϵ_v on V_s at $\Phi = $ const. According to (7.66a) and (7.66b) this dependence is represented by a set of straight lines parallel to the straight line AA, with the region to the right of the vertical BB corresponding to adsorption of acceptor particles and that to the left to adsorption of donor particles (cf. Fig. 7.2). This set constitutes a family of equiphotoadsorption straight lines, on each of which Φ is constant. Different values of Φ correspond to different lines, which are numbered in order of decreasing Φ. The straight line AA corresponds to a zero photoadsorption effect, i.e., $\Phi = 0$. In the case of acceptor particles Φ is positive in the region below the straight line AA (lines 1 and 2) and is negative above AA (lines 3, 4, and 5), while in the case of donor particles Φ is positive above AA (lines 1 and 2) and negative below AA (lines 3, 4, and 5) (see Fig. 7.5). The rhombus $CCCC$ outlines the range of values of ϵ_v and V_s within which the theory is valid.

Sample treatment of any kind always leads to a change in ϵ_v and V_s (one of them or both simultaneously) and, hence, to a shift from one of the points in Fig. 7.5 to another. This implies, via Fig. 7.5, that the magnitude (and, possibly, the sign) of the photoadsorption effect changes. This enables giving an interpretation to all the experimentally observed changes of the magnitude and sign of the effect arising from sample treatment (see Section 7.1.3) via Fig. 7.5. We will do this later in our exposition.

7.4. ADSORPTION CENTERS IN PHOTOADSORPTION

7.4.1. The Nature of Adsorption Centers

Up till now we considered photoadsorption processes as occurring on an ideal surface, i.e., a surface with strict periodicity of its structure. We will now turn to the case of a real surface, i.e., a surface with defects, or local imperfections in the above-mentioned periodicity. Discussing photoadsorption at a real surface, we have in mind the case where these defects play the role of adsorption centers. These may be defects of one type (a homogeneous surface) or of several types corresponding to the different values of the chemisorption binding energy (an inhomogeneous surface). For simplicity we will consider here the case of a homogeneous surface.

Adsorption at regions where the surface may be considered ideal, at the regular atoms (ions) of the lattice, may be ignored if the concentration of the defects, which serve as adsorption centers, is not too small when compared to the surface concentration of the lattice sites and yet small enough so that the interaction between defects can be neglected, and if the energy of binding of a chemisorbed particle to an adsorption center is sufficiently high compared to the binding energy on an ideal surface. This will be assumed throughout in what follows.

Since a defect of a given type may be either an acceptor or a donor (we will omit the general case where a defect is an acceptor and a donor simultaneously), i.e., is able to capture a free electron or hole, and since free carriers play the role of free (unsaturated) valences [17] (we will return to this question later), the defects in structure of the surface may be divided into two classes with respect to their adsorptive properties:

(1) Valence-saturated defects. These provide a chemisorption bond similar to the "weak" bond found in the case of adsorption at an ideal surface.
(2) Defects possessing free (unsaturated) valences, i.e., acting as surface radicals or ion radicals. In this case we have a "strong" donor acceptor bond.

Evidently, localization or delocalization of an electron or hole may lead a defect of one of these classes to be transformed into that of the other class. Since the chemisorption act may be viewed as formation of a valence bond between an adsorption center and the molecule to be adsorbed, only defects of the second class may be considered as true adsorption centers, while the adsorptive role of defects of the first class can be considered negligible in the first approximation. This means that the role of adsorption centers is now played by electrons and holes localized at surface defects.

Such an approximation amounts to neglecting the "weak" bond when considering adsorption at defects. This will be shown to be sufficient to understand many features of the photoadsorption effect. Within this approximation the neutral F- or V-centers (metalloid or metal vacancies in a binary lattice built from ions of the metal and metalloid with electrons or holes localized nearby) serve as adsorption centers, while ionized F- or V-centers (metalloid or metal vacancies) are not such centers. Likewise, a chemisorbed oxygen atom "weakly" bound to the surface (valence-saturated form of chemisorption; see [17] and [18]) is not an adsorption center, while, at the same time, an oxygen atom in the "strongly" bound state (the ion radical state) may serve as an adsorption center, i.e., could acquire other atoms or molecules.

The experimental data does indeed show that charge carriers (electrons or holes) localized at surface defects play the role of adsorption centers. For instance, Kohn and Taylor [23, 24] have established that for hydrogen at SiO_2 the holes localized at aluminum impurities serve as adsorption centers. The same conclusion was reached by Boreskov *et al.* [25, 26]. Muha [27] discovered that two types of centers are responsible for adsorption of hydrogen, deuterium, and methanol at the surface of SiO_2. The centers were identified as holes localized at surface defects. According to Stamiers and Turkevich [28], the adsorption centers for oxygen at synthetic zeolites are charge carriers localized at surface impurities. Lunsford and Jayne [29] have shown that electrons localized at anion vacancies (F-centers) serve as adsorption centers for O_2 and CO_2 at MgO. Adsorption at F-centers has also been studied by Bauer and Staude [30].

We will now consider some properties of such adsorption centers and the peculiarities of adsorption at them [32].

7.4.2. The Concentration of Adsorption Centers

Let the surface contain structural defects of only one type and let their concentration be X^* (the particular structure of the defects need not be specified here). We assume they are acceptors, i.e., that electrically they may be either neutral or negatively charged. We denote the corresponding surface concentration by X^0 or X^-. Further, in accordance with Section 7.4.1, we will assume that only charged defects act as adsorption centers, while neutral defects do not possess this property. In other words, electrons localized at the defects act as adsorption centers. Similarly, holes localized at donor defects and serving as adsorption centers constitute an alternative model. Let the surface concentration of adsorbed particles be N. Then

$$X^0 + X^- + N = X^*, \quad X^- + N = X, \tag{7.67}$$

where X is the total number of adsorption centers (both vacant and occupied) per unit surface area.

We wish to find the time variations of X and N, i.e., how the number of adsorption centers and the number of particles adsorbed at these centers vary with time. To this end we use the equations

$$\frac{dX^-}{dt} = (a_1 X^0 - a_2 X^-) - (b_1 X^- - b_2 N),$$

$$\frac{dN}{dt} = b_1 X^- - b_2 N. \tag{7.68}$$

Here the first and second terms in the first parentheses correspond to number of electron localization and delocalization acts per unit time per unit surface area, while the first and second term in the second parentheses give the adsorption and desorption rates per unit surface area. The factors a_j and b_j (with $j = 1, 2$) in (7.68) have the form

$$a_1 = \alpha_1 n_s + \gamma_1 \exp\left(-\frac{\epsilon_i^+ + v^+}{kT}\right),$$

$$b_1 = \beta_1 P \exp\left(-\frac{E}{kT}\right),$$

$$a_2 = \alpha_2 p_s + \gamma_2 \exp\left(-\frac{\epsilon_i^- - v^-}{kT}\right), \tag{7.69}$$

$$b_2 = \beta_2 \exp\left(-\frac{q - E}{kT}\right).$$

where P is the pressure in the gaseous phase, E the adsorption activation energy, q the bonding energy of the adsorbed particle and the adsorption center, v^- and v^+ the distances from the acceptor and donor levels, respectively, to the intrinsic Fermi level (see Fig. 7.1), and n_s and p_s the concentrations of free electrons and holes in the surface plane. Equations (7.68) will be solved on the assumption that

$$n_s = \text{const}, \qquad p_s = \text{const}. \tag{7.70}$$

which means that the surface charge is assumed to be practically constant during adsorption.

It is convenient to rewrite Eqs. (7.68) by replacing the first by the sum of the first and second equations and allowing for (7.67). This yields

$$\frac{dX}{dt} = a_1 X^0 - a_2 X^-,$$

$$\frac{dN}{dt} = b_1 X^- - b_2 N. \tag{7.71}$$

We will return to these equations in Section 7.4.3. If in (7.71) we assume that dX/dt is much less than $a_1 X^0$ or $a_2 X^-$, which means that electron equilibrium is maintained at the surface, then the first equation in (7.71) yields

$$\frac{X^-}{X^0} = \frac{a_1}{a_2} = \exp\left(-\frac{v^- - \epsilon_s}{kT}\right). \tag{7.72}$$

Combining this with (7.67), we obtain

$$X^- = \frac{X^* - N}{1 + \exp\left[(v^- - \epsilon_s)/kT\right]},$$

$$X = \frac{X^* + N \exp\left[(v^- - \epsilon_s)/kT\right]}{1 + \exp\left[(v^- - \epsilon_s)/kT\right]}. \tag{7.73}$$

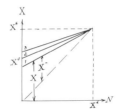

Fig. 7.6. Concentration of adsorption
centers vs. the surface concentration
of adsorbed particles.

The dependence of X on N given by (7.73) is depicted in Fig. 7.6, where

$$X_0 = \frac{X^*}{1 + \exp\left[(v^- - \epsilon_s)/kT\right]}. \tag{7.74}$$

We see that the total concentration of adsorption centers (both vacant and occupied) does not remain constant but increases during adsorption. New adsorption centers are constantly produced as those already present are filled by the adsorbed molecules. Figure 7.6 is drawn on the assumption that

$$\epsilon_s = \text{const} \tag{7.75}$$

[note that this condition is equivalent to (7.70)]. Different curves in Fig. 7.6 correspond to different values of ϵ_s (the numbers are in the order of increasing ϵ_s). We see that the number of adsorption centers X_0 available at the surface prior to adsorption depends on the position of the Fermi level and decreases as the level is lowered. Thus, the concentration of centers can be increased or diminished by doping the crystal sample (on its surface) with foreign impurities that shift the Fermi level but take no part in the adsorption process.

The concentration of the centers may also be changed by illumination, which changes the electron occupancy of the defect local levels. Some consequences of this will be considered in Section 7.4.3.

Consider again Eqs. (7.71). If we substitute (7.73) into the second equation and integrate the result with the initial condition that $N = 0$ at $t = 0$, we obtain

$$N(t) = N_\infty [1 - \exp(-t/\tau)]. \tag{7.76}$$

Combining this with (7.73), we find that

$$X(t) - X_0 = (X_\infty - X_0)[1 - \exp(-t/\tau)]. \tag{7.77}$$

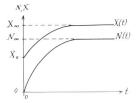

Fig. 7.7. Time dependence of the concentration of adsorption centers and the surface concentration of adsorbed particles in the course of adsorption.

where

$$\tau = \frac{1 + \exp\left[-(v^- - \epsilon_s)/kT\right]}{b_1 + b_2\{1 + \exp\left[(v^- - \epsilon_s)/kT\right]\}},$$

$$N_\infty = N(\infty) = \frac{b_1 X^*}{b_1 + b_2\{1 + \exp\left[(v^- - \epsilon_s)/kT\right]\}}, \tag{7.78}$$

$$X_\infty = X(\infty) = \frac{X^* + N_\infty \exp\left[(v^- - \epsilon_s)/kT\right]}{\{1 + \exp\left[(v^- - \epsilon_s)/kT\right]\}}.$$

Figure 7.7 depicts the N vs. t and X vs. t curves, which describe the kinetics of the increase in the number of adsorption centers during adsorption and the adsorption kinetics at such centers.

Consider first the initial stage of adsorption $t \ll \tau$. If the Fermi level lies far below the defect levels, so that

$$\exp\left(\frac{v^- - \epsilon_s}{kT}\right) \gg 1, \tag{7.79}$$

the adsorption rate is given, according to (7.76), (7.77), and (7.79), by the following formula:

$$\frac{dN}{dt} = \frac{N_\infty}{\tau} = b_1 X^* \exp\left(-\frac{v^- - \epsilon_s}{kT}\right). \tag{7.80}$$

We have a typical activation adsorption process with the activation energy depending on the position of the Fermi level, ϵ_s.

At $t \gg \tau$, i.e., when adsorption equilibrium is achieved, we have [according to (7.78) and (7.69)]

$$N = \frac{X^*}{1 + (\gamma/P)\exp(-q/kT)\{1 + \exp\left[(v^- - \epsilon_s)/kT\right]\}}, \tag{7.81}$$

where $\gamma = \beta_2/\beta_1$. The adsorptivity of the surface depends, as we see, on the position of the Fermi level and decreases as the level moves downward. Equation

(7.81) coincides with that obtained for the concentration of the "strongly" bound particles (charged form of chemisorption) in the theory of adsorption at an ideal surface (see Section 3.2.1).

The results obtained differ markedly from those of classical adsorption theory, which considers a fixed number of adsorption centers.

7.4.3. Variation of Adsorption Center Concentration under Illumination

Let us consider the time evolution of the concentration X of adsorption centers after the light is turned on or off [32]. To this end we turn to Eqs. (7.71). Using (7.67), we can rewrite these equations as

$$\frac{dX}{dt} = a_1 X^* - (a_1 + a_2)X + a_2 N, \tag{7.82a}$$

$$\frac{dN}{dt} = b_1 X - (b_1 + b_2)N, \tag{7.82b}$$

where the coefficients a_1, a_2, b_1, and b_2 are given by (7.69). Here

$$n_s = n_{s0} + \Delta n_s, \qquad p_s = p_{s0} + \Delta p_s, \tag{7.83}$$

with n_{s0} and p_{s0} the free carrier concentrations in the surface plane in the dark, and Δn_s and Δp_s the light-induced contributions.

Let us now turn to Eq. (7.82a) and assume that

$$\nu N \ll N^*, \text{ where } \nu = \frac{a_2}{a_1}. \tag{7.84}$$

In this case the last term on the right-hand side of (7.82a) can be ignored. The solution of Eq. (7.82a) subject to (7.84) and the initial condition $X = X_i$ at $t = t_i$ is

$$X(t) = \mu X^* - (\mu X^* - X_i) \exp\left[-a(t - t_i)\right], \tag{7.85}$$

where

$$\mu = \frac{1}{1 + \nu}, \qquad a = a_1 + a_2. \tag{7.86}$$

Solution (7.85) describes the time evolution of the adsorption center concentration under illumination.

We go over to Eq. (7.82b). Substituting (7.85) into it, we obtain

$$\frac{dN}{dt} + bN - b_1 \mu X^* + b_1(\mu X^* - X_i) \exp\left[-a(t - t_i)\right] = 0, \tag{7.87}$$

where

$$b = b_1 + b_2 . \tag{7.88}$$

The solution of Eq. (7.87) subject to the initial condition $N = N_i$ at $t = t_i$ has the form (see [31])

$$N(t) = A \exp[-a(t - t_i)] + B \exp[-b(t - t_i)] + C, \tag{7.89}$$

where

$$A = \frac{b_1}{a - b}(\mu X^* - X_i),$$

$$B = N_i - \frac{b_1}{a - b}(\mu X^* - X_i) - \frac{b_1}{b}\mu X^*, \tag{7.90}$$

$$C = \frac{b_1}{b}\mu X^* .$$

Solution (7.89) describes the adsorption kinetics when the concentration of adsorption centers varies in time according to (7.85). We will use this solution in Section 7.5.1.

We will now investigate solution (7.85). Let the light be turned on at $t = t_0$. We wish to determine X_i in (7.85) on the assumption that electron equilibrium is maintained prior to illumination. In the absence of light we have $\mu = \mu_0$, where, according to (7.86) and (7.84),

$$\mu_0 = \frac{1}{1 + \nu_0} , \qquad \nu_0 = \frac{a_{20}}{a_{10}} . \tag{7.91}$$

Here a_{10} and a_{20} are the values of the parameters a_1 and a_2 in the dark, i.e., at $\Delta n_s = 0$ and $\Delta p_s = 0$ [see (7.69) and (7.83)]. According to (7.72) and (7.91),

$$\nu_0 = \frac{X_0^0}{X_0^-} , \tag{7.92}$$

with X_0^0 and X_0^- the concentrations of the neutral and charged defects of the type considered prior to illumination. This together with (7.91) and (7.67) yields

$$\mu_0 = \frac{X_0 - N}{X^* - N}$$

and, hence,

$$X_0 = \mu_0 X^* \left(1 + \nu_0 \frac{N}{X^*}\right),$$

where, according to (7.67), $X_0 = X_0^- + N$. Allowing for (7.84) and assuming that

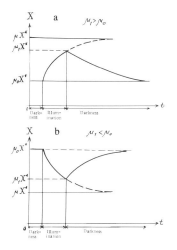

Fig. 7.8. Time variation of the concentration of adsorption centers with the semiconductor surface illuminated: a) $\mu_1 > \mu_0$; b) $\mu_1 < \mu_0$.

$X_i = X_0$ (electron equilibrium being maintained prior to illumination), we obtain

$$X_i = \mu_0 X^*. \qquad (7.93)$$

With this in mind and setting $t = t_0$ we can rewrite the solution (7.85) as

$$X(t) = \mu X \left\{ 1 - \frac{\mu - \mu_0}{\mu} \exp[-a(t - t_0)] \right\}. \qquad (7.94)$$

We see that when the light is turned on, the concentration of adsorption centers

$$\begin{cases} \text{increases if } \mu > \mu_0, \\ \text{decreases if } \mu < \mu_0, \end{cases}$$

and tends to the steady-state value $X = \mu X^*$ at $t \to \infty$.

Now suppose the light is turned off at $t = t_1$. The concentration of adsorption centers will still vary according to (7.85) but with the following changes:

$$t_i = t_1, \quad a = a_i, \quad \mu = \mu_0, \quad X_1 = \mu_1 X^*,$$

where, according to (7.86) and (7.94),

$$a_0 = a_{10} + a_{20}$$

and

$$\mu_1 = \mu \left\{ 1 - \frac{\mu - \mu_0}{\mu} \exp[-a(t_1 - t_0)] \right\}. \qquad (7.95)$$

Solution (7.85) then takes the form

$$X(t) = \mu_0 X^* \left\{ 1 - \frac{\mu - \mu_1}{\mu_0} \exp[-a(t - t_1)] \right\}. \tag{7.96}$$

which is the relaxation law. As $t_1 - t_0$ tends to ∞, $\mu_1 \to \mu$, as seen from (7.95). As $t \to \infty$, the adsorption center concentration tends to its initial value $X = \mu_0 X^*$ [see (7.96)], which it had prior to illumination. We can easily see that after the light is turned off, relaxation of the number of centers proceeds more slowly than the growth (or decay) of this number under illumination. Indeed, according to (7.69) and (7.83), $a_1 > a_{10}$ and $a_2 > a_{20}$ and hence, $a > a_0$. Figure 7.8 depicts the time variation of the concentration of adsorption centers when the light is turned on and off, as described by (7.94) and (7.96).

7.5. THE PHOTOADSORPTION EFFECT AT A REAL SURFACE

7.5.1. Adsorption after Illumination

We assume that adsorption takes place at a surface that has been illuminated. Then both adsorption in the dark and photoadsorption acquire some special features. Adsorption proceeds differently depending on the duration of the previous illumination and on the time interval between the cessation of illumination and the beginning of adsorption.

(1) Consider first adsorption in the dark. Let us assume that at $t = t_0$ the light is switched on in vacuum and at $t = t_1$ it is switched off. After this, at $t = t_2$ the gas is let into the adsorption volume and adsorption begins. The adsorption kinetics is described by solution (7.89), where, however, we must put

$$t_1 = t_2, \quad a = a_0, \quad \mu = \mu_0, \quad N_i = 0, \quad X_i = \mu_2 X^*, \tag{7.97}$$

where, according to (7.96),

$$\mu_2 = \mu_0 \left\{ 1 - \frac{\mu_0 - \mu_1}{\mu_0} \exp[-a_0(t_2 - t_1)] \right\} \tag{7.98}$$

and μ_1 is given by (7.95), which, according to (7.95) and (7.98), behaves in the following manner:

$$\begin{aligned}
\mu_1 &\to \mu & \text{for} \quad t_1 - t_0 &\to \infty, \\
\mu_1 &\to \mu_0 & \text{for} \quad t_1 - t_0 &\to 0, \\
\mu_2 &\to \mu_1 & \text{for} \quad t_2 - t_1 &\to 0, \\
\mu_2 &\to \mu_0 & \text{for} \quad t_2 - t_1 &\to \infty,
\end{aligned} \tag{7.99}$$

Here, obviously, $t_1 - t_0$ is the duration of the preliminary illumination, and $t_2 - t_1$ the time interval during which the sample is left in the dark before adsorption

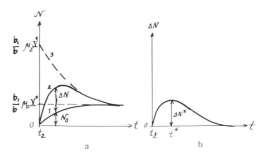

Fig. 7.9. Kinetic curves for adsorption after illumination: a) time dependence of the surface concentration of adsorbed particles; b) time variation of the contribution to the adsorptivity due to preliminary illumination.

begins. In this section we take $\mu_1 > \mu_0$, which means that the illumination creates new adsorption centers instead of destroying them (see Fig. 7.8a). Hence, in view of (7.98), $\mu_2 > \mu_0$.

Combining solution (7.89) with (7.97) and (7.90), we have

$$N(t) = N_0(t) + \Delta N(t),$$ (7.100)

where

$$N_0(t) = \frac{b_1}{b} \mu_0 X^* \{1 - \exp[-b(t - t_2)]\},$$

$$\Delta N(t) = \frac{b_1}{a_0 - b} (\mu_2 - \mu_0) X^* \{ \exp[-b(t - t_2)] - \exp[-a_0(t - t_2)]\}.$$ (7.101a)

According to (7.99), at $t_1 - t_0 = 0$ we have $\Delta N = 0$, and (7.100) becomes the kinetics equation in the absence of preliminary illumination, (7.101a). Thus, $\Delta N(t)$ is the contribution to the adsorptivity due to the preliminary illumination.

Figure 7.9a shows the kinetic curve N vs. t for the sample in the dark (curve 1) and for the sample that has been preilluminated (curve 2), calculated via (7.100), (7.101a), and (7.101b) (on the assumption that $\mu_2 > \mu_0$). It clearly indicates a memory effect, viz., preliminary illumination leads to additional adsorption in the dark. Curve 2 possesses a maximum, while both curves, as $t \to \infty$, tend to a common limit:

$$N = \frac{b_1}{b} \mu_0 X^*,$$ (7.102)

corresponding to adsorption equilibrium.

Figure 7.9b depicts the ΔN vs. t curve, as given by (7.101b). This curve has a maximum at $t = t^*$. For $t < t^*$ there is additional adsorption at the centers stored by the light, while at $t > t^*$ the additional adsorption begins to fall off due to the gradual disappearance of the light-induced centers. We can easily find that at the maximal point

$$\Delta N^* = \Delta N(t^*) = \frac{b_1}{b} (\mu_2 - \mu_0)X^* \exp[-a_0(t^* - t_2)], \qquad (7.103)$$

where, according to (7.98) and (7.95),

$$\mu_2 - \mu_0 = (\mu - \mu_0)\{1 - \exp[-a(t_1 - t_0)]\} \exp[-a_0(t_2 - t_1)]. \qquad (7.104)$$

Note that if

$$a_0 \ll b, \qquad (7.105)$$

and if, in addition,

$$\exp[-b(t - t_2)] \ll \exp[-a_0(t - t_2)], \qquad (7.106)$$

then (7.100) takes the form

$$N(t) = \frac{b_1}{b} \mu_0 X^* \left\{ 1 - \frac{\mu_0 - \mu_2}{\mu_0} \exp[-a_0(t - t_2)] \right\} = \frac{b_1}{b} X(t), \qquad (7.107)$$

where, in view of (7.98),

$$X(t) = \mu_0 X^* \left\{ 1 - \frac{\mu_0 - \mu_2}{\mu_0} \exp[-a_0(t - t_2)] \right\} = \mu_0 X^* \left\{ 1 - \frac{\mu_0 - \mu_1}{\mu_0} \exp[-a_0(t - t_1)] \right\}. \qquad (7.108)$$

This, as we see, is clearly (7.96). Equation (7.107) describes adsorption equilibrium in the case where the concentration of adsorption centers varies. It is shown in Fig. 7.9a by a dashed curve (curve 3).

(2) We now turn to photoadsorption at a surface that has been preilluminated. Let the sample be illuminated in a vacuum from time $t = t_0$ to time $t = t_1$. Then at $t = t_2$ (where $t_2 > t_1$) the gas is let into the adsorption volume, while the light is turned on again at $t = t_3$ (where $t_3 > t_2$). The sequence is shown in Fig. 7.10. We are interested in the photoadsorption kinetics at $t > t_3$ and the effect of the preliminary illumination. Finally, we adopt condition (7.105).

The kinetics of photoadsorption is described by (7.89), where, according to (7.108) and (7.107),

$$t_i = t_3, \qquad X_i = \mu_3 X^*, \qquad N_i = \frac{b_1}{b} \mu_3 X^*, \qquad (7.109)$$

Fig. 7.10. Sequence of events in the photoadsorption at a surface that has been illuminated.

and where we have introduced the notation

$$\mu_3 = \mu_0 \left\{ 1 - \frac{\mu_0 - \mu_2}{\mu_0} \exp[-a_0(t_3 - t_2)] \right\}. \tag{7.110}$$

Substituting (7.109) into (7.90) and the result into (7.89), we obtain

$$N(t) = \frac{ab}{a-b}(\mu - \mu_3)X^* \left\{ \frac{1}{a} \exp[-a(t - t_3)] - \frac{1}{b} \exp[-b(t - t_3)] \right\} + \frac{b_1}{b} \mu X^*. \tag{7.111}$$

We will confine the discussion to two limiting cases where (7.111) is simplified considerably:

$$\text{when} \quad a \ll b, \tag{7.112a}$$
$$\text{when} \quad a \gg b. \tag{7.112b}$$

Then (7.111) takes the form

$$N(t) = \frac{b_1}{b} \mu X^* \left\{ 1 - \frac{1}{1-\delta} \frac{\mu - \mu_3}{\mu} \exp[-k(t - t_3)] \right\}, \tag{7.113}$$

or

$$N(t) = \frac{b_1}{b} \mu X^* \left\{ 1 - \frac{\mu - \mu_3}{\mu} \exp[-k(t - t_3)] \right\}, \tag{7.114}$$

where k and δ are given by

a) $k = a, \quad \delta = \frac{a}{b} \qquad \text{when} \quad \delta \ll 1 \quad (\text{case } (7.112a)$

b) $k = b, \quad \delta = \frac{b}{a} \qquad \text{when} \quad \delta \gg 1 \quad (\text{case } (7.112b)$

$$\tag{7.115}$$

Here, in view of (7.110) and (7.105),

$$\mu - \mu_3 = (\mu - \mu_0) \{1 - [1 - \exp[-a(t_1 - t_0)]]\} \exp[-a_0(t_0 - t_1)]. \tag{7.116}$$

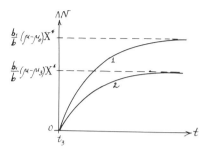

Fig. 7.11. Time variation of the contribution to the adsorptivity due to preliminary illumination.

Using (7.114) and (7.109), we obtain

$$\Delta N(t) = N(t) - N_i = \frac{b_1}{b}(\mu - \mu_3)X^* \{1 - \exp[-k(t - t_3)]\}. \qquad (7.117)$$

The curve ΔN vs. t is depicted in Fig. 7.11 for two cases, viz., $t_1 = t_0$ (photoadsorption without preliminary illumination; curve 1) and $t_1 > t_0$ (preliminary illumination during the time interval $t_1 - t_0$; curve 2). The quantity $\Delta N(\infty)$, obviously, characterizes the photoadsorptivity of the surface, which is the number of gas molecules additionally adsorbed at a unit surface area of the surface owing to the influence of illumination. According to (7.116) and (7.117),

$$\Delta N(\infty) = \frac{b_1}{b_0}(\mu - \mu_0)X^* \{1 - [1 - \exp[-a(t_1 - t_0)]]\exp[-a_0(t_1 - t_2)]\}. \quad (7.118)$$

We see that a surface subjected to light keeps a "memory" of this fact. This "memory" gradually decays in time. The origin of the memory is the following: by producing ionization of the sample followed by a subsequent localization of the charge carriers (electrons or holes) at the defects, the light transfers the system to a metastable state, which is an excited state separated from the ground state by an activation barrier. It is the presence of this barrier that makes the light-induced centers more or less stable.

The surface keeps its increased adsorptivity in the dark until the light-induced centers disappear. On the other hand, the photoadsorptivity proves to be lower than it was before the preliminary illumination. Indeed, the additional light can now add only a relatively small number of centers, since most of them have already been created during the preliminary illumination.

7.5.2. The Sign and Magnitude of the Photoadsorption Effect

We will consider the photoadsorption effect, again adopting the model of a real surface discussed above. The magnitude of the photoadsorption effect is defined by (7.14):

$$\Phi = \frac{N(\infty) - N_0(\infty)}{N_0(\infty)} , \qquad (7.119)$$

where $N(t)$ and $N_0(t)$ are the adsorptivities of the surface (i.e., the number of particles adsorbed on a unit surface area) at time t with and without illumination. As $t \rightarrow 0$, adsorption equilibrium is established.

According to (7.100), (7.101a), and (7.101b), if we put $t_1 = t_0$ (thermal adsorption without preillumination) and $t - t_2 = \infty$ (adsorption equilibrium), we have

$$N_0(\infty) = \frac{b_1}{b} \mu_0 X^* . \qquad (7.120)$$

On the other hand, if in (7.114) we put $t - t_3 = \infty$ (photoadsorption equilibrium), we have

$$N(\infty) = \frac{b_1}{b} \mu X^* . \qquad (7.121)$$

Substituting (7.120) and (7.121) into (7.119), we obtain

$$\Phi = \frac{\mu}{\mu_0} - 1, \qquad (7.122)$$

where, according to (7.91), (7.84), and (7.86),

$$\mu_0 = \frac{1}{1 + \nu_0} , \qquad \nu_0 = \frac{a_{20}}{a_{10}} ,$$

$$\mu = \frac{1}{1 + \nu} , \qquad \nu = \frac{a_2}{a_1} . \qquad (7.123)$$

Here the parameters a_1, a_2, a_{10}, and a_{20} have the same meaning as in (7.69), with the subscript 0 meaning the absence of light ($\Delta n_s = \Delta p_s = 0$). Thus, according to (7.69),

$$a_1 = a_{10} + \alpha_1 \Delta n_s, \qquad a_2 = a_{20} + \alpha_2 \Delta p_s,$$

and, hence,

$$\nu = \frac{a_{20} + \alpha_2 \Delta p_s}{a_{10} + \alpha_1 \Delta n_s} = \nu_0 \frac{1 + (\alpha_2/a_{20}) \Delta p_s}{1 + (\alpha_1/a_{10}) \Delta n_s} . \qquad (7.124)$$

We are now in a position to formulate the criteria for the positive and negative photoadsorption effects. According to (7.122), we have $\Phi \geq 0$ if $\mu \geq \mu_0$, i.e., $\nu \leq \nu_0$, and $\Phi \leq 0$ if $\mu \leq \mu_0$, i.e., $\nu \geq \nu_0$, or, allowing for (7.124),

$$\Phi \gtrless 0, \quad \text{or} \quad \frac{\alpha_1}{a_{10}} \Delta n_s \gtrless \frac{\alpha_2}{a_{20}} \Delta p_s.$$

In view of (7.123), this yields

$$\Phi \gtrless 0, \quad \text{or} \quad \nu_0 \frac{\alpha_1}{\alpha_2} \frac{\Delta n_s}{\Delta p_s} \gtrless 1. \tag{7.125}$$

Note that (7.72) yields

$$\nu_0 = \frac{a_{20}}{a_{10}} = \exp \frac{v - \epsilon_s}{kT}, \tag{7.126}$$

where ϵ_s and v are the distances from the intrinsic Fermi level to the Fermi level in the surface plane and, correspondingly, to the local level of the defect that serves as the adsorption center. Moreover, we note that according to (7.33) (low excitation) or (7.63) and (7.64) (high excitation),

$$\frac{\Delta n_s}{\Delta p_s} = \exp\left(-\frac{2V_s}{kT}\right), \tag{7.127}$$

where, as usual, V_s is the bending of the energy bands in the surface plane. Substituting (7.126) and (7.127) into (7.125), assuming that

$$\alpha_1 = \alpha_2 \tag{7.128}$$

and allowing for the fact that

$$\epsilon_s = \epsilon_v + V_s,$$

we have instead of (7.125) the following:

$$\Phi \gtrless 0, \quad \text{or} \quad (\epsilon_v + V_s - v) \gtrless 0. \tag{7.129}$$

which is exactly the same sign criterion as that obtained in the case of an ideal surface [cf. (7.47)]. In the present model the acceptor level of a defect adsorption center acts as the level of a chemisorbed particle.

We now turn to the problem of finding the magnitude of the effect. We assume that

$$\nu_0 \gg 1, \quad \nu \gg 1. \tag{7.130}$$

Note that, according to (7.126), the first condition means that the Fermi level is assumed to lie considerably lower than the levels of the adsorption centers; i.e., the latter levels, which in our model are assumed to be acceptor levels, are completely filled by electrons. In view of (7.130) and (7.122) we have, instead of (7.122), the following:

$$\Phi = \frac{\nu_0}{\nu} - 1. \tag{7.131}$$

Let the excitation be so high that

$$\Delta n_s \gg \frac{a_{10}}{\alpha_1}, \qquad \Delta p_s \gg \frac{a_{20}}{\alpha_2}. \tag{7.132}$$

Then, using (7.124), we can rewrite (7.131) in the following form:

$$\Phi = \frac{(\alpha_1/a_{10})\Delta n_s}{(\alpha_2/a_{20})\Delta p_s} - 1,$$

where, taking (7.126), (7.127), and (7.128) into account, we finally obtain

$$\Phi = \exp\left(-\frac{\epsilon_v + V_s - \upsilon}{kT}\right) - 1. \tag{7.133}$$

We have again arrived at a result obtained for an ideal surface. To see this, it suffices to insert (7.45) into (7.65a) and compare the result with (7.133). Thus, as far as the magnitude of the photoadsorption effect is concerned, it is immaterial in the model considered whether we are dealing with an ideal surface or a real surface.

7.5.3. "Direct" Photodesorption

Besides the mechanism described in Section 7.4.3 there is another mechanism of photodesorption, studied by Crucq and Coekelbergs [41] and by Molinari, Cramarosa, and Paniccia [42]. This mechanism is known as "direct" photodesorption, which is caused by absorption of a photon directly by an adsorbed particle (an atom or a molecule) with no participation of free charge carriers. The model adopted in Section 7.4.3 ignores the "weak" bond, or the electrically neutral form of chemisorption; i.e., the particle remains in the adsorbed state as long as it is bound to the charged carrier localized at or near it. As soon as the charge is delocalized, the particle is desorbed. This approximation constitutes the basis of the boundary layer theory (see [43, 44]) and is applicable as long as the Fermi level lies considerably higher (in comparison to kT) then the donor levels or considerably lower than the acceptor levels. Characteristically, the main regularities of photodesorption can be obtained (as we will see later) within the framework of this approximation, but at the same time it is this approximation that leaves direct desorption out of the realm of the theory.

Let us go outside the limits of the theory discussed in Section 7.4.3 and assume that the adsorbed particles (for the sake of simplicity we assume they are univalent atoms) are in the state of "weak" bonding with the surface. The electronic states of the system, i.e., the levels of the valence electron (we consider a

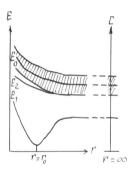

Fig. 7.12. Energy spectrum of the system of a univalent electropositive atom and a one-dimensional ionic crystal.

one-electron problem), are depicted in Fig. 7.12 (which is simply a repetition of Fig. 2.17). Two types of levels are present here: (1) levels corresponding to bound states (an example is the level E_1 with a characteristic minimum) and (2) levels corresponding to antibound (repulsive) states (level E_2 or the system of levels that form a band around level E_0). When the system is in its ground state, the electron lies at the bottom of the well at level E_1. An excitation due to the absorption of a quantum transfers the electron from E_1 (or from a lower lying level) to level E_2 (or a higher lying level), i.e., from a bound state to an antibound state. This transition, as seen from Fig. 7.12, leads to immediate desorption. This constitutes the essence of the direct desorption mechanism as developed by Crucq and Coekelbergs [41] and also independently by Molinari *et al.* [42].

The modification of the electronic theory of photodesorption consists in allowing, in addition to the two forms of chemisorption ("strong" and "weak"), for transitions between these two forms. This does not lead to any serious mathematical difficulties but results in a highly cumbersome mathematical apparatus.

7.5.4. The Aftereffect

Let us consider some specific features of photoadsorption (or photodesorption) kinetics at a real surface [32]. Suppose that a semiconductor sample is placed in a gaseous atmosphere of constant pressure P and there is equilibrium between the surface and the gas. At $t = t_0$ the light is turned on and photoadsorption (or photodesorption) begins.

The kinetics of this process obeys Eq. (7.89), where we assume, following (7.102) and (7.93), that

$$N(t) = \frac{ab_1}{a-b}(\mu - \mu_0)X^* \left\{ \frac{1}{a}\exp[-a(t-t_0)] - \frac{1}{b}\exp[-b(t-t_0)] \right\} + \frac{b_1}{b}\mu X^*. \quad (7.134)$$

Note that in the limiting cases (7.112a) and (7.112b), Eq.(7.134) is reduced to

$$N(t) = \frac{b_1}{b}\mu X^* \left\{ 1 - \frac{1}{1-\delta}\frac{\mu - \mu_0}{\mu}\exp[-k(t-t_0)] \right\}, \quad (7.135)$$

where k and δ are defined in (7.115).

As expected, Eqs. (7.134) and (7.135) coincide with Eqs. (7.111) and (7.113) if we substitute t_0 for t_3 and μ_0 for μ_3. Indeed, Eqs. (7.111) and (7.113) were obtained on the assumption that photoadsorption (or photodesorption) was preceded by illumination of the sample. If the duration of this preillumination is taken as zero, i.e., if in (7.116) we put $t_1 - t_0 = 0$, then (7.116) yields $\mu_3 = \mu_0$, and Eqs. (7.111) and (7.113) are reduced to (7.134) and (7.135), respectively, which is what we expected.

Now, let the light be turned off at $t = t_1$. We are interested in the behavior of the surface after the light has been turned off, i.e., at $t \gg t_1$. As before, we limit the discussion to the cases (7.112a) and (7.112b). According to (7.89), at $t \gg t_1$ the concentration of the chemisorbed particles, $N(t)$, should relax to the initial value it had prior to illumination. In the present case, using (7.135) (with $\delta \ll 1$), (7.94), and (7.95), we must rewrite Eq. (7.89) with

$$t = t_4, \qquad a = a_0, \qquad \mu = \mu_0,$$

$$N_i = N_1 = \frac{b_1}{b} \mu_1' X^*, \qquad X_i = X_1 = \mu_1 X^*, \tag{7.136}$$

where we have introduced the following notation:

$$\mu_1' = \mu\left\{1 - \frac{\mu - \mu_0}{\mu} \exp\left[-k(t_1 - t_0)\right]\right\},$$

$$\mu_1 = \mu\left\{1 - \frac{\mu - \mu_0}{\mu} \exp\left[-a(t_1 - t_0)\right]\right\}. \tag{7.137}$$

Let [see (7.105)] $a_0 \ll b$. Then,

$$N(t) = \frac{b_1}{b} \mu_0 X^*\left\{1 + \frac{\mu_1' - \mu_1}{\mu_0} \exp\left[-b(t - t_0)\right] - \frac{\mu_0 - \mu_1}{\mu_0} \exp\left[-a_0(t - t_1)\right]\right\}. \tag{7.138}$$

Consider first the case (7.112a). According to (7.115), $k = a$ and, hence, (7.137) yields $\mu_1' = \mu_1$. Equation (7.138) now takes the form

$$N(t) = \frac{b_1}{b} \mu_0 X^*\left\{1 - \frac{\mu_0 - \mu_1}{\mu_0} \exp\left[-a_0(t - t_1)\right]\right\} = \frac{b_1}{b} X(t), \tag{7.139}$$

where $X(t)$ is given by (7.94). Here, as in (7.107), we have the case of adsorption equilibrium at a varying (relaxing) number of adsorption centers. Equation (7.139) corresponds to the part BC of the curve shown in Fig. 7.13. [Note that the curve ABG corresponds to Eq. (7.135); cf. Fig. 7.11.]

We now turn to the case (7.112b), where $k = b$ and, consequently, $\mu_1' \neq \mu_1$. Equation (7.138) here corresponds to the curve $BDEF$ in Fig. 7.13, which passes through a maximum (minimum) at $t = t_1$. Expanding $N = N(t)$ in (7.138) in a Taylor series at $t = t_1$ and alloying for (7.106), we can easily show that the maximum (point D in Fig. 7.13) is at

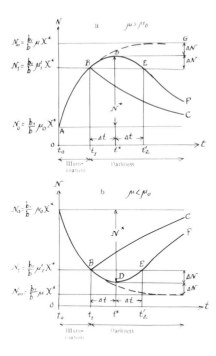

Fig. 7.13. Time variation of the concentration of adsorption centers after the surface has been illuminated: a) $\mu > \mu_0$; b) $\mu < \mu_0$.

$$t^* = t_1 + \Delta t, \qquad \text{where} \quad \Delta t = \frac{1}{b}, \qquad\qquad (7.140a)$$

$$N^* = N_1 + \Delta N, \qquad \text{where} \quad \Delta N = \frac{b_1}{2b}(\mu_1 - \mu_1')X^*. \qquad (7.140b)$$

Likewise, we can easily show that (see Fig. 7.13)

$$t_2' - t_1 = \frac{2}{b} = 2\,\Delta t,$$

$$N_\infty - N_1 = \frac{b_1}{b}(\mu_1 - \mu_1')X^* = \frac{b_1}{b}(\mu - \mu_0)X^* \exp[-b(t_1 - t_0)] = 2\,\Delta N.$$

We again encounter the semiconductor memory effect. Here it is reflected in the aftereffect, viz., adsorption (desorption) continues some time after the light has been turned off as if the light was present. The magnitude of the effect, ΔN, follows from (7.140b) and (7.137):

$$\Delta N = \frac{b_1}{2b}(\mu_1 - \mu_1')X^* = \frac{b_1}{2b}(\mu - \mu_0)X^* \exp[-b(t_1 - t_0)]. \qquad (7.141)$$

To sum up, semiconductors are seen to possess a certain "memory." It is revealed (a) in the influence of preillumination on subsequent adsorption both in the dark and under illumination (see Section 7.5.1), and (b) in the aftereffect

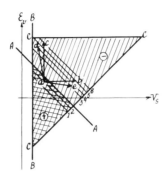

Fig. 7.14. Range of the position of the bulk Fermi level and the band bending at the surface in the event of adsorption of acceptor particles (oxygen; cf. Fig. 7.5).

(Section 7.5.4). The memory is not connected with the radiation damage occurring when the semiconductor is subjected to light (in fact, corpuscular radiation) and remaining afterwards. On the contrary, the memory is due to the nonequilibrium charge carriers localized at the intrinsic defects rather than radiation defects under illumination and remaining there for a long period. Such a memory originates in any semiconductor whose localized charge carriers act as adsorption centers.

7.6. COMPARISON OF THE THEORY OF THE PHOTOADSORPTION EFFECT WITH THE EXPERIMENTAL DATA

7.6.1. Influence of Illumination on the Adsorptivity of a Surface

We now turn to the experimental data reviewed in Section 7.1.3. At first glance the experimental findings seem to some extent to be contradictory. Let us consider them from the point of view of the theory described above.

To begin with, we will consider the influence of the experimental conditions and sample treatment on the sign and magnitude of the photoadsorption effect. For this purpose we will return to Fig. 7.5. The part of this figure that describes adsorption of acceptor particles (oxygen) is repeated in Fig. 7.14. Sample treatment or variation of the experimental conditions such as the temperature or the pressure leads to a change in V_s or ϵ_v (or both), which in Fig. 7.14 is reflected by a shift from one point to another. The straight lines parallel to AA in Fig. 7.14 constitute the set of points for which $\Phi = \text{const}$, the lines being numbered in order of decreasing values of Φ, i.e., the algebraic value of Φ is taken into account. In this way the straight line AA, which corresponds to $\Phi = 0$, divides the triangle CCC into regions of positive and negative effects (to the left and right of AA, respectively). All the experimental data can be divided into two groups:

(1) Studies in which the sample is treated in such a way that V_s changes but ϵ_v does not. This situation arises when external factors affect the surface properties but not the bulk of the sample.

An example is a study of the influence of the pressure of the gas, in whose atmosphere the sample is placed, on the photoadsorptive properties of the sample. If the gas is oxygen and the adsorbed particles are acceptors, i.e., also oxygen, then, as the pressure grows, the point (V_s, ϵ_v) moves to the right in Fig. 7.14, as shown by the horizontal arrow ab. We also see that the magnitude of the positive effect decreases, and the positive effect may be replaced by the negative.

Romero-Rossi and Stone [5, 6] observed this effect when they studied the system ZnO + O_2 at room temperature. They found that photoadsorption at low pressure was replaced by photodesorption at higher pressures, and photodesorption increased with pressure. Note that it is essential that the temperature not be too high, since at high temperatures the sample becomes oxidized and ϵ_v decreases as V_s grows (see below).

Another example is the increase of V_s as chemisorbed donor particles are removed from the surface. According to Fig. 7.14, photoadsorption should decrease in the process, which is in agreement with Stone's data [6, 8] obtained for the TiO_2 + O_2 system, viz., a decrease in photoadsorption was observed on removal of the bound water (a donor) from the surface.

Or take the case where the surface is hydroxylated. The negative charge of the surface increases, which leads to an increase in V_s and, as a result, a point in Fig. 7.14 moves horizontally from left to right. This, as shown by Fig. 7.14, must lead to an increase in photodesorption, which is observed in experiments (see [36]).

Take then the effect of chemisorbed water on the photoadsorption of O_2. When molecules of H_2O (donor) appear on a surface, a point in Fig. 7.14 moves horizontally from right to left, or stimulates photoadsorption, which is in agreement with Bickley and Jayanty's data [39].

Finally, we can now understand how an external electric field influences the photoadsorption effect. By increasing V_s (or bending the bands upward), or (in other words) by enriching the semiconductor surface with holes, a field may stimulate photodesorption, as seen from Fig. 7.14, but a field of opposite direction (decreasing V_s) may suppress photodesorption. This effect was observed by Bykova, Komolov, and Lazneva [37].

(2) The second group contains studies in which sample treatment leads to changes in ϵ_v, while V_s remains constant. Such seems to be the case of the influence of doping on photoadsorption at a "quasiisolated" surface. This case is often encountered; see [17, 18]. Doping by donor impurities causes the sample state to shift upward in Fig. 7.14, as shown by the vertical arrow ac, while doping by acceptor impurities shifts it downward. This agrees with Kwan's data [7] (O_2 at ZnO): a positive effect was observed in a Li (acceptor)-doped sample, while for an Al (donor)-doped sample the effect was negative, as in Fig. 7.14. This also agrees with observations described by Romero-Rossi and Stone [5, 6] (O_2 again at ZnO). These authors studied the positive effect range: doping with Li (acceptor) increased the effect while doping with Ga (donor) decreased the effect, as follows from Fig. 7.14.

(3) The third group is the most common. It contains all the studies in which sample treatment changes both V_s and ϵ_v. Numerous papers describe the influence of the degree of oxidation on the sign and magnitude of the photoadsorption.

To consider the oxidation mechanism we take the example of ZnO, which is the adsorbent most widely used in photoadsorption studies. At high temperatures (higher than 200°C) the adsorption of oxygen on ZnO is irreversible. Romero-Rossi, Stone, and Barry [5, 10, 33] explain this fact by a transformation of the adsorbed O^- ions into the O^{--} ions via the reaction

$$O_a^- + Zn\,_{in}^+ \rightarrow Zn^{++}\,O^{--}$$

as a result of which Zn_{in}^+ interstitial ions diffuse to the surface (here O_a^- refers to an adsorbed ion of oxygen and Zn_{in}^+ to interstitial zinc). The reaction becomes important only at temperatures so high that the Zn_{in}^+ interstitial ions diffuse from the bulk to the surface under the influence of the electric field of the chemisorbed O^- ions. According to this approach, which is shared by many authors, the high-temperature oxygen adsorption on ZnO is nothing more than crystal growth accompanied by a decrease in superstoichiometric zinc, which means a decrease in ϵ_v.

Thus, when the sample is oxidized, ϵ_v decreases while V_s increases. Hence, the system is shifted, say, from point d to point a in Fig. 7.14 or, say, from point a to point e. In the first case (transition $d \rightarrow a$) the negative effect is replaced by the positive, while in the second (transition $a \rightarrow e$) the positive effect is replaced by the negative.

The first case was observed by Fujita and Kwan [9], Barry [10], and Terenin and Solonitsyn [11] in their studies of the $ZnO + O_2$ system, and also by Kennedy *et al.* [12] and Kazanskii *et al.* [13], who studied the $TiO_2 + O_2$ system; viz., when the sample was oxidized, photodesorption was replaced by photoadsorption. The same situation was encountered by Romero-Rossi and Stone [5, 6] (the $ZnO + O_2$ system) and by Kwan [7] (the $TiO_2 + O_2$ system): at high temperatures (higher than 500°C) photodesorption was replaced by photoadsorption when the oxygen pressure was raised, i.e., when the sample became more oxidized.

The second case was apparently encountered by Haber and Kowalska [14] and by Romero-Rossi and Stone [5] ($ZnO + O_2$): photoadsorption was replaced by photodesorption on sample oxidation.

7.6.2. "Memory" Effects in Photoadsorption

We will now turn to the problem of the semiconductor's "memory," i.e., to the question of how long light-induced changes are maintained in a sample (for one, at its surface) after illumination is turned off. If the word "illumination" is understood in the broad sense, having in mind not only electromagnetic radiation (such as visible and ultraviolet light, x rays, and γ rays) but corpuscular radiation (bombardment of the sample by neutrons, protons, α particles, and electrons) as well, then the answer to the question proves to be different in different cases (see [32, 34]).

(a) There is the limiting case where the semiconductor once subjected to illumination remembers the event forever. This often happens when the sample is irradiated by heavy particles producing nuclear reactions in the material. Illumination here acts as a factor that ensures the doping of the semiconductor with impurities, in which process the regular lattice atoms are replaced by foreign atoms. This constitutes an irreversible process.

(b) The more general case is when the semiconductor remembers illumination for some time and then gradually "forgets" the event. This happens when the radiation does not change the chemical composition of the semiconductor but produces structural (radiation) defects, i.e., introduces additional disorder into the lattice. Such a process is also equivalent to doping, since, in the language of semiconductor physics, the words "defect" and "impurity" are synonyms. All the structure-sensitive properties of the semiconductor immediately react to this process. Radiation-induced disorder gradually relaxes after the illumination has been turned off, and the higher the temperature the greater the relaxation rate. Thus, heating proves to be a factor helping to "erase" the memory of illumination. The "impurities" introduced into the sample by the effect of illumination gradually disappear. In this respect they differ from those produced by nuclear reactions.

(c) Finally, there is another limiting case involving electromagnetic radiation, which is of major interest to us. Illumination is the ionization agent and directly produces nonequilibrium charge carriers, generating neither structural defects nor chemically foreign impurities. As a rule, the sample then remembers the illumination as long as it is under the light. After the light is turned off, the sample relaxes quickly (sometimes almost instantaneously) to its initial equilibrium state. Sometimes, however, this rule is broken and the material remembers the fact of illumination for a comparatively long time. It is this memory of a nontrivial origin (connected to no radiation defects) that we have to deal with in the photoadsorption effect (we studied this type of memory in the previous sections).

We will now return to some points mentioned in Section 7.1.3, where the memory effects of a semiconductor surface manifest themselves after illumination.

(1) The surface adsorptivity in the dark is often increased after preillumination. The additional adsorptivity, ΔN, is given, as we already know, by (7.101b), viz.,

$$\Delta N = \frac{b_1}{a_0 - b} (\mu_2 - \mu_0) X^* \{\exp[-b(t - t_2)] - \exp[-a(t - t_2)]\}, \qquad (7.142)$$

where [see (7.104)]

$$\mu_2 - \mu_0 = (\mu - \mu_0)\{1 - \exp[-a(t_1 - t_0)]\} \exp[-a_0(t_2 - t_1)]. \qquad (7.143)$$

Here t_0, t_1, and t_2 ($t_0 < t_1 < t_2$) are the times when the light was turned on, turned off, and the gas let into the adsorption volume, respectively; all the other notations are the same as in Section 7.5.1.

It follows from (7.142) and (7.143) that the additional (preillumination-induced) adsorptivity ΔN possesses the following properties:

(i) It increases with the exposure time $t_1 - t_0$ and tends to the saturation value.
(ii) It decreases as the time during which the sample is left in the dark following the exposure, $t_2 - t_1$, increases and tend to zero.

These results are in complete agreement with the experimental data of Solonitsyn [25] (see Section 7.1.3).

(2) The photoadsorptivity $\Delta N(t)$, i.e., the surface adsorptivity in the presence of light, may also be changed if preillumination was involved. For $\Delta N(\infty)$, which is the photoadsorptivity under adsorption equilibrium conditions, we have Eq. (7.118), viz.,

$$\Delta N(\infty) = \frac{b_1}{b}(\mu - \mu_0)X^* \{1 - [1 - \exp[-a(t_1 - t_0)]]\exp[-a_0(t_3 - t_1)]\}, \qquad (7.144)$$

where t_0, t_1, and t_3 $(t_0 < t_1 < t_3)$ are the times when the preillumination was turned on, turned off, and photoadsorption began, respectively.

It follows from (7.114) that the photoadsorptivity $\Delta N(\infty)$ possesses the following properties:

(iii) It decreases as the time of preillumination exposure, $t_1 - t_0$, increases.
(iv) It increases with the time during which the sample is left in the dark following the exposure, $t_3 - t_1$.
(v) It disappears completely in the limit of large $t_1 - t_0$ and small $t_3 - t_1$.

These results are again in good agreement with the experimental data (see [16] and Section 7.1.3).

(3) The memory effects appearing in photoadsorption also reveal themselves in the aftereffect; i.e., photoadsorption continues for some time after the light is switched off. The magnitude of the effect is characterized by the number of molecules additionally adsorbed by a unit surface area after the light is turned off, ΔN. This quantity is given by Eq. (7.141), but we will not write it here.

Considering (7.141) and returning to the notation employed in Section 7.5.4, we come to the following conclusions:

(vi) The aftereffect manifests itself only at $a \gg b$ (it is only in this case that $\mu_1' \neq \mu_1$).
(vii) The aftereffect is smaller the greater $t_1 - t_0$, i.e., the greater the exposure time.
(viii) The temperature dependence of the aftereffect is determined by the factor b in (7.141) and, hence, according to (7.88) and (7.69), the aftereffect ΔN falls sharply as the temperature grows.

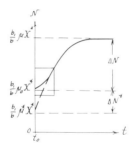

Fig. 7.15. Kinetic curves given by (7.134)
(solid curve) and by (7.135) (dashed curve).

Note that for $t \gg t_1$ Eq. (7.138) is reduced to (7.139) for both $a \ll b$ and $a \gg b$. Thus, a long time after the light is switched off, the chemisorbed particle concentration N relaxes according to the same law (7.139) irrespective of whether we are dealing with case (7.112a) or case (7.112b). The rate of relaxation is greater the higher a_0 and, hence, increases sharply with temperature, according to (7.86) and (7.69).

All these regularities are in agreement with the experimental data (see Section 7.6.1). The first to discover the aftereffect in photoadsorption and to study it were Coekelbergs and his collaborators [16].

Finally, we will consider certain peculiarities of photoadsorption (or photo-desorption) at a real surface which follow from the theory. According to the results of Section 7.54, the rate of photoadsorption (or photodesorption) is determined, provided that $a \ll b$, by the rate with which the surface defects, which play the role of adsorption centers, recharge. This rate is characterized by the time constant a^{-1}. When the light is turned off, the concentrations of adsorption centers and adsorbed particles relax to their equilibrium values in the dark with a time constant $a_0^{-1} > a^{-1}$ (see Fig. 7.8 and the curves ABC in Fig. 7.13). Under certain conditions, fixed by the prehistory of the adsorbent and the intensity of the light, we have $a_0 \ll a$. In such cases the adsorptive properties of the surface recover in the dark much more slowly than they change under illumination. The observer perceives this as the irreversibility of photoadsorption or photodesorption. If the temperature increases, the constant a_0 grows and the relaxation of the adsorptive properties increases its rate. The discussed regularities of photoadsorption or photodesorption relaxation agree with the experimental data cited in [6, 8, 9, 94]. Note that the irreversibility of photoadsorption follows from the theory in the case of an ideal surface, too, since the charged form of chemisorption may be considered as irreversible, as follows from [17, 18] and as we noted in Section 3.4.4.

Now let us examine another peculiar feature of photoadsorption that follows from the results of Section 7.5.4. Photoadsorption kinetics, as we know, is described by Eq. (7.134) or, in the limiting cases when $a \ll b$ and $a \gg b$, by Eq. (7.135), in which equation t_0 is the time when the light is turned on. The kinetic curves given by (7.134) and (7.135) for the case $\mu > \mu_0$ are depicted in Fig. 7.15

by a solid and dashed curve, respectively. We are interested in the initial adsorption period, just after the light is turned on.

Expanding the exponentials in (7.134) and (7.135) in Taylor series at $t = t_0$, we obtain for (7.134)

$$N(t) = \frac{b_1}{b} \mu_0 X^* + \frac{ab_1}{2} (\mu - \mu_0) X^* (t - t_0)^2,$$ (7.145)

and for (7.135)

$$N(t) = \frac{b_1}{b} X^* \left\{ \mu^* + \frac{\mu - \mu_0}{1 - \delta} k (t - t_0) \right\},$$ (7.146)

where we have introduced the notation

$$\mu^* = \mu - (\mu - \mu_0)/(1 - \delta).$$ (7.147)

Equation (7.145) shows that there is an "induction period" in the initial part of the kinetic curve. During this period the adsorption rate does not decrease with the passage of the time. On the contrary, it increases, i.e.,

$$\frac{d^2N}{dt^2} > 0.$$

The corresponding part of the kinetic curves is boxed in in Fig. 7.15. Such an induction period is indeed sometimes observed (see Section 7.6.1).

Adopting the notations (see Fig. 7.15)

$$\Delta N = \frac{b_1}{b} (\mu - \mu_0) X^*, \qquad \Delta N^* = \frac{b}{b+c} (\mu - \mu^*) X^*,$$

we find from (7.146) that

$$\Delta N^* = \frac{b}{b+c} (\mu - \mu_0) X^* \frac{\delta}{1-\delta} = \Delta N \frac{\delta}{1-\delta}.$$

Thus, ΔN^* tends to zero as $\delta \to 0$, i.e., the induction period disappears as $\delta \to 0$.

7.6.3. Some Theoretical Predictions

In this chapter we have developed the general theory of the photoadsorption effect both at an ideal and at a real semiconductor surface. The theory attempts to give a unified explanation of all the experimental findings, which at times seem to contradict each other. The theory of photoadsorption, like any physical theory, serves not only to explain the experimental data and regularities that are already known but also new facts and regularities that follow from the theory and may (and should) be verified experimentally. In this way the theory both follows and precedes experiment.

In conclusion, we point out some predictions that follow from the theory and are awaiting experimental verification.

(1) Both the magnitude and the sign of the photoadsorption effect are determined not only by the experimental conditions but also by the prehistory of the sample, i.e., they depend on the treatment the sample was subjected to before illumination. The prehistory of the sample enters the theory through the parameters ϵ_v and V_s, where the first describes the position of the Fermi level in the bulk and the second gives the bending of the bands near the surface.

Since the photoadsorption effect Φ depends on ϵ_v, there must be a correlation between the value of the photoadsorption effect and the initial electrical conductivity of the semiconductor (in the dark). In the case of acceptor particles, the smaller the electronic component of the conductivity and the higher the hole component at a given temperature, i.e., the greater the ϵ_v (we neglect the effect which the surface has on the conductivity; this is justified for a massive semiconductor sample), the higher the absolute value of Φ. In the case of donor particles the situation is just the opposite. An experimental verification of this theoretical prediction might be of interest.

(2) Since the photoadsorption effect Φ depends on V_s, both the magnitude and the sign of the effect must depend on the strength and direction of an external electric field applied at right angles to the surface of the sample. Therefore, it might be of interest to study the photoadsorption effect in conjunction with the field effect, i.e., the combined effect of illumination and an external electric field. Indeed, by varying the electric field strength we can vary the bending of the bands, V_s, within a broad range. In terms of Figs. 7.2, 7.3, 7.5, and 7.14, varying the field strength means shifting the corresponding points in these figures horizontally, which leads to a change in the magnitude of the photoadsorption effect. This theoretical prediction, as noted in Section 7.6.1, has been verified experimentally in [37, 38]. Moreover, an increase in the field strength may lead to a change in the sign of the photoadsorption effect. In other words, both positive and negative photoadsorption effects may be observed with one and the same sample under the same conditions, the only difference being in the direction and strength of the applied electric field. Experimental verification of this theoretical prediction could also be of interest.

(3) Since Φ depends on $\epsilon_s = \epsilon_v - V_s$ there must be a correlation between the photoadsorption effect and the work function of the initial sample. Indeed, the work function, to within a constant term, is equal to $\epsilon_i^- - \epsilon_s$, which is the distance between the Fermi level and the conduction band edge in the surface plane. For samples treated differently (neglecting the changes in the dipole contribution to the work function), the changes in the work function and the magnitude of the photoadsorption effect should be symbatic. It might also be interesting to check this theoretical prediction experimentally.

(4) We have seen that in certain conditions there is an aftereffect in photoadsorption, i.e., adsorption proceeds for some time after the light has been turned

off. This effect, which the theory predicts, was indeed observed experimentally (we noted this fact earlier). Note, however, that the theory predicts that this effect is possible not only in photoadsorption (Fig. 7.13a) but also in photodesorption (Fig. 7.13b). The latter possibility has not yet been studied experimentally and it might be of interest to search for this effect. It would consist of light-induced desorption going on for some time after the light is turned off.

(5) One more theoretical prediction is connected with the resonance action of monochromatic light on a semiconductor surface. Such action enables, at least in principle, carrying out selective recharging or excitation of surface active centers or adsorbed molecules. This, in turn, opens the way to controlled selective adsorption, desorption, and catalytic activity. Most promising possibilities here are connected with employing layers. This question was analyzed in [40].

The photoadsorption effect has been attracting and will continue to attract the attention of both experimenters and theoreticians since it provides the most direct means of studying the electronic mechanism of chemisorption processes.

7.7. THE PHOTOCATALYTIC EFFECT

7.7.1. The Mechanism of the Photocatalytic Effect

Light changes not only the adsorptive properties of semiconductors, which was shown in the previous sections, but the catalytic properties as well. We call this phenomenon the photocatalytic effect. As we have seen in Chapter 5, the expression for the catalytic reaction rate g incorporates the relative contents of the various forms of chemisorption, η^0, η^-, and η^+, for particles of each given type that take part in a reaction. These quantities change under illumination (see Section 7.2.1), which implies that the reaction rate g changes although the reaction mechanism remains the same. In other words, the catalytic activity of the surface with respect to the reaction changes.

Let g and g_0 be the reaction rates with and without illumination, respectively. The photocatalytic effect of a given reaction can be characterized by the fractional change in the reaction rate under illumination:

$$k = \frac{g - g_0}{g_0}. \tag{7.148}$$

Illumination may cause the reaction to increase its rate ($g > g^0$, or $k > 0$). In this case we are dealing with a positive photocatalytic effect. If, on the other hand, illumination slows the reaction, we are dealing with a negative photocatalytic effect ($g < g_0$, or $k < 0$). Finally, if illumination does not influence the catalytic activity of the catalyst with respect to the reaction, we are dealing with photocatalytically inactive light absorption ($g = g_0$, or $k \equiv 0$).

The absolute value and sign of the catalytic effect depend on the type of reaction involved, on the experimental conditions (temperature, intensity of the light, spectral composition of the light, etc.), and on the prehistory of the sample that serves as the catalyst (in particular, the concentration and nature of the impurities). We note, in passing, that the same impurity can serve as a poison for the reaction in the dark but promote the photocatalytic reaction. On the other hand, promoters that stimulate the reaction in the dark may suppress the photocatalytic effect. Examples of this are given below.

Note that in some cases the reaction does not proceed in the dark but proceeds violently under illumination. It is convenient then to characterize the photocatalytic effect by the rate of the photocatalytic reaction, g.

In catalysis one often has to deal with two or more parallel reactions. For instance, the reaction of decomposition of alcohol may lead both to dehydrogenation (discharge of H_2) and to dehydration (discharge of H_2O; see Section 5.3.2). The rate with which a catalyst drives a reaction usually depends on the direction of the reaction. This property, as we know, is called the selectivity of the catalyst. Under illumination the rates of the various parallel reactions change differently. Thus, light generally changes not only the activity of the catalyst with respect to every given reaction but the selectivity of the catalyst with respect to the reactions that occur simultaneously.

Just as in the case of the photoadsorption effect not all the photons absorbed by the semiconductor are photocatalytically active. Only those that generate photoconductivity in it can be considered photocatalytically active. This aspect was studied by Lyashenko [45] when he investigated the SiO_2, MgO, ZrO_2, ZnO, TiO_2, SnO_2, WO_3, and Al_2O_3 oxides. The first three proved to be photocatalytically inactive with respect to oxidation of CO and photoelectrically inactive as well. The others proved to be photoelectrically active and at the same time photocatalytically active.

An important characteristic of the photocatalytic effect is the quantum yield γ of a photoreaction. We define γ as the ratio of the number of reacting molecules to the number of quanta absorbed per unit time:

$$\gamma = \frac{g}{I}, \tag{7.149}$$

where I is the intensity of light of frequency ν, and g depends both on ν and I.

Note that while a surface in the dark is not quasiisolated, under illumination the same surface may become quasiisolated and remain such as long as the light remains turned on. Indeed, let us assume that under illumination some of the electrons move from donors to acceptors, which is the way in which illumination sustains the enhanced concentration of electrons and holes at the surface, so that

$$n_s = n_{s0} + \Delta n, \qquad p_s = p_{s0} + \Delta n,$$

where n_s, p_s and n_{s0}, p_{s0} are the surface concentrations of electrons and holes with and without illumination, respectively. Then, in the absence of illumination we have

$$\delta_0 = \left| \frac{p_{s0} - n_{s0}}{p_{s0} + n_{s0}} \right|,$$

while in the presence of illumination we have

$$\delta = \left| \frac{p_s - n_s}{p_s + n_s} \right| = \left| \frac{p_{s0} - n_{s0}}{p_{s0} + n_{s0} + 2\,\Delta n} \right|.$$

This may be combined with the following condition:

$$\delta \ll \delta_0 < 1,$$

i.e., illumination may make the surface quasiisolated, with the surface charge remaining unchanged. In order to avoid misunderstanding we note that by a quasiisolated surface we mean that $\delta \ll 1$ and not that the surface and bulk Fermi levels are independent of each other, since under illumination, i.e., when the electronic equilibrium is violated, the very notion of a Fermi level loses its meaning.

Below we study the mechanism of the following photoreactions:

1) Hydrogen–deuterium exchange (see Section 5.3.4)

$$H_2 + D_2 \rightarrow 2\,HD.$$

2) Oxidation of carbon monoxide (see Section 5.3.3)

$$2\,CO + O_2 \rightarrow 2\,CO_2.$$

3) Synthesis of hydrogen peroxide

$$2\,H_2O + O_2 \rightarrow 2\,H_2O_2.$$

To these reactions we will add those that have been studied experimentally:

4) Photooxidation of hydrogen on MgO (studied by Lisachenko and Vilesov [46]).
5) Photooxidation of methane on MgO (studied by the authors of [46], too).
6) Photooxidation of ethylene and propylene on TiO_2 (studied by McLintock and Ritchie [47]).
7) Photooxidation of methyl alcohol on ZnO (studied by Filimonov [48] and Noller, Schwab, Steinbach, and Venugopalan [49, 50]).
8) Oxidation of alkanes on TiO_2 (studied by Djedhri, Formenti, Juvllet, and Techner [51]).
9) Photooxidation of isopropyl alcohol on TiO and the role of adsorbed water (studied by Bickley and Jayanty [52]).
10) Photodecomposition of nitrous oxide on MgO (studied by Lisachenko and Vilesov [53]).
11) Photodecomposition of methyl alcohol on silica gels (studied by Bobrovskaya and Kholmogorov [54]).

12) Photodecomposition of hydrazine on Ge (studied by Lyashenko and Gorokhvatskii [55]).

13) Photoreduction of methyl blue on ZnO (studied by Borshchevskii and Nikolaev [56]).

This is not an exhaustive list of photocatalytic reactions occurring at semiconductor surfaces.

7.7.2. Hydrogen–Deuterium Exchange

The reaction

$$H_2 + D_2 \rightarrow 2\,HD$$

has been extensively studied (see Section 5.3.4). It was found that under illumination its rate changes considerably. Below we give a summary of the experimental results in this field.

(1) Illumination in some cases accelerates the reaction (a positive photocatalytic effect), while in other cases it inhibits the reaction (a negative photocatalytic effect). The sign and magnitude of the effect depends on the experimental conditions and the sample's prehistory. For instance, Kohn and Taylor [57] note that when zinc oxide is subjected to gamma radiation, which accelerates hydrogen–deuterium exchange, the magnitude of the effect diminishes if a donor impurity is introduced into the sample. The same authors observed a positive photocatalytic effect in silica gel samples (see [58]). The introduction of an acceptor impurity into the catalyst increased the rate of the reaction when gamma radiation was applied.

Lunsford and Leland [59] studied hydrogen–deuterium exchange on MgO crystals with V-centers. As we know, a V-center in ionic crystals is a cation vacancy with a hole localized in the neighborhood and acts as an acceptor. The two authors found, in agreement with the experiments of Kohn and Taylor, an increase in the magnitude of the photocatalytic effect with the concentrations of V-centers in the crystal.

(2) Kohn and Taylor [57] also studied the effect of light on hydrogen–deuterium exchange in barium hydride, calcium hydride, lithium hydride, and sodium hydride. For samples that had been annealed in an atmosphere of hydrogen the photocatalytic effect proved to be positive, while for the same samples annealed in a vacuum the illumination inhibited the reaction.

(3) The experimental studies conducted by Lunsford and Leland [59] and Shipman [60] have shown that the dependence of the reaction rate on the pressure in the reaction mixture is the same irrespective of whether illumination is present or not. Illumination did not change the order of the reaction.

(4) Freund [61] has studied the effect of ultraviolet radiation on the catalytic activity of zinc oxide with respect to hydrogen–deuterium exchange. The experimenter notes (see [61]) that the photocatalytic effect was positive and decreased as the temperature grew.

(5) Boreskov with collaborators [62] pointed out that the specific catalytic activity of silica gel with respect to hydrogen–deuterium exchange under gamma radiation first grows with the radiation dosage but then achieves saturation when the dosage is high.

Obviously, the theory of the photoreaction of hydrogen–deuterium exchange must explain all the regularities cited above.

Next we turn to the reaction mechanism. We will assume that the rate of the reaction in the dark, g_0, is given by (5.26):

$$g_0 = \eta_0^0 \alpha PN^*.$$

For the rate of the photoreaction, g, we have

$$g = \eta^0 \alpha PN^*, \tag{7.150}$$

where η^0 has the form (7.10). We will restrict our discussion to the case where H and D atoms at the semiconductor surface act as donors. We can then assume that

$$\eta_0^- = 0, \qquad \eta_0^+ = 1,$$

and, hence, Eq. (7.10) yields

$$\eta^0 = \frac{1}{\mu^+} \eta_0^0, \tag{7.151}$$

where μ^+ is defined in (7.65b) and (7.45).

According to (7.148), (5.26), (7.150), and (7.151), the photocatalytic effect K is given by the following formula:

$$K = \frac{\eta^0}{\eta_0^0} - 1 = \frac{1}{\mu^+} - 1. \tag{7.152}$$

Substituting μ^+ from (7.65b) and (7.45) into (7.152), we arrive at the following expression for the case of high excitation:

$$K = \exp\left(-\frac{\epsilon_v + V_s - v^+}{kT}\right) - 1. \tag{7.153}$$

Let us now compare the theoretical results with the experimental data. Below we will study the various factors in (7.153) that influence the photocatalytic effect.

7.7.2.1. Impurities

Figure 7.16 demonstrates the influence of sample treatment on the photocatalytic effect. This figure, similar to Fig. 7.15, gives the ϵ_v versus V_s curves at constant K for different values of K. The curves are numbered in order of increasing K:

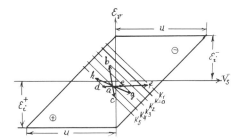

Fig. 7.16. Range of the position of the bulk Fermi level and the band bending at the surface in the event of sample treatment.

$$K_1 < 0 < K_2 < K_3 < K_4 < K_5 .$$

The parallelogram encloses the region in which (7.153) is valid. The straight line $K = 0$ divides this region into sections of positive and negative photocatalytic effects.

Introduction of an impurity into the sample (which alters both ϵ_v and V_s) moves a representative point in Fig. 7.16. Suppose that a donor impurity (which results in an increase in ϵ_v and a decrease in V_s) moves point a to point b. As Fig. 7.16 shows, K diminishes in the process. Such reduction in the photocatalytic effect resulting from introduction of donor impurities was observed by Kohn and Taylor, who studied the photoreaction of hydrogen–deuterium exchange on zinc oxide under gamma radiation.

Now suppose that an acceptor impurity (which decreases ϵ_v and increases V_s) moves point a to point c in Fig. 7.16. This, as we see, increases K, which agrees with the results of Kohn and Taylor [58], who observed an increase in the photocatalytic effect on silica gel when acceptor impurities were introduced into the catalyst, and also agrees with the results of Lunsford and Leland [59], who discovered an increase in the MgO when the concentration of V-centers (acceptors) increased.

7.7.2.2. The State of the Surface

Changes occurring in the state of the semiconductor surface accompanied by changes in V_s must also influence the photocatalytic effect. For instance, previous chemisorption of a foreign donor gas, leading to a decrease in V_s (at $\epsilon_v = $ const), must increase K (it moves point a to point d in Fig. 7.16). Contrary to this, chemisorption of an acceptor gas, leading to an increase in V_s (at $\epsilon_v = $ const), must decrease K (it moves point a to point e in Fig. 7.16).

If on a sample that has been annealed in an atmosphere of hydrogen we observe a positive photocatalytic effect, after annealing in a vacuum, accompanied by an increase in V_s, the positive effect may change to negative (point a moves to point f in Fig. 7.16). Such inversion (change of sign) of the photocatalytic effect as a result of annealing in a vacuum (with the sample having been previously annealed in hydrogen) was observed by Kohn and Taylor {57], who worked with hydrides of various metals.

7.7.2.3. Pressure

As Eqs. (5.26) and (7.150) show, the order of the reaction under illumination remains the same as in the dark. This fact agrees with the experimental data of [59, 60], which showed that illumination does not alter the order of the reaction. Thus, according to (7.148), K does not depend on P.

7.7.2.4. Temperature

Both ϵ_v and V_s can be considered as constants in a wide temperature range. As shown by (7.153), the positive photocatalytic effect $(\epsilon_v + V_s - v^+ < 0)$ must decrease and the negative effect $(\epsilon_v + V_s - v^+ > 0)$ must increase as the temperature grows. Indeed, Freund [61], who investigated the positive photocatalytic effect, found the effect to decrease as the temperature grew (in the case of hydrogen–deuterium exchange on zinc oxide under ultraviolet radiation).

7.7.2.5. The Intensity of Illumination

Substituting μ^+ from (7.8b) into the expression (7.152) for K and allowing for (7.41) and (7.42), we obtain

$$K = \frac{AI}{B + CI},$$ (7.154)

where I is the intensity of illumination. This formula agrees with the experimental data, according to which the catalytic effect K at small I's $(CI \ll B)$ grows with the intensity and for large I's $(CI \gg B)$ reaches saturation, i.e., becomes independent of I [62]. [Note that it is in the saturation range that we are dealing with high excitation levels for which formula (7.153) was found.]

7.7.3. Oxidation of Carbon Monoxide: The Experimental Data and the Reaction Mechanism

The reaction in the dark,

$$2\,CO + O_2 \rightarrow 2\,CO_2,$$ (7.155)

was studied extensively in Section 5.3.3. Illumination, as we have seen, in some cases accelerates the reaction, but in others slows it. The photoreaction (7.155) was studied in [64–66].

We will now turn to the experimental data pertinent to the photoreaction (7.155).

(1) A large number of works (e.g., see [67–77]) are devoted to the study of how the rate of the photoreaction (7.155) depends on the partial pressure of the reagents. The majority of researchers [45, 49, 50, 63, 67–69, 78–80] agree that the photooxidation of CO is of first order in CO and zero order in O_2. In particular, Dorfler and Hauffe [70] found this fact to be true in the case of a reaction mixture enriched with oxygen. The same authors found, however, that for a reaction mixture enriched with carbon monoxide the reaction is zero order in CO and first order in O_2. Steinbach [79] observed a first-order reaction in CO and a zero-order reaction in O_2 on samples of ZnO and NiO, while on samples of Co_3O_4 the reaction was first order in CO and 0.5 order in O_2. As noted by Steinbach, the order of the reaction in both reagents was the same as in the dark (which implies that illumination does not change the order of the reaction, just as in the case of hydrogen–deuterium exchange). Fujita [72], who worked with ZnO, found the reaction to be zero order in CO and 0.6 order in O_2. Lyashenko [45], who worked with various oxides, observed a zero-order reaction in both components (CO and O_2), both in the dark and under illumination.

(2) It was also found that illumination in the principal absorption band may accelerate the oxidation of CO (a positive photocatalytic effect [49, 50, 67–69, 78–81]) and may hinder this reaction (a negative photocatalytic effect [67, 81]). The magnitude and sign of the effect are determined by the experimental conditions. For instance, Romero-Rossi and Stone [67], who worked with ZnO, note that the magnitude and sign of the effect depend on the ratio of the partial pressures of O_2 and CO in the reaction mixture, P_{O_2}/P_{CO}. As this quantity increased, the magnitude of the positive effect diminished, and for a certain value of P_{O_2}/P_{CO} the researchers observed deceleration of the reaction under illumination. We note that Steinbach and Harborth [82], who also worked with ZnO, showed mass spectroscopically that in the course of the photooxidation of CO atomic oxygen is formed at the surface of ZnO.

(3) Some researchers have shown that the magnitude of the photocatalytic effect can change as a result of alloying of the sample. For instance, Romero-Rossi and Stone [67] observed an increase in the effect on ZnO when an acceptor dopant (Li) was added to the sample. An increase in the effect on Cu_2O with acceptor dopants (S and Sb) was observed by Ritchey and Calvert [71]. On the other hand, Romero-Rossi and Stone [67] found that a donor dopant added to ZnO lowered the effect. But Keier with collaborators [80] found that introduction of lithium into ZnO lowered the effect, in contrast to the data of Romero-Rossi and Stone [67]. Using aluminum as dopant lowers the effect even more.

(4) Dorfler and Hauffe [70] and Lyashenko and Gorokhvatskii [81], who studied the effect of visible and ultraviolet light on the rate of oxidation of CO on ZnO, observed a positive photocatalytic effect and found that the magnitude of the effect diminished as the temperature grew (at a temperature of about 250°C the light adsorption becomes practically inactive). Note that in [68, 70] the samples of zinc oxide were preannealed in an oxygen atmosphere, i.e., the catalyst's surface was enriched with oxygen.

(5) Lyashenko's data [45] shows that illumination does not change the reaction mechanism but only lowers the activation energy. This was found to be

true for WO_2 and other oxides. Keier and collaborators [80] arrived at the same conclusion after studying ZnO both pure and with Li and Al as dopants.

(6) Lyashenko and Gorokhvatskii [81] and Zakharenko, Cherkashin, and Keier [83] studied the quantum yield in the oxidation of CO. They found that at a given frequency of the incident light the quantum yield steadily decreased as the intensity of light increased.

Now we turn to the mechanism of reaction (7.155). For the reaction rate we have (see Section 5.3.3)

$$g = \frac{A\eta_{\bar{O}}}{B + (\eta_{\bar{C}O_2}/\eta_{CO_2}^0)} \frac{P_{CO}}{1 + (\eta_{\bar{O}}^0/C\sqrt{P_{O_2}})}, \tag{7.156}$$

where the subscripts O and CO_2 mean that the respective parameters refer to a chemisorbed oxygen atom and a chemisorbed CO_2 molecule, respectively. When no light is present, we arrive at the same formula for g_0, only now the thermal equilibrium values (3.5) must be substituted for $\eta_{\bar{O}}$, $\eta_{CO_2}^0$, and $\eta_{\bar{C}O_2}$.

We will study the case where P_{O_2} is so high (the surface is saturated with oxygen) that

$$\eta_{\bar{O}}^0 \ll C\sqrt{P_{O_2}}. \tag{7.157}$$

Under this condition (7.156) takes the form

$$g = \frac{A\eta_{\bar{O}}}{B + (\eta_{\bar{C}O_2}/\eta_{CO_2}^0)} P_{CO}, \tag{7.158}$$

where, according to (7.10),

$$\eta_{\bar{O}} = (\eta_{\bar{O}})_0 \frac{\mu_{\bar{O}}}{1 + (\eta_{\bar{O}})_0 (\mu_{\bar{O}} - 1)},$$

$$\frac{\eta_{\bar{C}O_2}}{\eta_{CO_2}^0} = \left(\frac{\eta_{\bar{C}O_2}}{\eta_{CO_2}^0}\right) \mu_{\bar{C}O_2}. \tag{7.159}$$

Here the subscript 0 means, just as before, that illumination is absent.

We will restrict our discussion, as in Section 5.3.3, to two limiting cases.

(a) The Fermi level at the surface of the sample in the dark lies high in the forbidden band between the valence band and the conduction band. We will assume that

$$(\eta_{\bar{O}}^0)_0 \ll (\eta_{\bar{O}})_0 \approx 1, \quad (\eta_{\bar{C}O_2}/\eta_{CO_2}^0)_0 \mu_{\bar{C}O_2} \gg B. \tag{7.160a}$$

This occurs when the limiting stage of the reaction is the desorption of CO_2. In this case, according to (7.159), Eq. (7.158) takes the form

$$g = A \frac{\eta_{CO_2}^0}{\eta_{\bar{C}O_2}} \frac{1}{\mu_{\bar{C}O_2}} P_{CO}. \tag{7.161a}$$

In the absence of light, we can put, according to (7.8a), $\mu_{CO_2^-} = 1$ in (7.161a). This yields

$$g_0 = A \frac{\eta^0_{CO_2}}{\bar{\eta}_{CO_2}} P_{CO}. \qquad (7.162a)$$

If we substitute (3.5) into (7.162a), we can see that g_0 increases as the Fermi level moves downward, i.e., we are dealing with the donor branch of the curve depicted in Fig. 5.12.

(b) The Fermi level at the surface of the sample in the dark lies low in the forbidden band. We will assume that

$$(\eta_{\bar{O}})_0 \ll (\eta^0_O)_0 \approx 1, \quad (\bar{\eta}_{CO_2}/\eta^0_{CO_2})_0 \, \bar{\mu}_{CO_2} \ll B. \qquad (7.160b)$$

This occurs when the limiting stage of the reaction is the adsorption of CO_2. In this case

$$g = a_2 N^*_O \, (\eta_{\bar{O}})_0 \, \mu_{\bar{O}} P_{CO}, \qquad (7.161b)$$

while in the dark

$$g_0 = a_2 N^*_O \, (\eta_{\bar{O}})_0 \, P_{CO}. \qquad (7.162b)$$

On the basis of (3.6b) we can see that g_0 decreases as the Fermi level moves downward, which means that we are dealing with the acceptor branch of the curve in Fig. 5.12.

On the donor branch, according to (7.147), (7.161a), and (7.162a), for the photocatalytic effect K we have

$$K = \frac{1}{\bar{\mu}_{CO_2}} - 1 \qquad (7.163a)$$

or, according to (7.65a) and (7.45),

$$K = \exp\left(\frac{\epsilon_v + V_s - \upsilon_{CO_2}}{kT}\right) - 1. \qquad (7.164a)$$

On the acceptor branch, according to (7.148), (7.161b), and (7.162b), we have

$$K = \bar{\mu}_{\bar{O}} - 1 \qquad (7.163b)$$

or, according to (7.65a) and (7.45),

$$K = \exp\left(-\frac{\epsilon_v + V_s - \upsilon_0}{kT}\right) - 1. \qquad (7.164b)$$

Here υ_{CO_2} and υ_O are the distances between the intrinsic Fermi level and the acceptor level of the chemisorbed CO_2 molecule and the chemisorbed O atom, respectively.

7.7.4. Oxidation of CO: Comparison of Theoretical Results with Experimental Data

We will now discuss the experimental data relevant to the photoreaction (7.155). The data will be considered in the light of the theory developed in the previous section.

7.7.4.1. Order of Reaction

If ϵ_s does not depend on the partial pressures of the reagents, the order of the reaction, as seen from (7.156), remains the same under illumination as in the dark. This fact agrees with the data of Steinbach [79] and other authors.

Under (7.157), we have [as shown by (7.158)] a first-order reaction in CO and a zero-order reaction in O_2, which agrees with the data obtained by the majority of researchers (see [49, 50, 67–69, 78–80]). Note that condition (7.157) is met with a large margin when the reaction mixture is sufficiently enriched with oxygen, as in the experiment of Dorfler and Hauffe [70].

If the reaction mixture is depleted of oxygen, then instead of (7.157) we can assume that

$$\eta_O^0 \gg C\sqrt{P_{O_2}}.$$

In this case, as shown by (7.156), we have a reaction that is first order in CO and 0.5 order in O_2. This fact was observed by Steinbach on Co_2O_3 samples. Note that when the mixture is depleted of oxygen, we can arrive at a zero-order reaction in CO and a first-order reaction in O_2, as reported by Dorfler and Hauffe [70]. It can be shown that this happens when the oxygen and carbon monoxide are adsorbed independently, with the oxygen adsorbed in the form of O_2 molecules and not dissociating into atoms (we did not consider this case). When the partial pressures of O_2 and CO are high, the result may be zero-order reaction in both components, which was observed by Lyashenko [45].

But if we wish to take into account the dependence of ϵ_s on the pressure of the reagents, then the order of the reaction in (7.156) cannot be assumed to remain constant.

7.7.4.2. Impurities

If the oxidation of CO is an acceptor reaction (region b in Fig. 5.12), the photocatalytic effect K is expressed either by (7.163b) or by (7.164b), which coincide with the appropriate formulas for hydrogen–deuterium exchange [cf., (7.164b) and (7.153)]. In this case the interpretation of the experimental data can be done with the aid of Fig. 7.16.

Suppose that our representative point for a catalyst without impurities is a in Fig. 7.16. Introduction of an acceptor impurity (ϵ_v decreases and V_s increases) moves point a into either point c or point g. This either weakens the effect or increases it, respectively. Probably, Ritchey and Calvert [71] had to do with the first case (S^+ and Sb in Cu_2O), as Romero-Rossi and Stone [67] did, while Keier and collaborators [80] dealt with the second case (Li in ZnO). Thus, the contradictory results obtained for different treatments of samples (i.e., many factors involved) may be incorporated into a single theoretical model.

Introduction of a donor impurity (ϵ_v increases and V_s decreases) moves point a to point b, which leads, as we see, to an increase in the effect. This case was studied by Romero-Rossi and Stone [67] (Cr in ZnO) and Keier and collaborators [80] (Al in ZnO).

Note that if the oxidation of CO belongs to the donor type (region a in Fig. 5.12), the experimental data can also be interpreted by employing a figure similar to Fig. 7.16.

7.7.4.3. State of Surface

As shown by (7.164a) and (7.164b) and Fig. 7.16, the photocatalytic properties of a surface at constant ϵ_v depend on V_s. This explains the dependence of the sign and magnitude of the effect on P_{O_2} and P_{CO}. Since oxygen acts as an acceptor and CO as a donor, an increase in P_{O_2}/P_{CO} increases the negative surface charge and V_s. If we assume that the acceptor branch of the reaction is involved and use Fig. 7.16, we find that as P_{O_2}/P_{CO} increases, the representative point in Fig. 7.16 moves horizontally from left to right (e.g., from point a into point e). In this case K remains positive but decreases, as reported by Romero-Rossi and Stone [67], for ZnO. For large values of P_{O_2}/P_{CO} the two researchers found that the effect changed its sign (point a moves to point f in Fig. 7.16), which was to be expected.

7.7.4.4. Temperature

As in the case of hydrogen–deuterium exchange, as long as K remains positive, its magnitude drops as the temperature grows, which is shown by (7.164a) and (7.164b) (since the parameters ϵ_v and V_s can be considered constant in a broad temperature range).

7.7.4.5. Activation Energy

According to (7.148), (7.163a), (7.163b), and (7.65a), we have

$$g - (K + 1) g_0 - g_0 \exp\left(\frac{\mp \varphi^-}{kT} \right), \tag{7.165}$$

where φ^- has the form (7.45), and for a donor reaction we must take the upper sign and substitute $v_{CO_2}^-$ for v^- in (7.45), while for an acceptor reaction we must take the lower sign and substitute $v_{O_2}^-$ for v^- in (7.45). If the reaction rate in the dark, g_0, obeys the Arrhenius equation, then, as shown by (7.165), illumination changes the reaction activation energy.

Since ϵ_v and V_s in (7.45) may generally be temperature dependent, the preexponential factor in the Arrhenius equation may also vary with temperature. This fact agrees with the experimental data (see [45, 80]).

7.7.4.6. Quantum Yield

According to (7.149) and (7.148), the quantum yield

$$\gamma = \frac{g}{I} = (K + 1)\frac{g_0}{I} \ .$$

Substituting (7.154), we obtain

$$\gamma = \left(\frac{A}{B + CI} + \frac{1}{I}\right)g_0 .$$

We see that the quantum yield steadily drops as the intensity of the light increases, which qualitatively agrees with the experimental data [80].

7.7.5. Synthesis of Hydrogen Peroxide: The Experimental Data and the Reaction Mechanism

The reaction of hydrogen peroxide synthesis (oxidation of water),

$$2H_2O + O_2 \rightarrow 2H_2O_2 ,$$

is a typical photocatalytic reaction occurring at the surface of semiconductors. In the dark the reaction either does not occur at all [73–76] or proceeds very slowly [77, 84–89]. As a result the regularities of this reaction in the dark have been little studied. Light stimulates this reaction considerably.

The photoreaction of water oxidation was discovered in 1927 by Baur and Neuweller [90] and later by a number of researchers. Korsunovskii [74–76] began with the excitonic mechanism of light absorption in analyzing the experimental data. Grossweiner [91] studied the kinetics of this reaction. Within the framework of the electronic theory of catalysis the reaction mechanism was studied in [66, 92].

Here are the experimental results pertinent to the photooxidation of water.

(1) Korsunovskii [73–76], Grossweiner [91], Stephens, Ke, and Trivich [77], and Marchem and Laidler [84] pointed out that the catalytic activity of semiconductor catalysts with respect to the oxidation of water illuminated by light from the principal absorption band first grows with the illumination period and then reaches saturation for sufficiently long periods.

(2) There are abundant data on the effect of adsorbed molecules on the photocatalytic activity of semiconductor catalysts with respect to oxidation of

water. These data show that acceptor molecules hinder the reaction [73 76, 85–87, 91], while donor molecules accelerate it [73–76].

For instance, according to Veselovskii's data [85, 86], adsorption of O_2 poisons the photoreaction. Korsunskii and Lebedev [73–76] found that the same is true of OH. Introduction into the liquid phase of an organic dye, C_6H_6, which clears the surface of hydroxyl groups (the reaction $C_6H_6 + 2OH \rightarrow C_6H_5OH + H_2O$), increases the activity of the catalyst. Adsorption of HCO_3 (acceptor molecules) hinders the reaction, as shown by Calvert et al. [87]. According to Grossweiner [91], who studied HgS samples, any negative charge at the surface hinders the reaction.

(3) Some researchers studied the dependence of the rate of water photooxidation on the conditions of preliminary treatment of the samples. Stephens, Ke, and Trivich [77] have established that preliminary heating of CdS samples in an atmosphere of sulfur (which results in CdS being enriched with an acceptor impurity) lowers the activity of the catalyst with respect to photooxidation of water.

Pamfilov, Mazurkevich, and Mushchii [93] found that heating of ZnO sample in air lowers the photocatalytic activity, while heating in vacuum, on the contrary, increases it.

On the other hand, Marchem and Laidler [84] and Veselovskii and Shub [85, 86] have shown that the photocatalytic activity of ZnO drops when the samples are annealed at high temperatures (about 1000°C) in a reducing atmosphere (such preliminary treatment leads to an increase in zinc concentration in the samples above the stoichiometric amount). In other words, the donor impurity (the superstoichiometric zinc) hinders the reaction.

Let us now turn to the possible mechanism of hydrogen peroxide synthesis. We will assume, as we did in the case of oxidation of CO, that the catalyst's surface holds chemisorbed atomic oxygen. Suppose that an H_2O molecule is adsorbed on this oxygen when the latter is in the ion-radical state. This disrupts the valence bond in the H_2O molecule. We will assume that the adsorption of H_2O proceeds as follows:

$$H_2O + \dot{O}eL \rightarrow HOOeL + \dot{H}L, \qquad (7.166a)$$

where, as usual, L stands for "lattice," eL designates a lattice electron, OeL a chemisorbed oxygen atom in the negatively charged state, and $\dot{H}L$ a chemisorbed hydrogen atom in the neutral state. Figure 7.17a depicts reaction (7.166a) by employing valence lines.

The atomic hydrogen appearing at the surface in the neutral state as a result of reaction (7.166a) may go over to a charged state and back. These transitions constitute electron transfers that result in localization or delocalization of an electron or hole at an H atom. The surface thus acquires a nonzero concentration of hydrogen atoms in a positively charged state (we will denote such atoms by HpL, where pL designates a lattice hole). Let us assume that the reaction of formation and desorption of an H_2O_2 molecule proceeds thus:

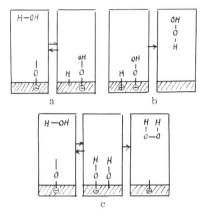

Fig. 7.17. Hydrogen peroxide synthesis;
a–c) various stages in the reaction.

$$HOOeL + HpL \rightarrow H_2O_2 + L. \qquad (7.166b)$$

The reaction is depicted in Fig. 7.17b. The mechanism of H_2O_2 synthesis [Eqs. (7.166a) and (7.166b)] discussed here is only one of the possible mechanisms. For instance, another mechanism is

$$H_2O + \dot{O}eL \rightarrow HOeL + \dot{H}OL, \qquad (7.167a)$$
$$HOeL + \dot{H}OL \rightarrow H_2O_2 + eL. \qquad (7.167b)$$

Here adsorption of an H_2O molecule is accompanied by the appearance at the surface of two hydroxyl groups OH, which recombine and form an H_2O_2 molecule, as depicted in Fig. 7.17c.

To be definite we will take the mechanism given by (7.166a) and (7.166b). [As we have seen, the mechanism given by (7.167a) and (7.157b) leads to the same result.] We will denote the partial pressures of oxygen and water vapor by P_{O_2} and P_{H_2O}. If the reaction proceeds in the liquid phase, the partial pressures must be replaced by their corresponding concentrations. Let the surface concentrations of the chemisorbed atoms O and H and groups HO_2 be N_O, N_H, and N_{HO_2}. Finally, let N_{H^+}, N_O^-, $N_{HO_2}^-$, N_H^0, N_O^0 be the surface concentrations of the respective particles in the charged and neutral states.

If we ignore the adsorption of H_2O_2 and assume that the surface coverage by HO_2 is low, (7.166a) and (7.166b) yield

$$\frac{dN_O}{dt} = a_1 P_{O_2} (N_O^* - N_O)^2 - b_1 (N_O^0)^2 - a_2 P_{H_2O} N_O^- + b_2 N_{HO_2}^- N_H^0, \qquad (7.168a)$$

$$\frac{dN_H}{dt} = a_2 P_{H_2O} N_O^- - b_2 N_{HO_2}^- N_H^0 - c N_{HO_2}^- N_H^+, \qquad (7.168b)$$

where $N_O{}^*$ is the surface concentration of the adsorption centers for O. The first term on the right-hand side of (7.168a) and (7.168b) is the number of O_2 and H_2O molecules adsorbed on a unit surface area per unit time. As in Section 7.7.3, we assume that both recombined O atoms are electrically neutral. The third term on the right-hand side of (7.168b) is the number of forming and desorbed (per unit time and per unit surface area) H_2O_2 molecules, i.e., gives the reaction rate g:

$$g = c N_{HO_2}^- N_H^+. \tag{7.169}$$

In equilibrium we have [see (7.168a) and (7.168b)]

$$a_1 P_{O_2} (N_O^* - N_O)^2 = b_1 (N_O^0)^2 - c N_{HO_2}^- N_H^+,$$

$$a_2 P_{H_2O} N_O^- = b_2 N_{HO_2}^- N_H^0 + c N_{HO_2}^- N_H^+. \tag{7.170}$$

Let us take the case where

$$b_2 N_{HO_2}^- N_H^0 \ll c N_{HO_2}^- N_H^+ \ll b_1 (N_O^0)^2. \tag{7.171}$$

Then, assuming that notations (3.4) for the O atoms hold, we find that (7.170) and (7.169) yield

$$N_O = N_O^* (1 + \eta^0 \sqrt{b_1/a_1 P_{O_2}})^{-1}, \tag{7.172a}$$

$$g = a_2 P_{H_2O} \eta^- N_O \tag{7.172b}$$

where the subscript O on η^0 and η^- is dropped. Substituting (7.172a) into (7.172b), we obtain

$$g = \frac{a_2 N_O^* P_{H_2O} \eta^-}{1 + \eta^0 \sqrt{b_1/a_1 P_{O_2}}}. \tag{7.173}$$

If, as in Section 7.7.3, we assume that condition (7.157) is met, then

$$g = a_2 N_O^* P_{H_2O} \eta^-. \tag{7.174}$$

Reasoning as we did in Section 7.7.3, we arrive at Eq. (7.163b) for the photocatalytic effect:

$$K = \mu^- - 1, \tag{7.175}$$

or for high-intensity light,

$$K = \exp\left(-\frac{c_v + V_s - v_O}{kT}\right) - 1. \tag{7.176}$$

Thus, for oxidation of water we have the same formula as for oxidation of CO in the acceptor region [cf., (7.176) and (7.164b)].

7.7.6. Synthesis of Hydrogen Peroxide: Comparison of Theoretical Results with Experimental Data

We will now compare theory with experiment. Comparing (7.176) with (7.164b) and (7.153), we can see that the dependence of the photocatalytic effect K on the position of the Fermi level ϵ_v in the bulk of a nonirradiated sample and on the surface potential V_s in the oxidation of water is the same as in the oxidation of CO (in the acceptor region) or in hydrogen–deuterium exchange. For this reason such factors as introduction of an impurity into the sample, adsorption of gases on the sample's surface, and preliminary treatment of the sample will affect the photocatalytic effect in all the above reactions in a similar manner. The dependence of K on the intensity I of the excited light must also be the same in all three reactions.

7.7.6.1. Intensity of Light

Substituting (7.8a) into (7.175) and then substituting (7.41) into the result and allowing for (7.42), we arrive at (7.154), which qualitatively agrees with the data of a number of researchers (see [73–75, 77, 84, 91]), viz., the magnitude of K steadily grows with I and tends to the saturation value. As we have seen, a similar dependence is characteristic of hydrogen–deuterium exchange.

7.7.6.2. Sample Treatment

Sample treatment leads, as a rule, to simultaneous variations in ϵ_v and V_s and, hence, to a variation in K [as can be seen from (7.176)].

To interpret the experimental data, it is convenient to turn again to Fig. 7.16. Suppose that the representative point is a in Fig. 7.16. Heating a CdS sample in a sulfur atmosphere, as done by Stephens, Ke, and Trivich [77], we enrich the sample with sulfur, which means a decrease in ϵ_v and an increase in V_s, since sulfur is an acceptor. This takes point a to, say, point g in Fig. 7.16, which weakens the effect, as observed by Stephens, Ke, and Trivich [77].

Heating ZnO samples in air may also transfer point a to point g, i.e., may lead to the effect decreasing, which agrees with data of Pamfilov, Mazurkevich, and Mushchii [93]. On the contrary, heating in a vacuum, which means enriching the samples with superstoichiometric zinc (ϵ_v increases and V_s decreases) and moving from point a to point h, increases the effect; this was observed by the above-mentioned researchers.

An opposite result was obtained, as noted earlier, by Marchem and Laidler [84] and Veselovskii and Shub [85, 86], who found that the effect decreased when the sample was enriched with zinc by annealing in a reducing atmosphere. This result can be explained if we assume that annealing in a reducing atmosphere (ϵ_v increases and V_s decreases) moves us from point a to point b (Fig. 7.16) rather than to point h.

7.7.6.3. Adsorption of Gases

Any acceptor particles that appear at a surface and impart a negative charge to the surface and, hence, increase V_s, must weaken the photocatalytic effect, according to (7.176). This was observed in the photosynthesis of hydrogen peroxide (see [73–75, 85–87, 91]). In the process point a moved to point e in Fig. (7.16), i.e., from left to right horizontally.

To conclude this chapter, a few remarks are in order concerning not only the synthesis of hydrogen peroxide, considered in this section, but photocatalytic reactions in general. As shown in Section 7.7.1, the theory provides us with a general method for studying such reactions. This was the method employed in studying the hydrogen–deuterium exchange, oxidation of CO, and synthesis of H_2O_2, but it can be used for any other heterogeneous photocatalytic reaction. For this we must know the electronic mechanism of the respective reaction in the dark. This mechanism is not always unique and is chosen according to additional information. But the laws governing the photocatalytic effect prove to depend on the electronic mechanism of the reaction in the dark. By comparing theory with experiment, we can arrive at a specific electronic mechanism.

We see that the magnitude and value of the photocatalytic effect are determined not only by the experimental conditions but by the prehistory of the sample as well, i.e., by the type of sample treatment prior to illumination. The prehistory of the sample is defined in the theory by ϵ_v and V_s, which enter into all the final expressions. Note that the rate of reaction in the dark also depends on ϵ_v and V_s. For this reason the rate of reaction in the dark and the photoreaction rate often depend on the same factors (doping, violations of stoichiometry, adsorption of foreign gases, etc.). We must note, however, that the formulas reflecting this dependence for the reaction in the dark and for the photoreaction often differ, as already noted in Section 7.7.1 and as follows from a comparison of theoretical results with experimental data (see Sections 5.3, 7.7.3, and 7.7.5).

Chapter 8

ADSORPTION AND LUMINESCENCE

8.1. BASIC FACTS ON LUMINESCENCE OF CRYSTALS

8.1.1. The Various Types of Luminescence

Luminescence is always a sequence of two acts: in the first (endothermic) the system transforms from the ground state to an excited state (or from an excited state to a more excited state), and in the second (exothermic) it returns from the excited state to the ground state (or less excited state) accompanied by emission of a photon.

Depending on the source that supplies the energy for exciting the system, the energy which then transforms into the luminescent radiation, we must distinguish between the following four types of luminescence:

(a) Photoluminescence (excitation by light).
(b) Electroluminescence (excitation by electric field).
(c) Cathodoluminescence (excitation by electron flux).
(d) Chemiluminescence (excitation as a result of chemical reaction).

Many gases, liquids, and solids are capable of emitting radiation via luminescence. Such substances are called luminophors. In what follows we will discuss only crystalline luminophors, known as phosphors.

In photoluminescence we must distinguish between two mechanisms: the recombination mechanism and the excitonic mechanism. With the first we are dealing with an electron transition from a low-lying energy level to a high energy level (the excitation act), after which the electron goes back to a lower level (the luminescence act). With the second an exciton is created in the phosphor (the excitation act) and then it annihilates (the luminescence act).

367

Usually in the literature on the theory of luminescence of solids the phosphor is taken as an infinite (limitless) crystal. In this way the very idea of surface effects in luminescence is excluded from the start. Such a phosphor exists, of course, only in the mind of the theoreticians. Actually, a phosphor is a crystal bounded by a surface and in contact with the ambient (e.g., the gaseous) phase.

The role of the surface in luminescence is twofold. First, we are dealing with heterogeneous chemiluminescence, i.e., luminescence of a solid excited by chemical processes at the surface of the solid. Second, we must keep in mind the effect of these processes on the common luminescence. In accordance with this, the present chapter is devoted to the effect of adsorption on the intensity and spectral composition of the luminescent radiation as well as the following processes of heterogeneous chemiluminescence:

(1)   Recombination of the atoms or radicals chemisorbed by the surface. This process can be accompanied by emission of luminescent radiation (radical-recombination luminescence).

(2)   Chemisorption accompanied by luminescent radiation from the adsorbent (adsorption luminescence).

(3)   A heterogeneous catalytic reaction accompanied by luminescent radiation from the catalyst (catalytic luminescence).

The mechanism of the heterogeneous luminescence processes is straightforward, viz., the energy liberated in the chemical processes at the surface is used to excite the crystal and then, when the crystal returns to the ground state, is released in the form of light quanta.

8.1.2. Luminescence Centers

We will study phosphors whose ability to luminesce is due to the presence of impurities. These impurities, which are responsible for luminescence, are known as activators. The nature of the activator determines the spectral composition of the luminescent radiation, while the activator concentration determines the intensity of the radiation. Note that an activator is understood as an impurity in the broad sense of the word, i.e., in the meaning attributed to this term in semiconductor physics. These may not be chemically foreign particles impregnated into the crystal. Any structural defects, e.g., local imperfections in the periodicity of the lattice, may do. For instance, vacancies or atoms or ions of the lattice proper shifted from the lattice sites to the interstices may serve as activators. Structural defects responsible for luminescence are called activator atoms or luminescence centers.

The excitation of a phosphor may be achieved by (a) excitation of the luminescence centers and (b) ionization of these centers. In the first case the excitation is localized at separate points in the crystal and can be considered a local effect. In the second case the excitation is distributed over the entire crystal and constitutes a collective effect.

We will start with the first case (localized excitation). Let AL stand for a luminescence center in the lattice (L stands for "lattice") in the ground state and A^*L for the center in an excited state. For photoluminescence we have

$$AL + h\nu \rightarrow A^*L \quad \text{(the excitation act)},$$
$$A^*L \rightarrow AL + h\nu \quad \text{(the luminescence act)}.$$

When the luminescence centers are acceptor–donor pairs in the crystal (often the case), each pair consisting of an acceptor and a donor defect in close proximity, the excitation of such a pair means that an electron is being transferred from the acceptor particle to the donor particle, while luminescence corresponds to the opposite transition (from donor to acceptor).

The explanation of this process is the following. Let AL and DL be the acceptor and donor defects in the lattice, both being in the electrically neutral state, and let AeL and DpL be the same defects with an electron and hole, respectively, localized at them. If the ionization energy of the donor defect is lower than the electron affinity of the acceptor defect, the state AL + DL is metastable, while the state AeL + DpL serves as the ground state. A Coulomb attraction emerges between the AeL and DpL defects, which brings these defects closer together and makes them form the acceptor–donor pair AeL·DpL. Neutralization of the acceptor and the donor particle constituting such a pair means its excitation. For photoluminescence we then have

$$AeL \cdot DpL + h\nu \rightarrow AL \cdot DL \quad \text{(the excitation act)},$$
$$AL \cdot DL \rightarrow AeL \cdot DpL + h\nu \quad \text{(the luminescence act)}.$$

Here excitation and luminescence mean an electron transfer from one partner in the pair to the other and back.

Let us now examine the case where the phosphor is excited by ionizing the luminescence center. The crystalline structure of the phosphor enables us to describe the ionization and neutralization of luminescence centers in terms of the energy band structure used in semiconductor theory. Such a band diagram is given in Fig. 8.1, where C is the conduction band, V the valence band, and A or D the activator level. Depending on its nature, the activator may be either an acceptor (level A) or a donor (level D); i.e., in the ground state it can be either free of an electron or occupied by an electron. Ionization of the activator in the first case means localization of a negative charge (an electron), while in the second it means localization of a positive charge (a hole).

In photoluminescence, the ionization of an activator may be due to the absorption of a light quantum directly by a luminescence center (called extrinsic light absorption). If we denote a free electron and hole in a lattice (i.e., an electron in the conduction band and a hole in the valence band) by eL and pL, the reaction of activator ionization can be written in the following form:

$$DL + h\nu \rightarrow DpL + eL, \tag{8.1}$$

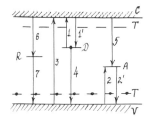

Fig. 8.1. Energy band structure for the ionization and neutralization of luminescence centers.

if the activator is a donor, or

$$AL + h\nu \to AeL + pL,\tag{8.2}$$

if the activator is a acceptor. The reverse reactions, which lead to neutralization of the ionized activator, are accompanied by the emission of photons, or luminescence:

$$DpL + eL \to DL + h\nu,\tag{8.1'}$$
$$AeL + pL \to AL + h\nu.\tag{8.2'}$$

The corresponding electron transitions are shown in Fig. 8.1 by vertical arrows 1, 1' and 2, 2'.

Activator ionization with subsequent luminescence may be achieved not only as a result of light absorption directly by luminescence centers but also by light absorption by the lattice proper, i.e., the regular lattice atoms or ions (called intrinsic light absorption). In this case an electron goes over from the valence band to the conduction band (a hole appears in the valence band and an electron in the conduction band):

$$L + h\nu \to eL + pL.\tag{8.3}$$

The endothermic act (8.3), which transfers the crystal into an excited state, is followed by two subsequent exothermic acts, which return the crystal to its ground state:

$$DL + pL \to DpL,\tag{8.4}$$
$$DpL + eL \to DL,\tag{8.1'}$$

if the activator is a donor, or

$$AL + eL \to AeL,\tag{8.5}$$
$$AeL + pL \to AL,\tag{8.2'}$$

if the activator is an acceptor. The corresponding electron transitions are depicted in Fig. 8.1 by the appropriate arrows. The transition 4 or 5 is the ionization of the activator, which is an exothermic transition 1 or 2, while the transition 1′ or 2′ is the neutralization of the activator. The transitions 1′ or 4 and 2′ or 5, or each of these, can be radiative, i.e., transitions in which electromagnetic radiation (luminescence) is emitted.

Note that a phosphor may have several types of activators rather than one type. This is reflected in Fig. 8.1 by the various levels in the forbidden gap between the bands (two types of such levels are shown in Fig. 8.1: levels A and D). In the luminescence spectrum of a given phosphor each activator has a characteristic band.* In the luminescence spectrum of a phosphor with several activators there are several bands, which may overlap to some extent. It is important to stress that we have bands instead of discrete spectral lines, since of the two states between which the radiative transition occurs one has a more or less broad energy band (the conduction band or the valence band) instead of a discrete level. Allowing for lattice vibrations leads to additional broadening of the spectral bands.

8.1.3. Traps and Quenching Centers

In addition to luminescence centers (of one or several types), a phosphor crystal may contain traps, which are defects that can trap electrons or holes released in the excitation process. These traps do not take direct part in the luminescence, but nevertheless play an important role in such processes. Each trap has a corresponding trap level, which may be either an acceptor level (a trap for an electron) or a donor level (a trap for a hole). The first type always lies near the conduction band, and the second near the valence band (levels T in Fig. 8.1).

An electron that finds itself in the conduction band or a hole created in the valence band as a result of excitation of the phosphor crystal may be trapped even before recombination. The buildup of electrons or holes on the trap levels means that the system shifts to a metastable state, and to leave this state it requires a certain activation energy. Indeed, an electron can return to the conduction band or a hole to the valence band as a result of heating or other external action, such as irradiation with infrared light. The position of the trap levels in the energy spectrum determines the duration of the afterglow of the phosphor crystal, i.e., the luminescence after the excitation has ceased. The concentration of these levels determines the maximal value of the integrated luminescence, which is the energy accumulated in the phosphor under excitation.

In addition to luminescence centers and traps, phosphors contain, as a rule, defects that are centers of radiationless recombination. In the energy spectrum such centers are represented by levels that, depending on the nature of the centers, may be either acceptors or donors (level R in Fig. 8.1). The recombination level R

---

*There are cases where one type of activator has several absorption and emission bands instead of one.

is the intermediary through which an electron from the conduction band recombines with a hole in the valence band (transitions 6 and 7).

Thus, there are two ways in which free electrons and holes, created as a result of transitions 3 (see Fig. 8.1), can recombine, viz., radiative recombinations through activator levels A or D (transitions 1′ and 4 or 2′ and 5), and radiationless recombination through the levels R (transitions 6 and 7). Both channels compete with each other. Obviously, recombination through R leads to quenching of luminescence (external quenching).

Apparently, luminescence quenching caused by adsorbed particles, often observed, is due to the presence of radiationless recombination centers that appear at the surface of phosphor crystal in the adsorption process. Note that recombination through A or D is not always due to radiation. There is always a (nonzero) probability of such recombination being radiationless. In other words, among the various types of recombination that occur through luminescence center levels some fraction may be radiationless (internal quenching).

The effectiveness of a phosphor crystal can be characterized by a quantity called the energy yield of luminescence, which is defined as the ratio of the amount of energy released during luminescence to the energy needed for excitation. Often the quantum yield is used instead of the energy yield. In photoluminescence the quantum yield is the ratio of the emitted photons (luminescence photons) to the number of adsorbed photons (excitation photons). In chemiluminescence the quantum yield is defined as the number of photons emitted per single act of chemical transformation.

For a given phosphor crystal the luminescence energy yield depends on the external conditions (temperature, pressure of the gas in whose atmosphere the phosphor is placed, etc.), the mode of excitation, and the external factors that are determined by the prehistory of the sample. The intensity of luminescence, therefore, depends on all the above-mentioned factors. For instance, every phosphor has a characteristic temperature called the luminescence quenching temperature. If the phosphor is heated above this temperature, the luminescence intensity begins to decrease. This is caused by the fact that the probability of radiationless transitions increases with temperature.

We have devoted this section solely to electronic processes in the bulk of a phosphor on the assumption that the phosphor's lattice is limitless, i.e., we have ignored surface effects. Actually, the surface of a phosphor, as noted earlier, is in contact with the ambient (e.g., with the gaseous phase) and physicochemical processes at the surface affect the luminescent properties. Only recently has the study of the role of the surface on luminescent properties gained impetus, but the prospects are highly promising. The next section deals with the effect of the surface on the luminescence of semiconductors.

8.2. THE EFFECT OF ADSORPTION ON LUMINESCENCE

8.2.1. The Various Mechanisms of Influence of Adsorption on Photoluminescence

The role that adsorption plays in the luminescence of semiconductors is as yet to be determined. A review of the scant work is given in [1–4]. The majority of researchers have established that adsorption leads to quenching of luminescence. In some cases, however, it was found that adsorption leads to a rise of luminescence emission. Sometimes adsorption resulted in the appearance of a new luminescence band, while the principal band was quenched (to a greater or lesser degree).

Let us first start with recombination luminescence. In this case there are two ways in which adsorption may affect luminescence.

(1) As we know, when chemisorbed particles appear on a surface, the surface becomes charged. Thus, the luminescence centers find themselves in the electric field of the surface charge. At the same time it is well known that an external electric field changes the intensity of luminescence (the luminescent field effect [5, 6]). With this mechanism by which adsorption influences luminescence, chemisorbed acceptor and donor particles must act in the opposite directions, i.e., if one type of particle increases luminescence, the other must decrease it. This mechanism will be called the field mechanism.

(2) Chemisorbed particles may act as surface recombination centers, with the recombination either radiationless or otherwise. When we are dealing with radiative recombination, the appearance of chemisorbed particles at the surface leads to a new luminescence band, while in radiationless recombination this does not happen. In both cases the intensity of the principal band drops. In this mechanism of adsorption influencing luminescence, the quenching of the principal band occurs via both acceptor and donor particles. We will call this mechanism the recombination mechanism.

Generally the two mechanisms operate simultaneously. In some cases one mechanism is predominant, while the other is obscured. In Section 8.2.2 we will find the criteria for the field and recombination mechanisms, while in Section 8.2.3 we will study the relative roles of these mechanisms using red and infrared luminescence emissions of CdS as examples.

We will now turn to excitonic luminescence. In Section 8.2.5 we will see what effect adsorption has on the intensity of such luminescence. The effect of an external electric field on excitonic luminescence has been studied in [7–10]. The first to study excitonic luminescence (the infrared luminescence of Cu_2O may serve as an example) were Karkhanin and Lashkarev [11].

According to Karkhanin and Lashkarev [11], luminescence emission is the result of the annihilation of the excitons generated by the light. This annihilation occurs at the ionized acceptor centers in the bulk of the crystal. In the case of Cu_2O the copper vacancy with a hole localized at it serves as such a center. The annihilation of excitons at ionized acceptor centers is assumed to be radiationless. In addition, in [7] it was assumed that radiationless annihilation of excitons occurs also at surface centers, whose nature, however, was not specified. In what follows we will remain within the scope of this mechanism.

There are three ways in which adsorption may influence excitonic luminescence.

(1) The variation of band bending, which is the result of the surface being charged in chemisorption, changes the relationship between the number of neutral and ionized annihilation centers in the bulk of the crystal. This in turn changes the ratio between the number of radiative and radiationless excitonic annihilation acts. The adsorption of donors and that of acceptors change the luminescence emission intensity in opposite directions.

(2) The variation of band bending in chemisorption changes the conditions for radiationless surface annihilation of excitons at the intrinsic defects of the surface, i.e., changes the external quenching of luminescence. As we will see below, this quenching may be either increased or decreased, depending on whether acceptor or donor particles are adsorbed.

(3) Centers created during adsorption may serve as additional annihilation centers for the excitons. If annihilation at such centers is radiationless, then according to this mechanism of adsorption influencing the luminescence process, luminescence is quenched irrespective of whether the adsorbed particles are donors or acceptors.

8.2.2. Recombination Luminescence: Statement of the Problem

Let us see how adsorption affects the intensity of recombination luminescence. We will take a phosphor crystal with two types of recombination centers: luminescence centers and external quenching centers. The centers of both types are assumed to be concentrated in a thin surface layer, so that we can replace them by certain effective surface recombination centers. We will assume that we are dealing with adsorbed particles of only one type.

The luminescence intensity I_L can be defined as the number of radiative transitions per unit time per unit surface area. Then, according to the well-known formula of Stevenson and Keyes [12],

$$I_L = N^A \frac{C_n^A C_p^A (n_s p_s - n_{0s} p_{0s})}{C_n^A (n_s + n_1^A) + C_p^A (n_s + p_1^A)} \ . \tag{8.6}$$

where N^A is the concentration of surface luminescence centers, C_n^A and C_p^A are the cross section of capture of an electron and hole by an A center, n_s and p_s are the free electron and hole concentrations at the surface under illumination, n_{0s} and p_{0s} are the same quantities in the dark, and n_1^A and p_1^A are the constants whose exact value is of no interest here. The excitation is assumed to be so high that

$$\begin{cases} n_s \gg n_{0s}, & p_s \gg p_{0s}, \\ n_s \gg n_1^A, & p_s \gg p_1^A. \end{cases}$$

Then instead of (8.6) we will have

$$I_l = N^A \; \frac{C_n^A C_p^A}{C_n^A p_s^{-1} + C_p^A n_s^{-1}} \; . \tag{8.7a}$$

Similarly, for the intensity of external quenching, I_R, we have

$$I_R = N^R \; \frac{C_n^R C_p^R}{C_n^R p_s^{-1} + C_p^R n_s^{-1}} \; , \tag{8.7b}$$

where N^R is the surface concentration of quenching centers, and $C_n{}^R$ and $C_p{}^R$ are the R-center capture cross section for electrons and holes, respectively. Obviously, I_L and I_R are related in the following manner:

$$I = I_L + I_R, \tag{8.8}$$

where I is the number of photons impinging every second on a unit surface area of the crystal.

If the recombination flux I_R through the quenching center is accompanied by radiation, there appears an additional luminescence band due to the adsorbed particles. In this case Eq. (8.7b) gives the intensity of the additional band, while Eq. (8.7a) gives the intensity of the principal band. In what follows we will study the effect of adsorption on the principal band.

Substituting (8.7a) and (8.7b) into Eq. (8.8) and solving this equation for p_s, we obtain

$$p_s = p_s(n_s; N^A; N^R), \tag{8.9}$$

which after substituting (8.9) into (8.6) yields

$$I_L = I_L(n_s; N^A; N^R). \tag{8.10}$$

The luminescence intensity I_L is affected by an external electric field through changes in n_s (the luminescent field effect), while adsorption acts both through changes in n_s (the field mechanism) and in N^R (the recombination mechanism). We see that the effects of adsorption and of an external electric field on I_L are the same for equal band bending (i.e., equal n_s) only if in (8.10) we ignore the dependence of I_L on N^R.

8.2.3. Recombination Luminescence: Limiting Cases

We must distinguish between two limiting cases, viz., when the luminescence quantum yield is low ($I_L \ll I_R$) and when the luminescence quantum yield is about unity ($I_R \ll I_L$). We start with the first case.

(1) We put $I_L \ll I_R$. Then (8.8) and (8.7b) yield

$$p_s^{-1} = \frac{N^R C_p^R}{I} - \frac{C_p^{'R}}{C_n^R} n_s^{-1} . \tag{8.11}$$

Substituting (8.11) into (8.7a) and assuming that

$$\frac{C_p^A}{C_n^A} \gg \frac{C_p^A}{C_n^A} , \tag{8.12}$$

which, according to [13], occurs for CdS, say, we then obtain

$$I_L = \frac{I_s N^A C_n^A C_p^A}{N^R C_p^R C_n^R + I C_n^R n_s^{-1}} \tag{8.13}$$

Note that the first term in the denominator on the right-hand side of Eq. (8.13), which appears due to adsorption, causes luminescence to quench. Since $C_n{}^A$, which depends on the origin of the luminescence centers, is present in this term, adsorbed particles of the same nature (i.e., having the same values of N^R) and of the same concentrations will produce different quenching for different luminescence bands.

Note, further, that the rise of luminescence emission caused by an electron field (owing to the increase in n_s) is always more pronounced than the rise caused by adsorption (for the same band bending) because the quenching term in (8.13) contains N^R. On the other hand, the quenching caused by a field (owing to the decrease in n_s) is always less pronounced than the quenching caused by adsorption (for the same band bending and for the same reasons).

Let us consider two cases where formula (8.13) simplifies considerably.

(a) Suppose that the electron concentration in the surface plane, n_s, is so low or that the surface coverage of the adsorbed particles, N^R, is so small or, finally, that the illumination intensity I is so large that

$$I C_p^A n_s^{-1} \gg N^R C_n^A C_p^R . \tag{8.14}$$

Then, according to (8.13),

$$I_L = N^A C_n^A n_s . \tag{8.15}$$

Here the luminescence intensity does not explicitly depend on N^R, and the effect of adsorption on luminescence is reduced to the effect of the electric field that emerges in adsorption (the field mechanism).

In this case adsorption and external electric fields produce the same changes in I_L (for the same band bending), with the adsorption of donors and of acceptors changing I_L in the opposite directions (donors increase I_L while acceptors decrease I_L).

(b) Let us now assume that the surface electron concentration n_s is so high or the surface coverage of adsorbed particles, N^R, is so large or, finally, the illumination intensity I is so small that

$$IC_p^A n_s^{-1} \ll N^R C_n^A C_p^R . \tag{8.16}$$

Then (8.13) yields

$$I_L = \frac{IN^A C_p^A}{N^R C_p^R} . \tag{8.17}$$

Here the adsorbed particles act as recombination centers (adsorption affects luminescence via the recombination mechanism). In this case adsorption always results in quenching, irrespective of whether the adsorbed particles are donors or acceptors.

(2) We put $I_R \ll I_L$. Substituting (8.5) into (8.8), we arrive at an equation for determining p_s:

$$I = \frac{N^A C_n^A C_p^A}{C_n^A p_s^{-1} + C_p^A n_s^{-1}} , \tag{8.18}$$

where

$$p_s^{-1} = \frac{N^A C_p^A}{I} - \frac{C_p^A}{C_n^A} n_s^{-1} \tag{8.19}$$

Substituting (8.19) into the expression (8.7) for I_R and taking into account condition (8.12) (the case with CdS), we obtain

$$I_R = N^R C_p^R p_s . \tag{8.20}$$

In the case at hand, I_R constitutes only a small correction to I_L and the effect of adsorption on the luminescence intensity can be found via the formula

$$I_L = I - I_R, \tag{8.21}$$

where I_R is given by (8.20) (here we have used a perturbation theory expansion, assuming that I_R is a first-order correction). On the basis of (8.20) and (8.21) we can say that the adsorption of both acceptor and donor particles leads to the quenching of luminescence; since p_s is larger for acceptor particles than for donor particles, acceptor particles lead to stronger quenching than donor particles, all other conditions being the same (the same values of N^R and C_p^R).

Thus, the effect that adsorption has on the luminescence intensity in this case is determined, as we have seen, by the free hole concentration in the surface plane, p_s, and the surface coverage of the adsorbed particles, N^R.

8.2.4. Recombination Luminescence: Experimental Data

The experiments that we will be studying here were conducted on plate-like single crystals of CdS grown from the gaseous phase by synthesis [3]. The adsorbates were water vapor, atmospheric air, oxygen, and ozone.

Luminescence studies were carried out in the red ($\lambda_{max} = 0.76\text{--}0.78\,\mu$m) and infrared ($\lambda_{max} = 1.03\,\mu$m) bands, where luminescence is sure to be of the recombination origin. The measurements usually involved single crystals of CdS with intense luminescence emission in both bands.

The following measurements were carried out:

(a) The variations in the luminescence intensity caused by an external electric field (the luminescent field effect). This involved a capacitor consisting of the sample, mica, and a semitransparent field electrode. The illumination of the same and the recording of the luminescent radiation were measured on the side where the field electrode was. Measurements were carried out for the same field strengths in air and in a vacuum (as low as 10^{-5} Torr) in a range of field strengths up to 10^5 V/cm. The magnitude of the effect was characterized by the fractional change in luminescence intensity at the maximum of the band when the field was switched on.

(b) The variations in the luminescence intensity caused by adsorption. Measurements were carried out in a vacuum (as low as 10^{-1} to 10^{-5} Torr) and in atmospheres of various gases. The magnitude of the effect was characterized by the change in luminescence intensity at the maximum of the band.

(c) The variations in the sample's conductivity caused by adsorption. The sensitivity of the circuit enabled measuring 1% variations.

These three types of measurements made it possible to compare the variations in luminescence intensity caused by an electric field with those caused by adsorption when variations of conductivity were the same. When no surface recombination centers emerge during adsorption (the field mechanism), equal variations in the conductivity imply equal variations in band bending.

The results of the measurements were as follows.

(1) All adsorbates investigated caused a reduction in the conductivity of CdS crystals, i.e., acted as acceptors (since CdS crystals are n-type semiconductors).

(2) All adsorbates investigated caused quenching of luminescence. An example is shown in Fig. 8.2, which shows the luminescence spectra of a CdS crystal taken in a 10^{-5} Torr vacuum (curve 1) and an oxygen atmosphere (curve 2). As shown by this figure, the red band was quenched much more strongly than the infrared band (quenching in the red band for some samples reached 70%).

(3) Under adsorption the luminescence spectrum did not change. No new bands in the spectral interval under investigation were found.

(4) Under an electric field and under adsorption the quenching of luminescence proved to be the same for the variations in conductivity for some samples, while for others samples the quenching under adsorption proved to be stronger than under a field.

These results lead to the following conclusions. First, the adsorbed particles do not serve as luminescence centers and under adsorption of the substances studied the surface of CdS acquires acceptor levels which in a number of cases act as radiationless recombination centers. Second, for some samples the adsorbed particles influence the luminescence intensity via the field mechanism [see (8.17)], while for others both the field and recombination mechanisms contribute

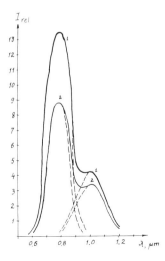

Fig. 8.2. Luminescence spectra of a CdS crystal taken in a 10^{-5}-Torr vacuum (curve 1) and an oxygen atmosphere (curve 2).

[see (8.13)]. Third, the fact that adsorption influences the red band more strongly than the infrared band is connected with the difference in the capture cross sections for electrons, C_n^A, on red and infrared luminescence centers. As shown by Shteikman, Ermolovich, and Belen'kii [13], the value of C_n^A for red luminescence is much larger than for infrared luminescence, which implies that the red band, according to (8.15), must be quenched more strongly than the infrared band.

8.2.5. Excitonic Luminescence: Statement of the Problem

Let us take a semiinfinite crystal that occupies the half space $0 \leq x \leq \infty$. Let the concentration of acceptor defects, which serve as luminescence centers, be N, and the concentrations of the neutral and charged centers under electronic equilibrium be N^0 and N^-, respectively. Obviously,

$$N = N^0 + N^-. \tag{8.22}$$

We will also assume that n is the exciton concentration in the crystal, and β^0 and β^- are the probabilities of exciton annihilation at a neutral and a charged center, respectively (in other words, β^0 and β^- are the radiative and radiationless annihilation probabilities).

Assuming, as in [11], that light absorption does not violate the equilibrium distribution of electrons on the levels, we can write the following formula for the luminescence intensity:

$$I_L = \beta^0 \int_0^\infty N^0(x)n(x)dx, \tag{8.23}$$

where both factors, $N^0(x)$ and $n(x)$, change under adsorption. In this way adsorption influences I_L.

Let us study the role of $N^0(x)$ in Eq. (8.23). Adsorption of acceptor particles bends the bands upward and, hence, makes $N^0(x)$ grow, while adsorption of donor particles makes $N^0(x)$ fall. Thus, if $n(x)$ does not change or changes very little under adsorption, the adsorption of an acceptor gas leads to a rise of luminescence emission, while the adsorption of a donor gas quenches luminescence.

We will now turn to the role of $n(x)$ in (8.23). To this end we will study the continuity equation for excitons:

$$g(x) - \{\beta^0 N^0(x) + \beta^- N^-(x)\} \, n(x) - \operatorname{div} j(x) = 0. \tag{8.24}$$

The first term on the left-hand side represents the number of excitons created per unit time per unit volume, the second term is the number of excitons annihilated per unit time per unit volume, and $j(x)$ is the diffusion flux of the excitons. Obviously,

$$g(x) = \kappa I \exp(-\kappa x), \tag{8.25}$$

$$j(x) = -D \frac{dn}{dx}, \tag{8.26}$$

where κ is the light absorption coefficient, I the intensity of the incident light (the number of incident photons per unit surface area per unit time), and D the diffusion coefficient for the excitons. We will assume that

$$\beta^- = \beta^0, \tag{8.27}$$

which, as shown in [8], agrees with the experimental data. On the basis of (8.27), (8.26), (8.25), and (8.22) we can write the continuity equation (8.24) in the following simple form:

$$D \frac{d^2 n}{dx^2} - \beta^0 N n + \kappa I \exp(-\kappa x) = 0. \tag{8.28}$$

The boundary conditions for this problem are

$$n(\infty) = 0, \qquad -j(0) = s n(0), \tag{8.29}$$

where s is the excitonic surface annihilation rate. Solving Eq. (8.28) with the boundary conditions (8.29) yields

$$n(x) = \frac{\kappa I}{D(\kappa^2 - \mu^2)} \left[\frac{D\kappa + s}{D\mu + s} \exp(-\mu x) - \exp(-\kappa x) \right], \tag{8.30}$$

with $\mu = (\beta^0 N/D)^{1/2}$. This implies that

$$\frac{dn(x)}{ds} < 0 \tag{8.31}$$

for all κ and μ. We see that $n(x)$, which is the exciton concentration in any given plane x, decreases as the excitonic surface annihilation rate s increases. Thus, if $N^0(x)$ in (8.23) does not change or changes little under adsorption, the effect of adsorption on luminescence intensity is reduced to the effect of adsorption on s. The value of s, therefore, will characterize the external quenching due to the surface.

Let us clarify the meaning of s. We write $s = s_A + s_B$, where s_A reflects the annihilation on the adsorbed particles, and s_B on the intrinsic defects of the surface. In the absence of adsorbed particles, $s_B = 0$ and $s_B = s_B^0$. If we assume that Δs is the variation of s under adsorption, then we have

$$\Delta s = s_A + \Delta s_B, \tag{8.32}$$

where $\Delta s_B = s_B - s_B^0$ is the variation in the annihilation rate on intrinsic defects under adsorption. The quantity Δs_B is due to the variation in the relative contents of neutral and charged intrinsic defects in the field created by the adsorbed particles and is present on the luminescent field effect. According to [8–10],

$$\begin{cases} \Delta s_B < 0 & \text{for adsorption of acceptors,} \\ \Delta s_B > 0 & \text{for adsorption of donors.} \end{cases} \tag{8.33}$$

As for s_A, it is always positive:

$$s_A > 0. \tag{8.34}$$

Thus, if $s_A \ll |\Delta s_B|$, then, according to (8.33), (8.32), and (8.31), the adsorption of acceptors increases $n(x)$, while the adsorption of donors decreases it. But if $s_A \gg |\Delta s_B|$, the adsorption of both acceptors and donors decreases $n(x)$.

8.2.6. Excitonic Luminescence: A Discussion

Summarizing the results concerning the effect of adsorption on $N^0(x)$ and $n(x)$ in Eq. (8.23), we arrive at the following conclusions:

(1) If s_A is large, the adsorption of any gas quenches luminescence. This is due to an additional channel for the surface radiationless annihilation of excitons (annihilation on adsorbed particles).

(2) If s_A is so small that $s_A \ll |\Delta s_B|$, then adsorption of an acceptor gas leads to a rise of luminescence emission, while adsorption of a donor gas leads to the quenching of luminescence. This is due to the simultaneous increase in the values of $N^0(x)$ and $n(x)$ in (8.23) under adsorption of acceptors and their simultaneous decrease under adsorption of donors.

(3) In the intermediate case, where Δs_B is comparable (in magnitude) to s_A, we have relatively strong quenching under adsorption of donors and a relatively weak rise of luminescence emission or weak quenching (all other things being equal) under adsorption of acceptors. This is due to the contribution which the

radiationless annihilation of excitons at the adsorbed particles makes to the quenching directly.

Let us now turn to the experimental data on the subject (see [4]). Macrocrystalline samples of Cu_2O were used in these studies, and the adsorbates were water vapor, atmospheric air, oxygen, and ozone. Measurements were done in the infrared luminescence band of Cu_2O ($\lambda_{max} = 0.96\,\mu$m). The experimental device and measurement technique are described in [7].

The following measurements were carried out:

(a) Variations in the luminescence intensity caused by an external electric field. Variations in conductivity were measured simultaneously.

(b) Variations in the luminescence intensity caused by adsorption. Variations in conductivity under adsorption were measured simultaneously.

Thus, it was possible to compare the variations in luminescence intensity caused by an electric field with those caused by adsorption when the variations in band bending were the same.

The measurements revealed the following:

(1) Adsorption of oxygen and ozone causes an increase in conductivity, while adsorption of water vapor causes a decrease. Since Cu_2O is a p-type semiconductor, this means that oxygen and ozone act as acceptors, while molecules of water act as donors.

(2) All adsorbates investigated (both donors and acceptors) proved to quench luminescence. It must be noted here that when the adsorbates were evacuated, the initial values of conductivity and luminescence intensity were restored. Stationary values were established in a vacuum of 5×10^{-2} Torr. This proves that in this case adsorption was reversible.

(3) Adsorption practically did not change the spectral distribution of the luminescence emission, changing only the luminescence intensity, as shown, for instance, in Fig. 8.3. The figure depicts the luminescence spectrum of H_2O in a vacuum (curve 1) and in an atmosphere of water vapor (curve 2). No new luminescence bands were observed in the range 0.5–1.3 μm.

(4) Quenching caused by adsorption proved to be much stronger than quenching caused by an external electric field (for the same values of band bending).

Combining these results with those of Section 8.2.5, we can conclude that the main way in which adsorption affects the excitonic luminescence of Cu_2O is to increase the surface excitonic annihilation rate by producing radiationless annihilation centers of an adsorption origin.

8.3. THE BASIC LAWS OF RADICAL-RECOMBINATION LUMINESCENCE

8.3.1. The Spectral Composition of Radical-Recombination Luminescence Emission

Radical-recombination luminescence (RRL) belongs to the class of heterogeneous chemiluminescence. The term "heterogeneous chemiluminescence," in-

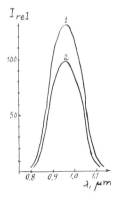

Fig. 8.3. Luminescence spectra of H_2O in a vacuum (curve 1) and in an atmosphere of water vapor (curve 2).

troduced by V. V. Styrov, means luminescence that accompanies a chemical reaction taking place at an interface (e.g., at the boundary between a solid and a gas). In the event of radical-recombination luminescence the atom or radical R adsorbed at the surface meets with a similar atom or radical coming from the gaseous phase or migrating along the surface, recombines with it, and forms a saturated molecule R_2, which then is desorbed. This process is accompanied by emission of a light quantum (luminescence). A similar mechanism lies at the base of candoluminescence, which means luminescence in flames, which always contain gas particles (atoms or radicals) with unsaturated valences [2].

Experimental study of radical-recombination luminescence spectra has revealed that in most cases they are composed of the same bands as other types of luminescence (photoluminescence and cathodoluminescence, to name some), although the energy distribution in RRL spectra with several bands may differ from that of the other types [1, 14–17]. Such bands will be called principal. Note that often there is a long-wave shift in these bands accompanied by broadening compared to the corresponding bands in photoluminescence emission of the same phosphors [18]. This can be explained by the special nature of the surface centers and by the local heating of the phosphor surface in RRL emission. In some cases the principal bands in RRL emission shift to the short-wave region as well.

But one of the remarkable features of RRL spectra is that there may be new, additional bands that do not appear in other luminescence spectra with traditional types of excitation [1, 20, 27, 29]. Two types of such additional bands were observed. The first is of a chemisorption origin; i.e., the bands are related to new adsorption centers at the phosphor surface or to gas-phase chemiluminescent reactions initiated by catalysts. Such reactions can also be realized via adsorption of gas particles at the surface of the solid [1, 17, 19].

The other type of additional bands is caused by centers of intrinsic origin, centers that are pesent in the phosphor but do not manifest themselves in other types of luminescence [20, 21].

Figures 8.4 and 8.5 show the RRL spectrum (curve 1) and the photoluminescence spectrum (curve 2) of self-activated zinc oxide (Fig. 8.4) and self-activated zinc sulfide (Fig. 8.5) [17]. Figure 8.6 depicts the spectra of RRL emission

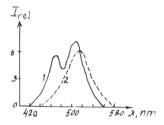

Fig. 8.4. RRL spectrum (curve 1) and photoluminescence spectrum (curve 2) of self-activated zinc oxide.

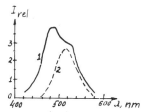

Fig. 8.5. RRL spectrum (curve 1) and photoluminescence spectrum (curve 2) of self-activated zinc sulfide.

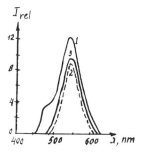

Fig. 8.6. RRL spectrum (curve 1), photoluminescence spectrum (curve 2), and cathodoluminescence spectrum (curve 3) of the ZnS·CdS·Cu-phosphor.

(curve 1), photoluminescence (curve 2), and cathodoluminescence of the ZnS·CdS·Cu-phosphor [17]. In all three cases the RRL spectra were excited by atomic hydrogen. We can easily see that compared to the photoluminescence and cathodoluminescence spectra there is an additional band in the RRL spectra with a

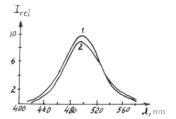

Fig. 8.7. RRL spectra of KCl crystals (curve 1) and NaCl crystals (curve 2).

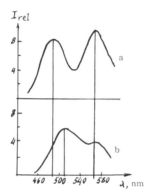

Fig. 8.8. Radical luminescence spectra of the ZnS·Sm-phosphor: a) excited by hydrogen atoms; b) excited by oxygen.

maximum in the 480–490-nm range. The same band was observed in RRL spectra of other phosphors excited by atomic hydrogen [1, 17, 22]. Morever, this band was excited even in activationless crystals, which do not emit other types of luminescence. Figure 8.7 shows the RRL spectra of KCl crystals (curve 1) and NaCl crystals (curve 2); in both cases the crystals exhibited neither photoluminescence nor cathodoluminescence [1]. Thus, under hydrogen excitation the RRL spectrum of these crystals had only one additional band with a maximum in the 480–490-nm range.

Often the position of the additional band does not depend (or depends very weakly) on the nature of the phosphor's base and on the type of activator. But its position in the spectrum depends on the type of atoms that initiate radical luminescence emission of the given substance. Figure 8.8 presents the radical luminescence spectra of the ZnS·Sm-phosphor under excitation by atoms of hydrogen (the upper curve) and oxygen (the lower curve) [23]. We can see that one maximum of the additional band for hydrogen excitation lies at 490 nm.

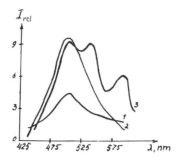

Fig. 8.9. Spectra of photoluminescence (curve 1) and radical-recombination luminescence (curves 2 and 3) of ZnO initiated by atomic nitrogen.

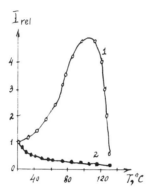

Fig. 8.10. RRL spectrum (curve 1) and photoluminescence spectrum (curve 2) of a ZnO film as functions of temperature.

Figure 8.9 shows the spectra of photoluminescence (curve 1) and radical-recombination luminescence (curves 2 and 3) of ZnO initiated by atomic nitrogen. Curve 2 corresponds to a nitrogen pressure of 5×10^{-2} Torr, while curve 3 corresponds to 1.0 Torr. In the latter case, as we can see, there appear two additional bands (curve 3) if compared to the case of photoluminescence, at 540 and 580 nm [11].

8.3.2. The Effect of Temperature on Intensity of RRL Spectra

Radical-recombination luminescence is characterized not only by the described peculiarities of its spectra but also by the special temperature dependence of the luminescence intensity. The intensity of RRL spectra (in the principal bands) usually passes through a maximum when the temperature varies. This feature is not present in any of the other types of luminescence.

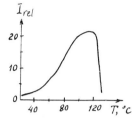

Fig. 8.11. Temperature dependence of the reduced RRL intensity for a ZnO film excited by radicals and obtained in a low-frequency discharge in water vapor.

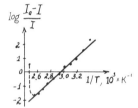

Fig. 8.12. Same temperature dependence as in Fig. 8.11 but given in semilogarithmic coordinates.

Experimental work on the temperature dependence of the RRL spectrum intensity has been reported in [10, 24–27]. Kornich and Gorban [24, 25] studied this aspect of RRL using self-activated zinc oxide with the green luminescence band.

The temperature dependence of the RRL spectrum intensity for a ZnO film is shown in Fig. 8.10 (curve 1). We see that in contrast to photoluminescence (curve 2), radical-recombination luminescence exhibits a maximum at about 100°C. Similar maxima, but at other temperatures, are characteristic of many phosphors. Note that a ZnO film has the same RRL and photoluminescence spectra. This makes it possible to assume that in both cases the same luminescence centers are responsible for the emission, and hence for sufficient excitation the radiationless transition probability in both cases must be the same.

Using the temperature dependence of the RRL and photoluminescence spectra, we can take into account the variation of radical-recombination luminescence intensity due to the drop in the radiative transition probability. To this end we must take the ratio of radical luminescence intensity to photoluminescence intensity at fixed temperatures. The quantity $I = I_{RRL}/I_{PL}$ will be called the reduced (or relative) RRL intensity.

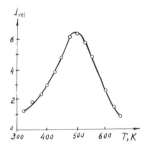

Fig. 8.13. Temperature dependence of the RRL intensity for the green (principal) band of the $Zn_2SiO_4 \cdot Mn$-phosphor.

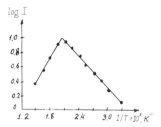

Fig. 8.14. Same temperature dependence as in Fig. 8.13 but given in semilogarithmic coordinates.

Figure 8.11 shows the temperature dependence of the reduced RRL intensity of ZnO film excited by radicals obtained in a low-frequency discharge in water vapor. Figure 8.12 shows the same dependence (for the ascending part of the curve) in $\log[(I_0 - I)/I]$ vs. T^{-1} coordinates, where I_0 is the maximal value of the reduced RRL intensity. In Section 8.4 we will find that such a dependence enables determination of the position of the levels of the chemisorbed radicals.

The temperature dependence of the RRL intensity represented by a curve with a maximum is characteristic of most phosphors. Figure 8.13 shows the temperature dependence of the time-independent RRL intensity for the green (principal) band of the $Zn_2SiO_4 \cdot Mn$-phosphor, while Fig. 8.14 gives the same dependence but in semilogarithmic coordinates [21]. We can see in the specific temperature interval from 300 to 700 K two clearly defined regions, one of which corresponds to a rapid rise of RRL emission with temperature, while the other corresponds to rapid quenching of RRL as the temperature rises still further. In other cases, however, the temperature dependence of RRL intensity may have a more complex form.

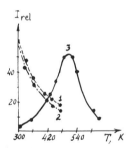

Fig. 8.15. Temperature dependence of the photoluminescence intensity for the AlN·Eu-phosphor (curve 1), of the RRL intensity for the AlN·Eu-phosphor (curve 2), and of the RRL intensity for the AlN·Mn-phosphor (curve 3).

Note that the principal feature in the temperature dependence of RRL intensity is the presence of a pronounced maximum in the principal bands, which is usually absent in other types of luminescence. An exception (for the principal bands) are some phosphors activated by rare earths. For instance, the temperature dependence of radical luminescence of the AlN·Eu-phosphor (the blue-green band caused by the presence of Eu^{2+} ions) does not have a maximum and for all practical purposes coincides with the temperature dependence of the blue-green line in the photoluminescence spectrum of the same phosphor. Figure 8.15 depicts the temperature dependence of RRL intensity for the AlN·Eu-phosphor (curve 2) and for the AlN·Mn-phosphor (curve 3) and the temperature dependence of the photoluminescence intensity for the AlN·Eu-phosphor (curve 1). The temperature dependence for such bands is usually represented by a descending curve.

The temperature dependence for an additional band whose origin is related to centers of intrinsic origin may be either monotonic or with maxima or minima [1, 21]. Morever, Naslednikov [21] describes a case where the temperature dependence of a fixed band varied. This was the case with the red (additional) band in radical luminescence for the Zn_2SiO_4·Mn-phosphor. It was found that when this phosphor is treated in an atmosphere of atomic hydrogen at relatively low temperatures (lower than 100–150°C), the temperature dependence of the band intensity is descendant, while with higher treatment temperatures and longer exposure of the phosphor to atomic hydrogen the curve acquires a maximum, which is characteristic of the principal RRL bands.

Without going into further detail, we will note that for additional bands of intrinsic origin the temperature dependence characteristic of principal bands appears to be only natural since both types of bands are caused by centers that were in the phosphor prior to luminescence.

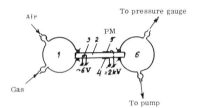

Fig. 8.16. Schematic of the equipment used in experiments on the effect of an electric field on the RRL intensity for the ZnS·CdS·Cu-phosphor.

In conclusion we note that an important characteristic of radical recombination luminescence is the RRL quantum yield, which, by definition, is the ratio of the number of photons emitted per unit time from a unit surface area to the number of recombination acts that occur in the same time interval and at the same surface area. There are many works (e.g., [28–31]) devoted to ascertaining the RRL quantum yield. It was found that this quantity is very small and does not exceed 2.5×10^{-5} photon per recombination act. In other words, out of every one hundred thousand recombination acts only one is accompanied by emission of a luminescence photon. The quantum yield is temperature dependent, and this temperature dependence is similar to that of the RRL intensity, i.e., is expressed by curves with maxima (as a rule), with the maxima on both curves corresponding approximately to the same temperature.

8.3.3. The Effect of Electric Field on Intensity of RRL Spectra

As we know, an external electric field applied at right angles to a semiconductor surface changes the adsorptivity of the latter (Section 4.3.2 was devoted to this aspect.) Since radical-recombination luminescence emerges during adsorption and recombination of active particles at the semiconductor surface, we can expect that an electric field must affect the RRL intensity. In some respects the effect of foreign impurities in the gaseous phase on the RRL intensity should be the same, since these impurities, when adsorbed, change the electrical state of the phosphor's surface.

The equipment used in the earlier experiments on the effect of an electric field on the intensity of RRL from the ZnS·CdS·Cu-phosphor is shown in Fig. 8.16. The radicals were created during combustion of a mixture of air and hydrogen [2]. The air and hydrogen were mixed in mixer 1, from which the gaseous mixture was sent to a quartz reaction tube, 2, at the beginning of which the mixture was ignited by a spiral, 3. The radicals created in the process (in the case at hand these were basically H atoms) were sent (under evacuation) to the phosphor on a substrate to which one of the electrodes, 4, was connected. The other electrode, 5, was placed above the quartz tube, while 6 was used as a buffer vessel. A high voltage of

Table 8.1

U, kV	−2,0	0	+2,0
I, rel. units	1.14	1.00	0.25

Table 8.2

U, kV	a) Hydrogen			b) Air		
	−5	0	+5	−5	0	+5
I, rel. units	1.11	1.0	0.30	0.97	1.0	1.16

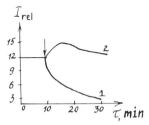

Fig. 8.17. Effect of electric field on RRL intensity.

about 2 kV was supplied to the electrodes. The luminescence emission produced by the radicals and the variations in its intensity under an electric field were registered by a photomultiplier.

The results of the experiment described above are presented in Table 8.1 and Fig. 8.17. The first line in Table 8.1 gives the values of the voltage U applied to the electrodes. The positive direction of the field is assumed to be from electrode 5 to electrode 4 in Fig. 8.16. Curve 1 in Fig. 8.17 corresponds to the negative direction of the field and curve 2 to the positive direction. The second line in Table 8.1 gives the luminescence intensity I in relative units. Both Table 8.1 and Fig. 8.17 show that a negative field leads to a small increase in the RRL intensity, while a positive field leads to a considerable drop in intensity.

Wolkenstein, Gorban, and Sokolov [32] conducted experiments to examine the effect of an electric field on RRL for a fixed phosphor but one excited first by hydrogen radicals and air radicals; the radicals were created by applying an electric discharge, with careful screening of the sample from the field created by the discharge. The source of radicals was a discharge tube filled with the appropriate gas. The radicals were drawn away by a pump into a side arm, which was the reaction vessel containing the glass substrate with the phosphor. The electrodes were applied externally and the luminescence emission was registered by a photomultiplier. The voltage across the electrodes was ±5 kV. The data (averaged over several experiments) is presented in Table 8.2.

The results show that when the phosphor is excited by hydrogen, the effect of the field is similar to that in the previous experiment (see Table 8.1), where a similar sample was treated largely by hydrogen atoms.   However, when the phosphor is excited by air, the effect is the opposite, namely, a negative voltage leads to a small decrease in the RRL intensity of the sample and a positive voltage to a small increase in the intensity.   The explanation, apparently, is that in the air dissociated under the action of the discharge there are radicals (atoms) of oxygen. In contrast to hydrogen, which is a donor gas, these radicals have acceptor properties.

The theory of this phenomenon will be discussed in Section 8.4.4, where we will see that the opposite effects of a field on phosphor excited by donor particles in one case and by acceptor particles in the other can be explained fairly well using the electron theory of chemisorption on semiconductors.

As for experiments to study the effect of an electric field on the RRL intensity, we must note that they are extremely difficult to reproduce and that the sample must be thoroughly screened from auxiliary fields.   The above-mentioned results can only be considered as preliminary, requiring further substantiation and verification in various experimental conditions.

8.3.4. The Effect of Illumination on RRL Intensity

Let us examine the effect of light on the RRL intensity and, in addition, consider the effect of radical-recombination processes taking place at the surface of phosphors on the intensity of photoluminescence (PL).   In both cases we are dealing with photoexcitation and radical-recombination excitation acting simultaneously.   The luminescence emission that emerges under the simultaneous action of these two types of excitation we will call radical photoluminescence (RPL).

We will focus our attention on the work of Naslednikov [21] and Kornich and Gorban [33] devoted to RPL.   Naslednikov used a film of self-activated zinc oxide as the phosphor.   The excitation of RRL emission was achieved by the products of water vapor dissociation in a glow discharge, primarily by hydrogen atoms [25]. Photoexcitation was carried out via a mercury–quartz lamp with a light filter.   The luminescence intensity was measured by a photomultiplier.

The results for not very high (room) temperatures are shown in Figs. 8.18 (weak photoexcitation) and 8.19 (strong photoexcitation).   Here I_{rel} is the relative luminescence intensity [1, 33].   The results suggest that the simultaneous action of the two types of excitation is not additive, i.e., the RPL intensity is lower than the sum of radical luminescence and photoluminescence when they are excited separately.   At elevated temperatures (about 110°C) we have the same result (nonadditivity) (see Figs. 8.20 and 8.21).   In this case the contribution of RRL to RPL is minor, i.e., the RPL intensity differs little from the photoluminescence intensity.   In Sections 8.4.5 and 8.4.6 we will continue with an analysis of Figs. 8.18–8.21.

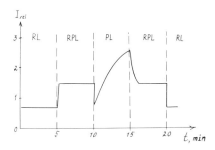

Fig. 8.18. Effect of illumination on RRL intensity (weak photoexcitation).

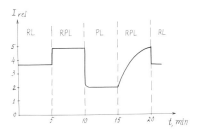

Fig. 8.19. Effect of illumination on RRL intensity (strong photoexcitation).

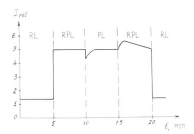

Fig. 8.20. Effect of illumination on RRL intensity at elevated temperatures (weak photoexcitation).

The radical photoluminescence of the phosphor $Zn_2SiO_4 \cdot Mn$ was investigated by Naslednikov [21]. The peculiar feature of the RRL spectrum of this phosphor is that it consists of two bands, the principal band (green) and an additional band (red), the latter manifesting itself in the RRL case as well. For other types of excitation the additional band can be observed only when BeO is added or for a high concentration of Mn. It was found that the RPL spectrum of this phosphor has several features that set it apart from the RPL spectra of the single-band phosphor considered earlier (ZnO).

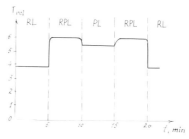

Fig. 8.21. Effect of illumination on RRL intensity at elevated temperatures (strong photoexcitation).

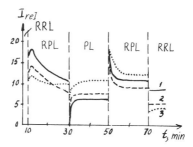

Fig. 8.22. Simultaneous radical excitation and photoexcitation for the principal (green) band of the $Zn_2SiO_4 \cdot Mn$-phosphor at sample temperatures of 600°C (curve 1), 400°C (curve 2), and 300°C (curve 3).

Figure 8.22 presents the results of simultaneous radical excitation and photoexcitation for the principal (green) band of the $Zn_2SiO_4 \cdot Mn$-phosphor at sample temperatures of 300°C (curve 3), 400°C (curve 2), and 600°C (curve 1). Photoexcitation was carried out in the impurity absorption band ($\lambda = 365$ nm), while radical excitation was carried out via hydrogen atoms.

As Fig. 8.22 shows, the stationary value of the RPL intensity for this band is always smaller than the sum of the stationary values of RRL and PL intensities. However, at the beginning of the period during which the light acts (Fig. 8.22, the interval between 10 and 12 min) there is a rise in luminescence for a constant level of excitation of RRL and the RPL intensity passes through a maximum, with the maximal RPL intensity being greater than the sum of the stationary values of radical luminescence and photoluminescence intensities. This is the first feature in which RPL of $Zn_2SiO_4 \cdot Mn$ differs from RPL of ZnO. This may be due to the spectral properties of the phosphor, since in radical-recombination exciation the $Zn_2SiO_4 \cdot Mn$-phosphor accumulates considerable light sums on the traps in the green (principal) band. When the sample is illuminated by the light of a mercury–

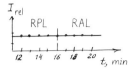

Fig. 8.23. Luminescence intensity for the red (additional) band of the $Zn_2SiO_4 \cdot Mn$-phosphor.

quartz lamp, the luminescence stimulation effect of ultraviolet radiation [34] leads to a rise in RPL immediately after the radiation is applied. The luminescence intensity then falls off to a stationary value.

Another feature of the $Zn_2SiO_4 \cdot Mn$-phosphor is that in the red (additional) band no rise in luminescence intensity is observed, and, in general, photoexcitation at small exposures (3 to 10 min) shows no effect on the intensity of this band (Fig. 8.23). The absence of the photoluminescence component in RPL can be explained by peculiarities of the red (additional) band of the $Zn_2SiO_4 \cdot Mn$-phosphor. As noted in Section 8.1, this band occurs only in radical-recombination luminescence and does not manifest itself in photoexcitation at concentrations of Mn present in the given phosphor.

In Section 8.4.6 we will return to the special features of RPL.

8.4. THE MECHANISM OF RADICAL-RECOMBINATION LUMINESCENCE

8.4.1. The Excitation Mechanism

The recombination of free atoms or radicals at the surface of a phosphor may in certain conditions lead to excitation of the crystal. This process consists in the creation of free electrons and holes at the crystal's surface. Under certain conditions each recombination act corresponds to the creation of a pair consisting of a free electron and a free hole. Subsequent recombination of such an electron with a hole, a process that can proceed along different routes (see below), leads to emission of one or serveral light quanta, i.e., the recombination of an electron with a hole may be accompanied by a luminescence act.

Let us assume, to be definite, that the atoms and radicals taking part in the recombination process (as usual we denote them by R) are acceptors. In the energy spectrum of the surface, shown in Fig. 8.24 (the y axis is parallel to the surface plane), they are depicted by acceptor levels that lie close to the valence band. If the recombination of atoms (radicals) leads to a pair consisting of a free electron and a free hole, the reaction may go through the following states [35, 36]:

$$R + L \to RL, \tag{8.35}$$
$$RL \to ReL + pL, \tag{8.36}$$

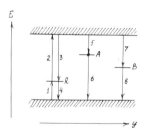

Fig. 8.24. Recombination of free atoms and radicals (acceptors) at the surface of a semiconductor.

$$R + ReL \rightarrow R_2 + eL. \qquad (8.37)$$

Let us consider each stage separately.

(1) The first stage is the chemisorption of the atom or radical R accompanied by the formation of a "weak" bond with the surface.

(2) The second stage is the transition of the "weak" bond into a "strong" bond. According to (8.36), this is achieved by a thermal jump of the electron from the valence band onto an acceptor local level of an adsorbed particle and the accompanying formation of a free hole in the valence band (the electron transition 1 in Fig. 8.24). This stage of "strengthening" the bond with the surface consists in the chemisorbed particle passing from the electrically neutral state RL into the negatively charged state ReL. Note that this transition may proceed either according to (8.36) or according to the following scheme:

$$RL + eL \rightarrow ReL \qquad (8.38)$$

(transition 3 in Fig. 8.24). In the latter case the strengthening of the band between the given chemisorbed particle and the surface is due to the electron that was previously shifted to the conduction band, e.g., in the recombination act involving another chemisorbed particle, i.e., as a result of stage (8.37).

Along with the strengthening of the bond we can often encounter the opposite process, the "weakening" of the bond, according to

$$ReL + pL \rightarrow RL, \qquad (8.39)$$

or

$$ReL \rightarrow RL + eL. \qquad (8.40)$$

The transition (8.39) is the recombination of an electron localized at the chemisorbed particle with a hole in the valence band (the electron transition 4 in Fig. 8.24). We will ignore the transition (8.40), which is a thermal jump of an electron into the conduction band (transition 2 in Fig. 8.24).

Thus, there is an exchange of electrons between a local level of a chemisorbed particle and the energy bands. We assume that the exchange with the valence band has a thermal origin, while that with the conduction band is ensured by recombination of atoms (radicals) via (8.37).

(3) The third stage [see (8.37)] is the recombination of an atom or radical R from the gaseous phase with a chemisorbed atom or a radical R bound (strongly) to the surface. This stage was examined in Section 2.4.4. The recombination act is accompanied by desorption of the molecule R_2 and creates an electron in the conduction band.

Recombination of free atoms or radicals R at the surface of a phosphor may proceed either via (8.37) or according to the following scheme:

$$R + RL \to R_2 + L, \tag{8.41}$$

i.e., the recombination act involves a chemisorbed atom (radical) that can be either "strongly" bound to the surface [the case of (8.37)] or "weakly" bound [the case of (8.41)]. The reader must bear in mind, however, that the act (8.41), in contrast to (8.37), is useless from the viewpoint of luminescence, since it does not lead to the liberation of an electron and leaves the phosphor unexcited. Thus, not all recombination acts can lead to luminescence emission of phosphors, but only those that proceed according to (8.37).

When we studied the recombination of two atoms or radicals, we assumed that one is chemisorbed while the other comes from the gaseous phase [see (8.37) and (8.41)]. Note, however, that there is another possibility, namely, when both atoms or radicals participating in the recombination act are in the chemisorbed state. Then instead of (8.37) and (8.41) we have, respectively,

$$RL + ReL \to R_2 + eL \tag{8.42}$$

and

$$RL + RL \to R_2 + L. \tag{8.43}$$

From the viewpoint of luminescence it is irrelevant whether recombination follows the mechanism (8.37) and (8.41) or the mechanism (8.42) and (8.43), since both (8.37) and (8.42) lead to the same result, i.e., an electron appears in the conduction band.

Up to this point we have assumed that the atoms or radicals R participating in the recombination process are acceptors. But suppose that they are donors. This means that the donor local levels R lie near the conduction band in Fig. 8.24. Instead of (8.36) and (8.37) we then have, respectively,

$$RL \to RpL + eL, \tag{8.44}$$

and

$$R + RpL \to R + pL. \tag{8.45}$$

In this case the "strengthening" of the bond, (8.44), is accompanied by formation of a free electron (transition 2 in Fig. 8.24) instead of a hole, and the recombination act (8.45) is accompanied by formation of a free hole instead of an electron. In both cases, irrespective of whether the chemisorbed particles R are acceptors or donors, the sequence of "strengthening" of the bond and the recombination act leads, as can be seen comparing (8.36) and (8.37) with (8.44) and (8.45), to the formation of a pair consisting of a free electron and a free hole; i.e., the phosphor passes into an excited state.

8.4.2. The Mechanism of Luminescence

Let us see what the fate is of this paired free electron and free hole, which appears, respectively, in the conduction band and the valence band as a result of the recombination of free atoms (radicals) at the surface of a phosphor. We again turn to Fig. 8.24. Level R depicts the level of a chemisorbed atom or radical R participating in the recombination act. Let A be a level of the activator. In Fig. 8.24 the activator levels are assume to lie in the surface plane. This assumption is not mandatory and is taken only to simplify the picture. Nothing will change in our line of reasoning if we assume that the activator atoms lie in the bulk of the crystal. To be definite we will assume that the levels A are donor levels. Besides R and A we have a level B in Fig. 8.24, which is the level of an intrinsic surface defect. Let us suppose that B and R are acceptor levels, but what we will discuss below can readily be applied to the case where either both levels or one is of donor origin. In some cases the intrinsic defects may be the particles of a foreign gas prechemisorbed at the surface. These may also be the particles R participating in the recombination process. In the latter case the level B in Fig. 8.24 must lie on the same horizontal line with R.

An electron that is sent to the conduction band can recombine with a hole in the valence band by the following schemes:

(1) Through the activator level A as a result of two successive transitions:

$$AL + pL \rightarrow ApL, \tag{8.46}$$

$$ApL + eL \rightarrow AL \tag{8.47}$$

(the transitions 6 and 5 in Fig. 8.24, respectively). The first transition is the ionization of the activator and the second is its neutralization. One transition (or both) may be accompanied by emission of a quantum (luminescence). As a result the phosphor returns to its ground state. This is the mechanism that forms the principal band of RRL [1, 2].

We see that the spectral composition of the principal band depends on the nature of the activator, i.e., the position of level A in the energy diagram in Fig. 8.24, and does not depend on the nature of the recombining atoms or radicals, i.e., the position of level R. These facts, as we have seen, agree with the experimental data. We also see that the principal band of RRL coincides with the photoluminescence band, which also agrees with the experimental data. Note that a phosphor may contain several types of activator rather than a single type, which explains why sometimes there are several principal bands of RRL instead of one.

Thus, the mechanism of principal band formation is depicted in Fig. 8.24 as the sequence of electron transitions 1-2-6-5. The entire mechanism is triggered by the recombination act, which ensures that transition 2 is present, and contains the luminescence act as a separate stage, depicted by transition 5 or 6.

(2) A free electron may recombine with a free hole not only through level A but through level B as well (as noted above, through level R, in particular):

$$BL + eL \rightarrow BeL, \tag{8.48}$$

$$BeL + pL \rightarrow BL. \tag{8.49}$$

The first transition is the capture of a free electron by an intrinsic surface defect, which leads to the defect acquiring charge. The second transition is the capture by the charged defect of a free hole, which leads to neutralization of the defect. One of these transitions (or both) may be radiative (luminescent). This explains the origin of the additional band of RRL [15]. In some cases these transitions may be radiationless. Then no additional line appears. In all cases the appearance of transitions (8.48)–(8.49), which constitute an additional channel of recombination, leads to a quenching of the principal band of RRL.

We see that the spectral composition of an additional band does not depend on the nature of the activator present in the phosphor. However, it does depend on the nature of the intrinsic defects on the surface. Why intrinsic surface defects do not manifest themselves in photoluminescence was discussed in Section 8.3.1. Note that there may be not one but several additional bands since the intrinsic defects present at the surface may be of various types. Note also that in some cases the luminescence spectrum may contain only additional bands, i.e., lack a principal band. This happens when the crystal produces radical-recombination luminescence but fails to produce photoluminescence. In particular, when the chemisorbed atoms or radicals R participating in the recombination act as intrinsic defects, the spectral composition of an additional band is determined by the nature of the reacting gas. Samples with different activators produce similar additional bands for the recombining atoms (radicals) of the same origin. At the same time, when the nature of the recombining particles changes, the additional band for a sample with the same activator shifts. The additional bands that are insensitive to the nature of the activator but are sensitive to the nature of the recombining atoms (radicals) can indeed be observed, as we have noted earlier in [15].

Thus, the mechanisms of formation of an additional band of RRL are depicted in Fig. 8.24 by the sequence of electron transitions 1-2-7-8. We see that the same mechanism works for both principal and additional bands.

Finally, we will touch on one more mechanism of formation of additional bands. Suppose that as a result of the recombination of an atom (radical) R from the gaseous phase with a chemisorbed atom (radical) R "weakly" bound to the surface we have a molecule R_2^* that is in the excited state:

$$R + RL \rightarrow R_2^* + L.$$

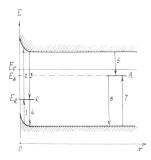

Fig. 8.25. Energy band structure of a phosphor used to study the dependence of the RRL intensity on the position of the Fermi level.

Transition of the excited molecule R_2^* into the ground state R_2 can be accompanied by emission of a photon, which constitutes a luminescence act:

$$R_2^* \to R_2 + h\nu.$$

This is another possibility for an additional band. Such bands of a chemiluminescent origin are sometimes observed in experiments. What gives off the luminescence emission is not the crystal but the gaseous phase in the region around the surface of the crystal [1, 19, 28, 37–39].

8.4.3. The Dependence of RRL Intensity on the Position of the Fermi Level

We will now determine the factors that determine the intensity of the luminescence that proceeds according to the radical-recombination mechanism described above [1]. We will restrict our discussion to the principal line. For the sake of simplicity we will assume that the surface does not contain intrinsic defects participating in the luminescence act.

Suppose that the half space $x \geq 0$ is occupied by a phosphor while the half space $x < 0$ is occupied by the gaseous phase. Figure 8.25 shows the energy band structure of the phosphor. We assume for the sake of definiteness that the surface is negatively charged (the bands are bent upward). Level A is the level of an activator that we assume to be a donor, and R is the level of the chemisorbed atom (radical) participating in the luminescence act. We assume that R is an acceptor level; the case of a donor level will be considered later.

Among the electron transitions, depicted in Fig. 8.25 by vertical arrows, there are transitions of two types: viz., transitions that incorporate local levels of chemisorbed particles (transitions 1, 2, 3, and 4) and transitions that incorporate activator levels (transitions 5, 6, and 7). We define s_i as the number of ith transitions ($i = 1, 2, 3, 4$) occurring per unit time per unit surface area, and $r_k(x)$ as the number of kth transitions ($k = 5, 6, 7$) occurring per unit time per unit volume and referred to the x plane (where $x \geq 0$).

In the steady-state case we have (for all $x \geq 0$)

$$s_1 - s_4 = s_2 - s_3,$$ (8.50)
$$r_6(x) - r_7(x) = r_5(x).$$ (8.51)

To these we add the continuity equation, which in the present case has the form

$$\frac{dj_n}{dx} = -\frac{dj_p}{dx} = r_5(x),$$ (8.52)

where $j_n(x)$ and $j_p(x)$ are the electron and hole fluxes in the x plane. Obviously

$$j_n(x) = j_p(x),$$

with

$$j_n(\infty) = j_p(\infty) = 0, \quad j_n(0) = j_p(0) = j_s,$$ (8.53)

where

$$j_s = s_2 - s_3.$$ (8.54)

If we integrate Eq. (8.52) over the entire volume of the crystal, we have, according to (8.53),

$$j_s = \int_0^\infty r_5(x)\, dx$$ (8.55)

and

$$I = (1 - \kappa)j_s,$$ (8.56)

where I is the luminescence intensity (the number of photons emitted by the phosphor per unit time from a unit surface area), and κ is the radiationless transition probability.

In what follows we will assume that

$$s_3 \ll s_2,$$ (8.57)

or, according to (8.54) and (8.55),

$$s_3 \ll I,$$

which means that we ignore the additional bands and deal only with the principal one. Here, as follows from (8.37) and (8.38),

$$s_2 = \alpha P N^-,$$
$$s_3 = \beta n N^0,$$

(8.58)

where P is the partial pressure of the gas consisting of the R particles, N^- and N^0 are the number of such particles chemisorbed on a unit surface area and are in charged and electrically neutral states, respectively, i.e., in states of "strong" and "weak" bonding with the surface (states ReL and RL), n is the conduction electron concentration in the surface plane (the $x = 0$ plane), and α and β are proportionality factors, with

$$\alpha = \alpha_0 \exp\left(-\frac{\epsilon}{kT}\right),$$

(8.59)

where ϵ is the activation energy in recombination.

According to (8.57), (8.55), (8.54), and (8.58) we have

$$I = (1 - \kappa)\alpha P N^-.$$

(8.60)

We will consider our system to be thermodynamically in quasiequilibrium, i.e.,

$$s_2 \ll s_1 + s_4,$$ (8.61)
$$r_5 \ll r_6 + r_7$$ (8.62)

(the latter condition is assumed to be met for all $x \geq 0$). According to (8.50), (8.57), and (8.51), we can rewrite conditions (8.61) and (8.62) in the following form:

$$\frac{s_1 - s_4}{s_1 + s_4} = \delta, \quad \text{where } \delta \ll 1,$$

$$\frac{r_6 - r_7}{r_6 + r_7} = \gamma, \quad \text{where } \gamma \ll 1,$$

which yield

$$s_1 = s_4, \qquad r_6 = r_7.$$

In this case, for N^- and N^0 we can employ the formulas obtained in the electronic theory of chemisorption for electron equilibrium conditions.

According to (3.6b), we have

$$\eta^- = \frac{N^-}{N} = \frac{1}{1 + \exp\dfrac{E_R - E_F}{kT}},$$

$$\eta^0 = \frac{N^0}{N} = \frac{1}{1 + \exp\dfrac{E_F - E_R}{kT}},$$

(8.63)

where $N = N^0 + N^-$ is the total number of particles chemisorbed on a unit surface area, and T is the absolute temperature; the meaning of E_F and E_R is clear from Fig. 8.25 (E_F represents the Fermi level). When there is an equilibrium between the surface and the gaseous phase, then, according to (3.9),

$$N = \frac{N^*}{1 + (b/P)\, \eta^0}\,,\tag{8.64}$$

where N^* is the surface concentration of the adsorption centers (in other words, the maximal number of particles capable of being chemisorbed on a unit surface area, or the number of particles chemisorbed on a unit surface area at $P = \infty$), and b is the adsorption coefficient (2.14).

Equations (8.60), (8.63), and (8.61) yield

$$I = (1 - \kappa)\,\alpha P \eta^- N = \frac{(1 - \kappa)\,\alpha N^* P}{(1/\eta^-) + (\eta^0/\eta^-)(b/P)}\,,$$

or, after substitution of (8.63),

$$I = \frac{A}{1 + B \exp\dfrac{E_R - E_F}{kT}}\,,\tag{8.65}$$

where

$$A = (1 - \kappa)\,\alpha N^* P, \quad B = 1 + b/P.\tag{8.66}$$

We see that for a given pressure and temperature (which means that A and B are fixed) the luminescence intensity is determined by the position of the Fermi level at the surface of the crystal (we are speaking of the Fermi level that characterizes the state of the system without allowance for recombination and luminescence acts).

We note, finally, that when we derived Eq. (8.65), we proceeded from the assumption that the chemisorbed particles R, which take part in recombination, are acceptors. But if they are donors, we arrive at the same equation, (8.65), the only difference being that E_R and E_F change places; parameters A and B are, as before, given by (8.66).

8.4.4. The Dependence of RRL Intensity on an External Electric Field

The I vs. E_F dependence, given by (8.65), is shown in Fig. 8.26. Figure 8.26a corresponds to the case when the particles R are acceptors, and Fig. 8.26b to the case where these particles are donors. Comparing the two, we can see that the dependence of RRL intensity on the position of the Fermi level in the first case is opposite to that in the second, namely, as the Fermi level moves downward (all other parameters remaining the same), luminescence is quenched in the first case (acceptor particles), while in the second (donor particles) it rises.

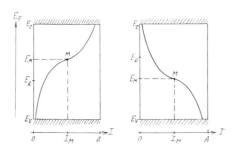

Fig. 8.26. Dependence of RRL intensity on an external electric field: a) R are acceptors; b) R are donors.

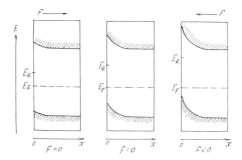

Fig. 8.27. Variation of band bending of a phosphor placed in a homogeneous electric field of strength F.

The two curves, in Figs. 8.26a and 8.26b, lie between two asymptotes, $I = 0$ and $I = A$, and have a point of inflection M, whose coordinates we denote by E_M and I_M. As can easily be shown,

$$E_M = E_R \mp kT \ln B, \quad I_M = \tfrac{1}{2} A, \tag{8.67}$$

where the upper sign is taken for donor particles and the lower for acceptor particles, and A is given by (8.66). We see that $E_M < E_R$ for donor particles and $E_M > E_R$ for acceptor particles, which means that $B > 1$ [see (8.66)].

Equation (8.65) shows that RRL intensity must depend on factors that shift the Fermi level. The most striking example is the influence of an external electric field applied to a phosphor.

Suppose that a phosphor is placed in a homogeneous electric field of strength F directed normally to the adsorbing surface. The presence of a field changes the bending of the bands, as a result of which the Fermi level at the crystal's surface proves to be shifted with respect to level E_R, as shown in Fig. 8.27 for $F > 0$ and $F < 0$.

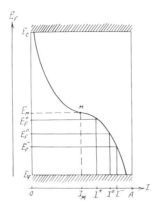

Fig. 8.28. Dependence of RRL intensity on
an external electric field when R are donors.

Thus, a field applied to a phosphor must result in a variation in the luminescence intensity. When the field changes sign, a decrease in luminescence emission changes to an increase and vice versa. In the final analysis, this is the result
of a change in the adsorptivity brought on by the field, an effect studied earlier
[40, 41] (see Section 4.4.2).

As we see, the sign of the effect of a field depends on whether the phosphor is
excited by donor particles or acceptor particles. This is observed in experiments
(see Section 8.3.3). For instance, if a field leads to an increase in luminescence
emission from a phosphor excited by hydrogen (donor), it leads to a decrease in
emission from the same phosphor but excited by oxygen (acceptor).

Let us return to the experimental data on the effect of a field on the RRL
intensity of the $ZnS \cdot CdS \cdot Cu$-phosphor excited by hydrogen atoms (see Section 8.3.3). If we interpret the data given in Tables 8.1 and 8.2 in terms of the
ideas just discussed, we arrive at the following conclusions:

(1) An increase in the luminescence intensity due to an electric field directed
negatively (i.e., when the Fermi level moves downward) and a decrease in
intensity due to an electric field directed positively (i.e., when the Fermi level
moves upward) indicate that the chemisorbed particles participating in recombination are donors. In the case of acceptors the effect would be reversed.

(2) The fact that the effect is not symmetric, i.e., a small increase in luminescence intensity when the field is negative (by 14% according to the data of
Table 8.1 and by 11% according to Table 8.2a) and a large decrease in intensity
(by 75% or by 70% according to Tables 8.1 and 8.2b) when the field is positive
(but of the same magnitude), indicates that the Fermi level in the initial state (i.e.,
with a zero field) lies below the point of inflection M in Fig. 8.28. If it were to lie
above this point, the effect would be reversed. This is clear from Fig. 8.28, which
repeats Fig. 8.26b and in which $E_F{}^0$, $E_F{}^+$, and $E_F{}^-$ stand for the Fermi levels in the
absence of a field, with a positive field, and with a negative field, respectively,
while I^0, I^+, and I^- are the corresponding luminescence intensities. We see that

$$(I^- - I^0) < (I^0 - I^+) \text{ при } (E_F^0 - E_F^-) = (E_F^+ - E_F^0).$$

In this case the Fermi level E_F lies under the point of inflection, i.e.,

$$E_F < E_M < E_R. \tag{8.68}$$

This condition is favorable from the viewpoint of radical-recombination luminescence, since then the majority of atoms (radicals) R chemisorbed at the phosphor's
surface are in a charged state instead of the neutral state, so that the reaction
proceeds according to Eq. (8.45),

$$R + RpL \to R_2 + pL,$$

and not according to Eq. (8.41),

$$R + RL \to R_2 + L,$$

the latter being useless from the point of luminescence.

The results given in Table 8.2b, which refer to the case where the phosphor is
excited by acceptor particles (oxygen), can be interpreted in a similar manner.

The studies concerning the effect of an electric field on RRL intensity can
serve as indirect proof of the theoretically predicted effect produced by an electric
field on the adsorptivity of a surface, which at present has direct experimental
verification (see Section 4.4.2).

8.4.5. The Temperature Dependence of RRL Intensity

The dependence of radical-recombination luminescence intensity I on temperature T is hidden in Eq. (8.65). Let us study this dependence. The reader will
recall that, according to (8.59) and (3.11),

$$\alpha = \alpha_0 \exp\left(-\frac{\epsilon}{kT}\right), \quad b = b_0 \exp\left(-\frac{q}{kT}\right), \tag{8.69}$$

where ϵ is the activation energy in the reaction of recombination, and q is the
energy of "weak" bonding in chemisorption (the adsorption heat). The factors α_0
and b_0 depend little on the temperature (compared to the dependence of the
exponential factors) and we will consider them constants.

On the basis of (8.66), (8.68), and (8.69) and assuming, for the sake of
definiteness, that the particles R participating in the recombination acts are
donors, we can rewrite Eq. (8.65) as follows:

$$I = \frac{(1 - \kappa)\alpha_0 N^* P \exp(-\epsilon/kT)}{1 + \exp[-(E_R - E_F)/kT] + (b_0/P)\exp[-(q + E_R - E_F)/kT]}, \tag{8.70}$$

where κ (the radiationless transition probability) is assumed to depend little on the temperature or to be small compared with unity, so that the factor $1 - \kappa$ can be considered constant. Note that the position of the Fermi level E_F depends on the temperature, too. Let us assume that in the temperature range we are studying here

$$\exp\left[-(E_R - E_F)/kT\right] \ll 1, \tag{8.71}$$

i.e., the Fermi level lies far below the E_R level. (Otherwise practically no luminescence takes place.)

The experimentally observed temperature dependence of I (see Fig. 8.14) can be found from Eq. (8.70) if we assume the following. Suppose that $q + E_R - E_F$ changes little with temperature (i.e., compared with kT), so that it can be considered to be independent of T. We will also assume that P is so small that

$$\frac{b_0}{P} \gg 1. \tag{8.72}$$

Here are two limiting cases.
(a) The case of "high" temperatures, when

$$(b_0/P)\exp\left[-(q + E_R - E_F)/kT\right] \gg 1. \tag{8.73}$$

Then, according to (8.71) and (8.73), instead of (8.70) we have

$$I = \frac{(1 - \kappa)\,\alpha_0 N^* P^2}{b_0}\, \exp\frac{E_R - E_F + q - \epsilon}{kT}. \tag{8.74}$$

When $E_R - E_F + q - \epsilon$ is positive, the $\ln I$ vs. T^{-1} dependence is represented by a straight line with a positive slope (the left branch in Fig. 8.29).
(b) The case of "low" temperatures, when

$$(b_0/P)\exp\left[-(q + E_R - E_F)/kT\right] \ll 1. \tag{8.75}$$

Then, according to (8.71) and (8.75), instead of (8.70) we have

$$I = (1 - \kappa)\alpha_0 N^* P \exp\left(-\frac{\epsilon}{kT}\right). \tag{8.76}$$

In this case the $\ln I$ vs. T^{-1} is represented by a straight line with a negative slope (the right branch in Fig. 8.29). The intersection of the straight line in Fig. 8.29 and, hence, the maximum on the I vs. T curve, are ensured by condition (8.72). Indeed, as can be seen from (8.76) and (8.74), we have (see Fig. 8.29)

$$\ln I_1 = \ln\left[(1 - \kappa)\alpha_0 N^* P\right],$$

$$\ln I_2 = \ln\left[(1 - \kappa)\alpha_0 N^* P\right] - \ln(b_0/P), \tag{8.77}$$

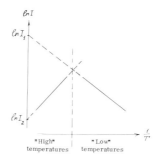

Fig. 8.29. Temperature dependence of RRL intensity.

which means that for the intersection to take place we must be sure that $\ln I_1 > \ln I_2$, i.e., $\ln(b_0/P) > 0$ and, hence, $b_0/P > 1$.

Note, finally, that when the recombining particles are acceptors instead of donors, the above reasoning remains valid, the only difference being that in all formulas E_R and E_F must be interchanged.

To establish the relation between the simple theory above and the facts we must have more experimental data.

8.4.6. Radical Photoluminescence

Let us return to radical photoluminescence (RPL), which we have discussed briefly in Section 8.3.4.

One could expect that luminescence emission emerging from the joint action of radical-recombination excitation and photoexcitation would consist of radical-recombination luminescence (RRL), which takes place in the absence of illumination, and photoluminescence (PL), observed in the absence of radical-recombination processes on the phosphor's surface. But, as we already know, this is not the case. We cannot simply add RRL and PL: the RPL intensity is always less than the sum of the RRL and PL intensities at the same excitation levels. This nonadditivity is due to the fact that photoexcitation to a certain extent quenches radical-recombination luminescence, and radical-recombination excitation quenches photoluminescence. In other words, each term (RRL or PL) decreases under the influence of the other term.

Let us now discuss the possible mechanisms of such an interrelationship between PL and RRL. We will start with the effect of photoexcitation on RRL.

As we already know, light usually changes the adsorptivity of a surface with respect to each given gas. This constitutes the so-called photoadsorption effect, which we have discussed at length in Chapter 7. In RRL we are dealing with atoms (or radicals) that are chemisorbed at the surface of a semiconductor. If illumination induces photoadsorption of these atoms (or radicals), which act in the recombination process, it will quench the luminescence emission. In the present case the quenching of RRL is due to a decrease in the number of recombination centers as a result of illumination.

Let us now turn to the effect produced by radical-recombination processes on PL. The problem can be reduced to chemisorption acting on PL. The mechanism of this influence is discussed in detail in Section 8.2. As we have seen, either the field mechanism or the recombination mechanism is responsible for this influence. In some cases, however, only one mechanism acts, while the other remains in the background.

Thus, the nonadditivity of PL and RRL is due to the decrease in one of the two components or both.

The case of photoluminescence being weakened by radical-recombination excitation is encountered when the RPL intensity is lower than the PL intensity. An example is the weak photoexcitation of self-activated zinc oxide that emits luminescence under the action of atomic hydrogen (see Figs. 8.18 and 8.19). These figures show that the photoluminescence intensity is restored after the radicals cease to act, i.e., after the discharge is switched off. In this case the luminescence intensity drops by a quantity that is exactly the contribution of the radical-recombination luminescence.

At the same temperature but with a more powerful source of exciting light (Figs. 8.20 and 8.21) the nonadditivity is due to the increase in both the photoluminescence intensity and the RRL intensity. In this case after the discharge is switched off the PL intensity is restored faster than in the previous case, while the RRL intensity is restored almost immediately after the light is turned off.

At elevated temperatures (see Figs. 8.22 and 8.23) we, obviously, have to deal with the case where illumination reduces the radical-recombination luminescence. Chemisorption does not have a strong influence on photoluminescence intensity in this case. Figures 8.22 and 8.23 show that photoluminescence does not change when the discharge is switched off.

An example of the case where both RRL and PL are reduced (mutual quenching of RRL and PL) is encountered in the studies of the principal (green) band in the luminescence emission from the $Zn_2SiO_4 \cdot Mn$ (see Fig. 8.23). The fact that quenching of photoluminescence takes place due to the chemisorption of the atoms (radicals) that participate in the recombination process is corroborated by the observed restoration of PL after radical-recombination excitation is switched off. On the other hand, the quenching of the green band of RRL in $Zn_2SiO_4 \cdot Mn$ due to photoexcitation is proved by the fact that the stationary intensity of radical photoluminescence is lower than the sum of the quenched PL and stationary RRL intensities.

We see that there are two interesting aspects in studying RRL: the mechanism by which illumination acts on RRL, and the mechanism by which chemisorption acts on PL.

8.5. ADSORPTION LUMINESCENCE

8.5.1. The Fundamentals of Adsorption Luminescence

In 1966, Rufov, Kadushin, and Roginskii [42–44] experimentally observed luminescence emission that accompanied the chemisorption of molecules at solid

Fig. 8.30. Typical signal from photomultiplier when O_2 was admitted into NiO.

surfaces. The same phenomenon, which was called adsorption luminescence, was observed later by Ivankiv *et al.* [45]. In 1971, Popov and Styrov [46] for the first time observed luminescence emission of phosphors during chemisorption of an atomic gas instead of a molecular gas passed on at the phosphor surface. In 1973, studies of adsorption luminescence were started in the USSR by Sokolov with collaborators [47, 48] and in France by Claudel with collaborators [49, 50].

In adsorption luminescence experiments [42, 43] the emission was observed in the form of a short burst (several tenths of a second) when O_2 was passed onto NiO, ZnO, Cr_2O_3, and Fe_2O_3. Figure 8.30 presents a typical picture of the signal from the photomultiplier when O_2 was passed onto NiO. Similar peaks have been observed when CO was passed onto NiO, ZnO, and Cr_2O_3, and when SO_2 and acetone were passed onto NiO. The luminescence emission was absent only when NO acted on NiO, CO_2 on ZnO, CO on Fe_2O_3, and O_2 on NiO with preadsorption of CO. Popov and Styrov [46] studied a large number of oxides, sulfides, and silicates from the viewpoint of adsorption luminescence. Luminescence emission in the form of bursts at the moment of gas release was observed for the majority of these substances. The character of the bursts was approximately the same in all cases. After the peak there was attenuation of the luminescence emission lasting from one to several seconds. A typical case of such attenuation is shown in Fig. 8.31 for oxygen adsorbed on MgO (at 475 K).

There is a close relationship between the ability of a substance to emit RRL and its ability to exhibit adsorption luminescence. For instance, all solids that emit RRL in atmospheric oxygen are capable of performing adsorption luminescence in molecular oxygen, and the stronger the RRL the more intensive the adsorption luminescence burst. On the other hand, compounds for which RRL emission is not characteristic do not exhibit adsorption luminescence.

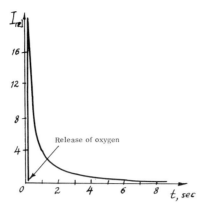

Fig. 8.31. Attenuation of luminescence emission for oxygen adsorbed on MgO (at 475 K).

Adsorption luminescence can be clearly observed under the initial action of the gas. A second release of gas leads either to a very weak burst or to no luminescence emission at all. However, after the sample is degassed in the preliminary processing mode, its ability to emit adsorption luminescence is restored almost completely. This signifies, apparently, that saturation of the adsorption centers is achieved during the very first release of gas.

Note that the results of various Soviet researchers agree with each other but differ from those of French authors. There are three features in which the two approaches differ.

(1) According to the Soviet authors, the adsorption luminescence spectrum coincides with the RRL spectrum (the principal band) and, hence, with the photoluminescence spectrum, i.e., depends not on the nature of the adsorbed gas but entirely on the nature of the adsorbent. On the other hand, according to Claudel and his coworkers, the spectral composition of the adsorption luminescence emission differs from that of the RRL and photoluminescence spectra for the same adsorbent. (Note that Claudel dealt with thorium dioxide, while the other authors dealt with a whole range of other oxides and sulfides.)

(2) According to the Soviet authors, in a broad range the luminescence kinetics obeys the hyperbolic law

$$I = \frac{A}{t + t_0},$$ (8.78)

where I is the luminescence intensity, t time, and A and t_0 parameters, while Claudel and his coworkers observed this hyperbolic law only for a limited number of cases; in other cases they found that for the same adsorbent–adsorbate system the following power law was valid at the beginning of the adsorption process:

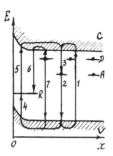

Fig. 8.32. Energy band structure of phosphor in the presence of donor levels responsible for conductivity.

$$I = \frac{B}{t^p}, \qquad (8.79)$$

where B and p are constants, with $p < 1$. Note that the same power law was observed by Roose and Offergald [51] during oxygen adsorption on magnesium.

Obviously, the empirical laws (8.77) and (8.78) work for not too high t's and reflect the attenuation of luminescence after the peak at $t = t^*$ is passed, i.e., for $t > t^*$.

(3) Finally, according to the Soviet researchers, the luminescence intensity is proportional to the adsorption rate:

$$I = \alpha \frac{dN}{dt}, \qquad (8.80)$$

where N is the surface concentration of the adsorbed particles; i.e., luminescence stops when adsorption equilibrium is established. But according to Claudel, luminescence continues even after such equilibrium sets in (i.e., after increase in weight due to adsorption is no longer observed).

In what follows we will study these three factors separately.

8.5.2. The Mechanism and Kinetics of Adsorption Luminescence

We will study the mechanism of adsorption luminescence for the case where the adsorbing particles are acceptors (oxygen) and the phosphor is an n-type semiconductor; for donor particles and a p-type semiconductor the results are similar.

Let us now turn to Fig. 8.32, where we have once more given the band structure of a phosphor (cf. Fig. 8.25). Figure 8.32 differs from Fig. 8.25 in the presence of donor levels D responsible for conductivity (in Fig. 8.25 these levels are not shown). Here C is the conduction band; V the valence band; A an activator level, which we assume occupied by an electron in the ground state; R a chemisorbed particle level, which we assume to be an acceptor level and to lie below level A; the x axis is directed normally to the surface plane (plane $x = 0$); and, for the sake of definiteness, the surface is assumed to be negatively charged (the bands are bent upward). In ordinary luminescence, when the phosphor is excited by the light from the intrinsic adsorption band, the act of luminescence is caused by the sequence of transitions 1-2-3, where transition 3 is assumed radiative.

When a chemisorbed particle R appears on the surface, the system goes over to a metastable state, since a vacant level R appears that is below the occupied level A and donor levels D. There are two ways in which the system can proceed from the metastable state to the ground state.

First, level R may capture a free electron from the conduction band (transition 6). If this transition is accompanied by emission of a quantum, we are dealing with adsorption luminescence. In this case the spectral composition of the adsorption luminescence depends on the nature of the adsorbed atoms. Apparently the French researchers encountered this case. But if we assume that transition 6 is radiationless, then there is another way by which the system can return to its ground state and emit a quantum (adsorption luminescence). If level R lies close to the valence band, then an electron from the valence band can thermally be shifted to level R (transition 4), with a radiationless transition 4 or transitions 2 and 3, the latter transition being radiative. Transition 4 does not require any preexcitation. As a result the electron is shifted from level D to level R, and the system, therefore, returns to its ground state. In this case the adsorption luminescence spectrum coincides with the photoluminescence spectrum. This was the case encountered by the Soviet researchers.

Let us now turn to the kinetics of adsorption luminescence. The best-known empirical laws in the theory of adsorption are the Roginskii–Zel'dovich law

$$N = \beta \ln \left(1 + \frac{t}{t_0} \right), \quad \text{or} \quad \frac{dN}{dt} = \frac{\beta}{t + t_0}, \tag{8.81}$$

and Bangham's law

$$N = \gamma\, t^n, \quad \text{or} \quad \frac{dN}{dt} = \frac{\gamma}{t^p}, \tag{8.82}$$

where N is the total surface concentration of the adsorbed particles, t time, and β, γ, t_0, n, and p parameters, with $p = 1 - n$ and $n < 1$ (as a rule). Note that

$$N = N^0 + N^-, \tag{8.83}$$

where N^0 and N^- are the surface concentrations of adsorbed particles in the electrically neutral and charged states, respectively.

Assuming that (as is often the case)

$$\frac{dN^0}{dt} \ll \frac{dN^-}{dt}, \tag{8.84}$$

and employing (8.80) and (8.83), we have

$$I = \alpha \frac{dN}{dt} = \frac{A}{t + t_0} \, , \qquad (8.85)$$

for (8.81), while for (8.82) we have

$$\frac{dN}{dt} = \frac{B}{t^p} \, , \qquad (8.86)$$

where $A = \alpha\beta$ and $B = \alpha\gamma$. We see that we have the empirical laws (8.77) and (8.78).

We can, therefore, assume that in [42–48] the authors were involved with the Roginskii–Zel'dovich kinetics, while the authors of [49, 50] were involved both with the Roginskii–Zel'dovich kinetics and with Bangham's kinetics, depending on the prehistory of the sample.

8.5.3. Adsorption Luminescence at Adsorption Equilibrium

Let us examine the chemisorption kinetics in the region close to saturation. We will restrict our discussion to two limiting cases:

$$\tau^- \ll \tau^0 \ll \tau \qquad (8.87)$$

and

$$\tau \ll \tau^- \ll \tau^0 , \qquad (8.88)$$

where τ is the lifetime of a particle in the adsorbed state, and τ^0 and τ^- are the lifetimes of adsorbed particles in the electrically neutral and charged states, respectively. As shown in Section 3.3, for the kinetics of the neutral and charged forms we have the following:

$$N^0(t) = N_\infty^0 \, [1 - \exp(-t/\tau)],$$
$$N^-(t) = N_0^- \, [1 - \exp(-t/\tau)] \qquad (8.89)$$

for (8.87), and

$$N^0(t) = N_\infty^0 \, [1 - \exp(-t/\tau)],$$
$$N^-(t) = \frac{N_\infty^-}{\tau^-} \, \{ \tau^- \, [1 - \exp(-t/\tau^-)] - \tau \, [1 - \exp(-t/\tau)] \} \, , \qquad (8.90)$$

where N_∞^0 and N_∞^- are the surface coverages of the neutral and charged forms, and N_∞ the total surface coverage under adsorption equilibrium (i.e., at $t = \infty$), with

$$\frac{N_\infty^0}{N_\infty^-} = \frac{\tau^0}{\tau^-} .$$

(8.91)

In the limiting case (8.87), with (8.89), (8.80), and (8.83) taken into account, we find that $I(t) \to 0$ and $N(t) \to N_\infty$ as $t \to \infty$, i.e., luminescence is quenched as we approach the adsorption equilibrium, while in the limiting case (8.88), with (8.90), (8.80), and (8.83) taken into account, we find that for fairly large values of t (but not very large), viz., $\tau \ll t \ll \tau^-$,

$$I(t) = \frac{\alpha N_\infty^-}{\tau^-}, \quad N(t) = N_\infty,$$

since in this case $N_\infty^- \ll N_\infty^0$, and the luminescence intensity remains nonzero, while the adsorption equilibrium is as much as reached.

Thus, there is reason to believe that the case (8.87) was realized in [42–48], while the case (8.88) was realized in [49, 50].

We can, therefore, say that the results of the Soviet researchers [42–48], on the one hand, and those of the French researchers [49, 50], on the other, can be understood within the framework of the general theory of adsorption lumines-cence as two separate limiting cases.

In conclusion we note that the quenching of luminescence characteristic of adsorption luminescence can be expected to be suspended and the adsorption luminescence made stationary if the charged chemisorbed particles are constantly removed from the surface and replaced by neutral particles. This process occurs in radical-recombination luminescence, whose mechanism was discussed in Section 8.4. Studying the RRL mechanism and the adsorption luminescence mechanism, we can discover a certain relationship between the two, viz., radical-recombination luminescence can be interpreted as adsorption luminescence prolonged.

8.5.4. Adsorption Luminescence and the Adsorption Emission of Electrons

Adsorption luminescence, i.e., luminescence accompanying chemisorption, is often accompanied, in turn, by emission of electrons from the surface of the phosphor. This phenomenon, which can be called adsorption emission, was discovered and studied experimentally in [52–55]. A theoretical discussion of this effect was given in [55, 56].

In Sections 8.5.1–8.5.3 we discussed electron transitions at the surface and in the bulk of a semiconductor that accompany chemisorption and cause lumines-cence to occur. There we assumed that the entire energy liberated during chemi-sorption is spent on heating the adsorbent and not used for electron excitation of the system. This is one possibility.

Another possibility is when the chemisorption energy is used entirely or partly for transferring the system from one electronic state to a more excited state, a tran-

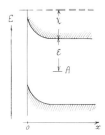

Fig. 8.33. Energy band structure of a semi-conductor at the semiconductor surface.

sition from which into the initial state is accompanied by luminescence. In this context the theory must include the distance r between the chemisorbed particle and the adsorbent surface as an independent parameter.

Figure 8.33 shows the band structure of a semiconductor. The plane $x = 0$ is the semiconductor's surface, which we assume, in order to be definite, to be negatively charged (the bands are bent upward). By A we denote an activator level, which we assume to be a donor level, while ϵ is the distance from this level to the conduction band above the level, and χ is the height of the surface potential barrier, i.e., the minimal energy required to bring a free electron from the bottom of the conduction band in the crystal's bulk outside the semiconductor.

Let AeL and AL be an activator atom in respectively, an electrically neutral and ionized state, eL an electron in the conduction band, and C an adsorbed particle. The ionization of the activator (the excitation act) and the neutralization (the luminescence act) have the following form:

$$\text{AeL} \rightleftarrows \text{AL} + \text{eL}.$$

We can write the adsorption and desorption reactions as follows:

$$\text{C} + \text{L} \rightleftarrows \text{CL},$$

where L denotes the lattice.

The right-hand side of Fig. 8.34 shows the energy levels corresponding to states of the system in which particle C lies at an infinite distance from the adsorbent surface. The lower level corresponds to the ground state of the system, with the electron localized at the activator level, AeL, while the upper level corresponds to an electron raised from the activator level to the bottom of the conduction band, AL + eL. The upper level, therefore, is the lower edge of the continuous spectrum, since a free electron can be raised to any level inside the conduction band.

The left-hand side of Fig. 8.34 shows the system's energy as a function of distance r. The direction from right to left in Fig. 8.34 corresponds to a gas molecule approaching the adsorbent surface. The minimum of each curve corresponds to the equilibrium distance ($r = r_0$) between the chemisorbed particle and the surface.

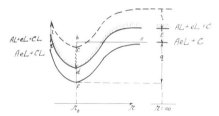

Fig. 8.34. Energy band structure of a semi-conductor in the event of adsorption luminescence accompanied by adsorption emission of electrons.

In chemisorption we are dealing with transition from point a to point b in Fig. 8.34, i.e., from the ground state to an excited state. The subsequent transition from point b to point d (the wavy arrow in Fig. 8.34) reflects the fact that the electron drops to the bottom of the conduction band, a process in which heat is released. Transition from point d to point e takes the system to its ground state, which is accompanied by emission of a photon $h\nu = \epsilon$ (the luminescence act).

Thus, the energy of chemisorption q (which is in the depth of the potential well in Fig. 8.34) is used to electronically excite the system (excitation of the activator). If point b lies higher than point c, as depicted in Fig. 8.34, then along with the transition of the system to its ground state (adsorption luminescence), an electron may leave the crystal (see Fig. 8.33). The latter effect constitutes adsorption emission.

Thus, if $q \geq \epsilon$, we are dealing with adsorption luminescence, while if $q \geq \epsilon + \chi$, we have adsorption emission. In both cases the energy liberated during the chemisorption of a gas particle is used to transfer the system to an excited state.

The processes of adsorption luminescence and adsorption emission are symbatic; i.e., a rise in luminescence increases electron emission, while a quenching of luminescence decreases electron emission. This is illustrated by Fig. 8.35, where the kinetic curves of adsorption luminescence (a) and adsorption emission of electrons (b) are shown for ZnO in molecular oxygen. In Figure 8.35a the vertical axis gives the number of photons emitted every second from 1 cm^2, while in Fig. 8.35b the vertical axis gives the number of electrons emitted every second from 1 cm^2. (Curves 1 correspond to 80 K and curves 2 to 185 K.) Figure 8.35 is taken from [55].

8.5.5. Luminescence Emission Accompanying Catalytic Reactions at Surfaces

In 1976, Claudel and his coworkers [57] discovered a new type of heterogeneous chemiluminescence: luminescence emission from a semiconductor involved in a catalytic reaction that takes place at the semiconductor's surface. The

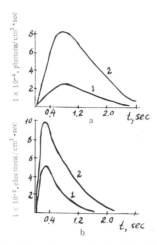

Fig. 8.35. Kinetic curves for ZnO in molecular oxygen: a) of adsorption luminescence; b) of adsorption emission of electrons.

name they chose was catalytic luminescence. Since each catalytic heterogeneous reaction includes adsorption and desorption, it is natural to assume that catalytic luminescence is nothing more than masked adsorption luminescence. But this is not always the case. As Claudel has shown, we have a new phenomenon, which cannot be reduced to adsorption luminescence. In catalytic luminescence the emission of radiation is observed only as long as the reaction takes place, and the luminescence intensity is proportional to the reaction rate.

Claudel and his coworkers studied catalytic luminescence on the catalyst ThO_2 in the reaction of oxidation of CO:

$$2CO + O_2 \rightarrow 2CO_2.$$

The researchers dealt with two samples, A and B, of ThO_2, which differed in preparation, viz., sample A was obtained from sample B as a result of annealing the latter in hydrogen. In other words, the Fermi level in sample A was higher than in sample B. This means that the surface of sample A had a higher relative content of charged oxygen (O^- or O_2^-) than the surface of B.

The oxidation of CO, according to [57], follows one of two routes:

$$COpL + OeL + CO_2 epL \rightarrow CO_2 + h\nu + L \tag{8.92}$$

or

$$COpL + OL \rightarrow CO_2 + pL. \tag{8.93}$$

Here we have used the notation employed in the electronic theory of catalysis,

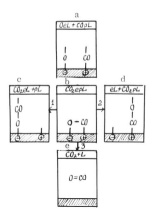

Fig. 8.36. Reaction of oxidation of CO:
a–e) various stages in the reaction.

viz., L is the lattice, eL and pL, respectively, an electron in the conduction band and a hole in the valence band, and COpL and OeL, respectively, an adsorbed CO molecule and an adsorbed O atom in the charged state (CO^+ and O^-).

There are many works devoted to oxidation of CO. This reaction was extensively studied within the framework of the electronic theory (see Section 5.3.3), where it is assumed to follow the radical mechanism. Note, however, that this mechanism automatically leads, as we will now see, to luminescence emission accompanying the reaction.

Reaction (8.92) can be thought of as two successive stages. The reaction starts with the formation of chemisorbed O^- and CO^+ particles at the surface, i.e., they formation of oxygen atoms and CO molecules "strongly" bound to the surface (OeL and COpL). The first are acceptors, i.e., they capture a free electron, while the second are donors, i.e., they localize a free hole around them. The clusters OeL and COpL are surface ion-radicals, as illustrated by Fig. 8.36a. When such clusters merge, they form a valence-saturated, electrically neutral cluster CO_2epL (depicted in Fig. 8.36b; cf. Fig. 5.13), which constitutes a "weak" form of chemisorption of the CO_2 molecules:

$$OeL + COpL \rightarrow CO_2 epL.$$

Here we are dealing with adsorption at Mott's exciton (an electron and a hole that are near each other and are coupled by the Coulomb interaction), which in this process is virtual, i.e., is not created beforehand but in the process of adsorption.

After this stage the reaction can follow either of the three routes, as depicted in Fig. 8.36:

$$1)\ CO_2\,epL \rightarrow CO_2\,eL + pL,$$
$$2)\ CO_2\,epL \rightarrow CO_2\,pL + eL,$$
$$3)\ CO_2\,epL \rightarrow CO_2 + epL \rightarrow CO_2 + h\nu + L,$$

where epL is the symbol of a free exciton. The first two routes lead to the creation of two different surface ion-radicals CO_2eL and CO_2pL and a free carrier (an electron or a hole). The third route leads to desorption of the CO_2 molecule and annihilation of the exciton at the surface, i.e., to emission of a photon:

$$epL \rightarrow h\nu + L.$$

Returning to Eqs. (8.92) and (8.93), we see that if the reaction follows Eq. (8.92), it is accompanied by luminescence emission, while if it follows Eq. (8.93), there is no room for luminescence. Reaction (8.92) is predominant on sample A, and reaction (8.93) on sample B. This means that sample A gives off luminescent radiation while sample B does not.

The experimental data on catalytic luminescence is so meager that building a detailed, quantitative theory must be considered premature.

Chapter 9

CONCLUSION

9.1. THE "LOCAL" AND "COLLECTIVE" EFFECTS IN CHEMISORPTION AND CATALYSIS

The problem of the interaction of the gaseous phase with the lattice of a solid incorporates two types of problems. The first group consists of problems dealing with the interaction of an adsorbed particle with an adsorption center. Here both the electrons of the particles being adsorbed and those of the particle at which the adsorption takes place are involved. This group of problems is commonly known as the local interaction problem. But there are also entirely different problems, which chemists call problems of collective interaction. Here the interaction of the adsorbed particle with the entire lattice is studied, and the entire collection of free electrons and holes of the lattice come into play.

According to Morrison's terminology (see [1]), in the first case we are dealing with the surface molecule model, while in the second we are dealing with the rigid band model.

The problems of the first type belong to quantum mechanics, precisely, quantum mechanics of a system of electrons. The problems of the second type belong to quantum statistics. Here the concept of a Fermi level comes into play, which is characteristic of quantum statistics.

An example of the first type of problems is the formation of the "weak" bond in chemisorption. Here the free electrons and holes of the crystal lattice play no part, and an adsorbed particle remains on the whole electrically neutral, although it may be polarized (see Section 1.4). The simplest problem of this type is that of the adsorption of a hydrogen atom at an ionic crystal. This problem can be considered to be a one-electron one (see [2, 3]). Generally, however, problems of the "weak" form of chemisorption are essentially multielectron problems. Figures 2.19b, 2.20a, and 2.21a depict the "weak" forms of chemisorption for

Fig. 9.1. Effect of O⁻ ions on a semiconductor surface.

H_2O, CO, and CO_2 molecules by valence lines. The electron and hole involved in the process act as free valences. Within the framework of quantum mechanics solution of any multielectron problem is extremely difficult.

An example of problems of the second type (collective interaction) is the problem of formation of a "strong" bond in chemisorption (see Section 2.3), in which a free electron or free hole of the crystal lattice acts in the bond, as a result of which the chemisorbed particle acquires an electric charge. The effect of adsorption on electrical conductivity and work function and the effect of an external electric field on adsorptivity and catalytic activity are typical examples of collective effects. From the viewpoint of collective interactions the entire individuality of a chemisorbed particle lies in a single parameter, the position of the local level of that particle. This parameter reflects the particle's chemical nature.

An example of a problem with two aspects, i.e., one which can be considered as a problem of local and collective interactions, is the effect of impurities on the adsorptive and catalytic properties. Indeed, there are two ways in which this influence can manifest itself:

First, through the Fermi level, whose position depends on the nature and concentration of the impurity. In turn, the position of the Fermi level determines the adsorptivity and catalytic activity of the surface. Here the impurity acts in the form of a typical collective effect.

Second, by direct participation of the impurity in the adsorption and catalytic act, since impurity atoms may manifest themselves as active centers in adsorption and catalysis or may, on the contrary, block such centers. This is a typical local effect.

The two mechanisms (local and collective) often act in opposite directions. The dominant role of each depends on the conditions of the experiment. For instance, lithium atoms implanted in a ZnO or NiO crystal as an impurity and forming a substitutional solution can be considered as adsorption centers for oxygen, since near each lithium atom there appears a free (nonsaturated) valence (Fig. 9.1). From this viewpoint introduction of lithium must increase the adsorptivity of the surface with respect to oxygen. On the hand, the same lithium atoms, forming an interstitial solution, act as donor impurities and lower the Fermi level (i.e., increase the work function). In this case, according to the electronic theory, lithium suppresses, rather then enhances, the adsorptivity of the surface with respect to such an acceptor gas as oxygen. Indeed, according to the data of Bielanski and Deren [4], lithium increases the work function and slows down oxygen adsorption, while according to Keier's data [5], lithium increases the work function *and* also accelerates oxygen adsorption. In the first case (Bielanski and

Deren's data) the collective effect manifests itself, while in the second (Keier's data) the collective effect is overshadowed by the local.

Photoadsorption and the photocatalytic effect are typical examples of collective effects. Indeed, both in the case of intrinsic light absorption and in the case where the absorption centers are the atoms of an impurity or the adsorbed particles proper, photoelectrically active absorption leads to the appearance of nonequilibrium carriers in the semiconductor. This changes the charge state of the chemisorbed particles participating in the reaction. This, in turn, influences the adsorptivity and the catalytic activity of the surface. Thus, in photoadsorption and the photocatalytic effect we are dealing not so much with chemisorbed particles as such as with their interaction with the entire crystal lattice.

The literature often debates what interaction (local or collective) is more important for catalysis (e.g., see [6]). Krylov [6] writes that "the question of collective and local effect has long been the central problem of catalysis theory." However, we believe there is no problem here at all. Both collective and local effects are present in catalysis. The very question of which effect is more important cannot be considered correct, since the answer depends on the phenomenon referred to.

There exists the false notion that all cases in which the local interactions mask the collective contradict the electronic theory of catalysis. Actually they are included in the theory. The electronic theory deals with local effects as well as with collective. Indeed, its significance lies in the fact that it embraces the collective effects as well and establishes their role in catalysis and chemisorption.

9.2. THE BASIC CONCEPTS OF THE ELECTRONIC THEORY OF CHEMISORPTION

In this book we have studied the interaction of a semiconductor with the gaseous medium surrounding it. We have focussed largely on the theory of chemisorption at semiconductors. This theory is usually called the electronic theory of chemisorption, since it studies the electronic processes at semiconductor surfaces accompanying chemisorption and determining its rate.

In concluding this book, let us restate the main concepts of this theory.

(1) The quantum-mechanical approach to the interaction of a foreign molecule with a crystal lattice shows that there are various possible forms of chemisorption, differing in the character of the bonding of the adsorbed particle with the adsorbent lattice and reflecting the ability of the chemisorbed particle to make free electrons and holes of the lattice participate in the bond. Here one must distinguish between the "weak" form of chemisorption, which does not require free electrons and holes of the lattice and in which the chemisorbed particle remains electrically neutral, and the "strong" from of chemisorption, in which a free electron or hole is localized at (or near) the chemisorbed particle and in which the chemisorbed particle, therefore, is charged (see Section 2.3).

(2)   Among the various forms of chemisorption (electrically neutral and positively and negatively charged) we must distinguish between valence saturated forms, on the one hand, and radical or ion-radical forms, on the other. The formation of radical or ion-radical forms is determined by the role that the free electrons and holes act as free surface valences (positive and negative, respectively). In radical forms of chemisorption the chemisorbed particle has an enhanced reactivity, which means a higher capability of entering into a chemical reaction with another chemisorbed particle or a particle coming from the gaseous phase. Thus, the various forms of chemisorption differ not only by the sign of the charge and the nature and strength of the bonds but also by the reactivity of the chemisorbed particles (see Section 2.4).

(3)   The various forms of chemisorption may transform into each other. In other words, a chemisorbed particle, while remaining in the adsorbed state, may change its type of bond with the surface. It may go over from a state with one type of bond to a state with a different type. Such transitions mean that a free electron or hole is being localized or delocalized at the chemisorbed particle (or near it) (see Section 3.1).

(4)   When electron equilibrium in a semiconductor has set in, each chemisorbed particle possesses a definite probability of being in one or another state characterized by the type of its bond with the surface. In other words, out of the total number of particles of a given type chemisorbed at a unit surface area, a certain fraction will be in the state of "weak" bonding, while the other will be in the state of "strong" bonding. The relative content of each form of chemisorption and, hence, the reactivity of the chemisorbed particles are determined by the position of the Fermi level at the crystal surface (see Section 3.5).

(5)   All other factors being the same, the position of the Fermi level determines the adsorptivity of the surface with respect to molecules of a definite type, i.e., the total number of molecules of this type that can be adsorbed under equilibrium with the gaseous phase. The position of the Fermi level at the surface also determines the magnitude and sign of the surface charge formed during chemisorption (see Section 3.2). In some cases it determines the fraction of reversibly (or irreversibly) chemisorbed particles (see Section 3.4). Finally, the position of the Fermi level, all other factors being the same, determines the catalytic activity of the semiconductor with respect to a given reaction (see Section 5.2).

(6)   Generally the position of the Fermi level at the surface depends on its position in the bulk. This establishes a relationship between bulk and surface properties of a crystal. In this way the factors that shift the Fermi level in the bulk influence the surface properties (see Section 4.1). But when the surface state density is high, this relationship breaks, and then the position of the Fermi level in the bulk does not influence the position of the Fermi level at the surface (see Section 4.3).

These basic results of the electronic theory of chemisorption lead to a number of consequences which may be checked experimentally. Comparison of theory and experimental data runs through this entire book.

9.3. THE ELECTRONIC THEORY OF CHEMISORPTION
AND EXPERIMENT

There are two aspects to the interrelationship between experiment and the electronic theory of chemisorption. First, the theory explains the known facts, i.e., follows the experiment. Second, the theory predicts certain phenomena, which are then verified by experiment, in which it precedes the experiment.

Here is a short summary of the main predictions.

(1) First we must point to the effect that chemisorption has on the work function and electrical conductivity of the semiconductor. This is because the semiconductor surface is charged under chemisorption and thereby shifts the Fermi level. The charge arises from the localization of free electrons or holes of the crystal lattice at the chemisorbed particles and was theoretically predicted in the late 1940s and early 1950s. The effect has since been studied in fairly great detail experimentally for various systems. It often serves the experimenter as a method for establishing whether the adsorbate is a donor or acceptor if the nature of the adsorbent is known (i.e., whether the semiconductor is of the n- or p-type). It can also be used as a method for determining the nature of the adsorbent if that of the adsorbate is known (see Section 4.2).

(2) Next we must point to the correlation between the catalytic activity of a semiconductor and its electrical conductivity. This correlation follows from the fact that both catalytic activity and the electrical conductivity are determined by the same factor, the position of the Fermi level. This correlation was predicted theoretically in 1950, when there was no experimental data on the subject. At present we have a broad range of works in which this correlation has been observed. Several authors measured the electrical conductivity and catalytic activity of samples that differed in their prehistory. They found that the variations of these two characteristics of the semiconductors studied were either symbatic or the opposite, depending on the type of sample (see Section 5.4).

(3) From the viewpoint of the electronic theory there must be a similar correlation between the adsorptivity and catalytic activity, on the one hand, and the electron work function in the semiconductor, on the other. This is again due to the fact that, all other factors being the same, the adsorptivity, catalytic activity, and work function are determined by the position of the Fermi level at the crystal's surface. The existence of such a correlation, predicted theoretically, has been verified in many experiments (see Sections 3.5 and 5.4).

(4) We must also note the role played by an external electric field. From the theory's viewpoint, a change can be expected in the adsorptivity and catalytic activity of a semiconductor when an electric field normal to the adsorbing surface is applied. This effect, caused by a shift in the position of the Fermi level under the electric field, was predicted before it became the subject of experimental investigation. It can now be considered as experimentally verified (see Sections 4.4 and 5.5).

(5) Finally, we must point to the special chemisorptive and catalytic properties of thin semiconductor films at metal surfaces. If the film's thickness is less than the screening length, the adsorptivity of the film and its catalytic activity

must depend on the thickness according to the theory. This follows from the fact that in a thin film the position of the Fermi level at the outer surface depends on the thickness of the film. This has been observed in experiments (see Section 4.6).

This list of experimental facts that follow from the theory and on which the theory is based could be continued. Experiments are the "nutrient" on which theory grows. Theory throws light on experiments. Without experiments theory has no meaning, and without theory experiments are blind. Of course, experiments, from which we always start and to which we always return, are the arbitrator in theory. The theoretician always follows the experimenter but he also shows him the way. All the material in this book bears this out. The reader must never forget, however, that nature is far more complex than our theories and never ceases to challenge us.

REFERENCES

CHAPTER 1

1. F. Bloch, *Z. Phys.*, **52**, 555 (1928).
2. N. F. Mott and W. Jones, *The Theory of Properties of Metals and Alloys*, Oxford University Press, Oxford (1958).
3. A. I. Ansel'm, *Introduction to Semiconductor Theory*, Mir Publishers, Moscow (1981).
4. D. Blokhintsev (Blochinzew), *Sow. Phys.*, **5**, 316 (1934).
5. A. H. Wilson, *Proc. R. Soc. London, Ser. A*, **133**, 458 (1931).
6. N. F. Mott and R. W. Gurney, *Electronic Processes in Ionic Crystals*, Clarendon Press, Oxford (1940).
7. I. E. Tamm, *Sow. Phys.*, **1**, 733 (1932); *Z. Phys.*, **76**, 849 (1932).
8. S. Ryzhanov (Rijanow), *Z. Phys.*, **89**, 806 (1934).
9. E. T. Goodwin, *Proc. Cambridge Philos. Soc.*, **35**, 221, 232 (1939).
10. W. Shockley, *Phys. Rev.*, **56**, 317 (1939).
11. J. J. Koutecky, *J. Phys. Chem. Solids*, **14**, 227 (1960).
12. Ya. I. Frenkel, *Vestn. Akad. Nauk SSSR*, No. 10, 61 (1946).
13. R. S. Mulliken, *Phys. Rev.*, **40,** 55 (1932).
14. F. Hund, *Z. Phys.*, **73**, 1, 565 (1931).
15. S. I. Pekar, *Zh. Éksp. Teor. Fiz.*, **18**, 525 (1948).
16. Th. Wolkenstein, *Usp. Fiz. Nauk*, **43**, 11 (1951).

CHAPTER 2

1. I. Langmuir, *J. Am. Chem. Soc.*, **38**, 2217 (1916).
2. S. Z. Roginskii (Roginsky) and Ya. B. Zel'dovich (Zeldovitch), *Acta Physicochim. URSS*, **1**, 554, 595 (1934).

3. H. S. Taylor and N. Thon, *J. Am. Chem. Soc.*, **74**, 4169 (1952).

4. F. S. Stone, in: *Chemistry of the Solid State*, ed. by W. E. Garner, Butterworths, London (1955), p. 367.

5. J. M. Thuillier, *Ann. Phys. (Paris)*, **5**, 865 (1960).

6. R. Schuttler and J. M. Thuillier, *C. R. Acad. Sci.*, **355**, 877 (1962).

7. D. H. Bangham and F. P. Burt, *Proc. R. Soc. London, Ser. A*, **105**, 481 (1924); D. H. Bangham and W. Sever, *Philos. Mag.*, **49**, 938 (1928).

8. H. Freundlich, *Kapillarchemie*, Academische Verlag, Leipzig (1922).

9. A. N. Frumkin and M. Shlygin, *Acta Physicochim. URSS*, **3**, 791 (1935).

10. F. V. Lenel, *Z. Phys. Chem., (Leipzig)* **B23**, 379 (1933).

11. W. J. C. Orr, *Trans. Faraday Soc.*, **35**, 1247 (1939).

12. J. E. Lennard-Jones, *Trans. Faraday Soc.*, **28**, 333 (1932).

13. Th. Wolkenstein, *Zh. Fiz. Khim.*, **23**, 917 (1949); *Probl. Kinet. Katal.*, **7**, 360 (1949).

14. Th. Wolkenstein, *Zh. Fiz. Khim*, **27**, 159, 167 (1958); *Usp. Fiz. Nauk.*, **50**, 253 (1953).

15. Th. Wolkenstein, *Zh. Fiz. Khim.*, **21** 163 (1947).

16. Th. Wolkenstein, *Zh. Fiz. Khim.*, **22**, 311 (1948).

17. Th. Wolkenstein, *Zh. Fiz. Khim.*, **26**, 1462 (1952).

18. Th. Wolkenstein, *Zh. Fiz. Khim.*, **28**, 422 (1954).

19. Th. Wolkenstein and S. Z. Roginskii, *Zh. Fiz. Khim.*, **29**, 485 (1955).

20. Th. Wolkenstein, *Usp. Fiz. Nauk*, **60**, 249 (1956).

21. Th. Wolkenstein, *Zh. Fiz. Khim.*, **21**, 1317 (1947).

22. V. L. Bonch-Bruevich, *Zh. Fiz. Khim.*, **25**, 1033 (1951).

23. Th. Wolkenstein, *Izv. Akad. Nauk SSSR, Otd. Khim. Nauk*, No. 8, 916 (1957); *J. Chim. Phys.*, **54**, 175 (1957).

24. T. B. Grimley, *Proc. Phys. Soc., London*, **72**, 103 (1958).

25. J. Koutecky, *Proc. Phys. Soc., London*, **73**, 323 (1959).

26. É. L. Nagaev, in: *Scientific Conference of Junior Scientists* [in Russian], Moscow University Press, Moscow (1959), p. 18.

27. Th. Wolkenstein, *Izv. Akad. Nauk. SSSR, Otd. Khim. Nauk*, No. 2, 147 (1957).

28. Th. Wolkenstein, *Probl. Kinet. Katal.*, **8**, 79 (1955).

29. Th. Wolkenstein, *Izv. Akad. Nauk. SSSR, Otd. Khim. Nauk*, No. 8, 924 (1957); *J. Chim. Phys.*, **54**, 181 (1957).

30. Th. Wolkenstein and V. L. Bonch-Bruevich, *Zh. Éksp. Teor. Fiz.*, **20**, 624 (1950).

31. Th. Wolkenstein, *Zh. Éksp. Teor. Fiz.*, **22**, 184 (1952).

32. Th. Wolkenstein, *Usp. Khim.*, **27**, 1304 (1958); *Chem. Tech.*, **11**, 8, 103 (1959).

33. V. L. Bonch-Bruevich and V. B. Glasko, *Vestn. Mosk. Gos. Univ.*, No. 5, 91 (1958); *Dokl. Akad. Nauk SSSR*, **124**, 1015 (1959).

34. Th. Wolkenstein, *Kinet. Katal.*, **19**, 90 (1978).

35. J. C. Slater, *Phys. Rev.*, **38**, 1109 (1931).

36. A. A. Balandin, *The Current State of the Multiplet Theory of Heterogeneous Catalysis* [in Russian], Nauka, Moscow (1968).

37. S. R. Morrison, *The Chemical Physics of Surfaces*, Plenum Press, New York (1977).
38. K. H. Johnson and R. P. Messmer, *J. Vac. Sci. Technol.*, **11**, 235 (1974).
39. A. P. Zeif, in: *Elementary Physicochemical Processes at the Surface of Single Crystal Semiconductors* [in Russian], Nauka, Novosibirsk (1975).
40. G. M. Zhidomirov, *Kinet. Katal.*, **18**, 1192 (1977).
41. R. P. Messmer, in: *Semiempirical Methods of Electronic Structure Calculation, Part B: Applications* (G. A. Segal, ed.), Plenum Press, New York (1977).
42. G. V. Gadiyak, A. A. Karpushin, and Yu. N. Morokov, in: *The Problems of the Physical Chemistry of Semiconductor Surfaces* [in Russian], Nauka, Novosibirsk (1978), p. 72.
43. H. Dunken and V. Lygin, *Quantenchemie der Adsorption an Festkörperoberflächen*, Deutscher Verlag, Leipzig (1978).
44. A. W. Goddard, and T. C. McGill, *J. Vac. Sci. Technol.*, **16**, 1308 (1979).
45. C. W. Bauschlicter, P. S. Bagus, and H. F. Schaefer, *IBM J. Res. Dev.*, **22**, 213 (1978).
46. C. Pisani and F. Ricca, *Surf. Sci.*, **92**, 481 (1980).
47. I. D. Mikheikin, I. A. Abronin, G. M. Zhidomirov, and V. B. Kazanskii (Kazansky), *J. Mol. Catal.*, **3**, 435 (1977/78).
48. I. D. Mikheikin, I. A. Abronin, G. M. Zhidomirov, and V. B. Kazanskii, *Kinet. Katal.*, **18**, 1580 (1977).
49. I. D. Mikheikin, A. I. Lumpov, G. M. Zhidomirov, and V. B. Kazanskii, *Kinet. Katal.*, **19**, 1053 (1978).
50. I. D. Mikheikin, A. I. Lumpov, and G. M. Zhidomirov, *Kinet Katal.*, **20**, 501 (1979).
51. A. G. Pel'menshchikov, I. N. Senchenya, G. M. Zhidomirov, and V. B. Kazanskii, *Kinet. Katal.*, **24**, 233 (1983).
52. V. A. Korsunov, N. D. Chuvylkin, G. M. Zhidomirov, and V. B. Kazanskii, *Kinet. Katal.*, **19**, 1152 (1978).
53. V. A. Korsunov, N. D. Chuvylkin, G. M. Zhidomirov, and V. B. Kazanskii, *Kinet. Katal.*, **21**, 402 (1980).

CHAPTER 3

1. Th. Wolkenstein, *Zh. Fiz. Khim.*, **22**, 311 (1948).
2. Th. Wolkenstein, *Zh. Fiz. Khim.*, **26**, 1462 (1952).
3. Th. Wolkenstein, *Zh. Fiz. Khim.*, **28**, 422 (1954).
4. Th. Wolkenstein and S. Z. Roginskii, *Zh. Fiz. Khim.*, **29**, 485 (1955).
5. Th. Wolkenstein, *Probl. Kinet. Katal.*, **8**, 79 (1955).
6. Th. Wolkenstein, *Vestn. Mosk. Gos. Univ.*, No. 4, 79 (1957); *Zh. Fiz. Khim.*, **32**, 2383 (1958).
7. Sh. M. Kogan, *Zh. Fiz. Khim.*, **33**, 156 (1959).
8. Yu. A. Zarif'yants, *Zh. Fiz. Khim.*, **52**, 3030 (1978).
9. C. G. B. Garrett, *J. Chem. Phys.*, **33**, 966 (1960).

10. Th. Wolkenstein and Sh. M. Kogan, *J. Chim. Phys.*, **55**, 483 (1958); *Izv. Akad. Nauk SSSR, Otd. Khim. Nauk*, No. 9, 1536 (1959).
11. S. Yu. Elovich and L. Ya. Margolis, *Izv. Akad. Nauk SSSR, Ser. Fiz.*, **21**, 206 (1957).
12. A. I. Gubanov, *The Theory of the Rectifying Effect of Semiconductors* [in Russian], Gostekhizdat, Moscow (1956), p. 32.
13. Th. Wolkenstein and O. Peshev, *Kinet. Katal.*, **6**, 95 (1965); *J. Catal.*, **4**, 301 (1965).
14. O. Peshev, *Usp. Khim.*, **35**, 1830 (1966).
15. Th. Wolkenstein, *Kinet. Katal.*, **12**, 179 (1971).
16. A. E. Bazhanov, Yu. A. Zarif'yants, and V. S. Kuznetsov, *Zh. Fiz. Khim.*, **49**, 1771 (1975).
17. N. F. Mott and R. W. Gurney, *Electronic Processes in Ionic Crystals*, Clarendon Press, Oxford (1940).
18. J. H. de Boer, *The Dynamical Character of Adsorption*, Oxford University Press, London (1953).
19. S. G. Kakashnikov, in: *Proc. VIth Int. Conf. on Phys. of Semiconductors*, Czech. Acad. Sci., Prague (1961), p. 241.
20. R. Kubo and Y. Toyozawa, *Prog. Theor. Phys.*, **13**, 160 (1955).
21. O. Peshev, *Kinet. Katal.*, **7**, 84 (1966).
22. Sh. M. Kogan and V. B. Sandomirskii, *Zh. Fiz. Khim.*, **33**, 1709 (1959).
23. Th. Wolkenstein, *Zh. Fiz. Khim.*, **24**, 1068 (1950).
24. P. Aigrain and C. R. Dugas, *Z. Elektrochem.*, **56**, 363 (1952).
25. K. Hauffe and H. J. Engell, *Z. Elektrochem.*, **56**, 366 (1952).
26. H. J. Engell, Halbleiterprobleme, Braunschweig, Vieweg 1, 249 (1954).
27. Sh. M. Kogan and V. B. Sandomirskii, *Zh. Fiz. Khim.*, **33**, 1129 (1959).
28. J. E. Germain, *C. R. Acad. Sci.*, **238**, 235, 345 (1954); *J. Chim. Phys.*, **51**, 263, 691 (1954).
29. J. E. Germain, *Catalyse hétérogène*, Dunod, Paris (1959).
30. K. Hauffe, *Angew. Chem.*, **67**, 189 (1955).
31. K. Hauffe, *Reaktionen in und an festen Stoffen*, Springer, Berlin (1955).
32. P. B. Weitz, *J. Chem. Phys.*, **21**, 1531 (1953).
33. K. Hauffe, *Reaktionen in und an festen Stoffen*, 2nd. ed., Springer, Berlin (1966).

CHAPTER 4

1. Th. Wolkenstein and Sh. M. Kogan, *Z. Phys. Chem.*, **211**, 282 (1959).
2. É. Kh. Enikeev, L. Ya. Margolis, and S. Z. Roginskii, *Dokl. Akad. Nauk SSSR*, **130**, 807 (1960).
3. A. Biclanski and J. Deren, in: *Symposium on Electronic Phenomena in Chemisorption and Catalysis on Semiconductors* (K. Hauffe and Th. Wolkenstein, eds.), De Gruyter, Berlin (1969), p. 156.
4. *Electronic Phenomena at Semiconductor Surfaces* [in Russian], (V. I. Lyashenko, ed.), Naukova Dumka, Kiev (1968).

5. G. P. Peka, *The Physics of Semiconductor Surfaces* [in Russian], Kiev Univ. Press, Kiev (1967).

6. F. I. Vilesov and A. N. Terenin, *Dokl. Akad. Nauk SSSR*, **125**, 1053 (1959).

7. V. I. Lyashenko and I. I. Stepko, *Probl. Kinet. Katal.*, **8**, 180 (1955).

8. N. P. Keier and L. N. Kutseva, *Dokl. Akad. Nauk SSSR*, **117**, 259 (1957).

9. V. I. Lyashenko and V. G. Litovchenko, *Zh. Tekh. Fiz.*, **28**, 447, 454 (1958).

10. G. Parravano and C. A. Domenicali, *J. Chem. Phys.*, **26**, 359 (1957).

11. S. Yu. Elovich, L. Ya. Margolis, and S. Z. Roginskii, *Dokl. Akad. Nauk SSSR*, **129**, 732 (1959).

12. É. Kh. Enikeev, *Probl. Kinet. Katal.*, **10**, 88 (1960).

13. É. Kh. Enikeev, L. Ya. Margolis, and S. Z. Roginskii, *Dokl. Akad. Nauk SSSR*, **129**, 372 (1959).

14. J. R. Schrieffer, *Phys. Rev.*, **97**, 641 (1955); in: *Semiconductor Surface Physics* (R. H. Kingston, ed.), University of Pennsylvania Press, Philadelphia (1957), p. 55.

15. V. A. Tyagai and A. V. Savchenko, *Fiz. Tverd. Tela (Leningrad)*, **7**, 3472 (1965).

16. Th. Wolkenstein, *Kinet. Katal.*, **11**, 395 (1970).

17. Th. Wolkenstein and S. F. Timashov, *Fiz. Tekh. Poluprovodn.*, **5**, 1460 (1971).

18. V. F. Kiselev, S. N. Kozlov, and Yu. A. Zarif'yants, in: *Problems of the Physical Chemistry of Semiconductor Surfaces* [in Russian], Nauka, Novosibirsk (1978), p. 200.

19. A. V. Rzhanov, *Radiotekh. Elektron.*, **1**, 1086 (1956).

20. A. V. Rzhanov, Yu. F. Novototskii-Vlasov, and I. G. Neizvestnyi, *Zh. Tekh. Fiz.*, **27**, 2440 (1957).

21. A. V. Rzhanov, Yu. F. Novototskii-Vlasov, and I. G. Neizvestnyi, *Fiz. Tverd. Tela (Leningrad)*, **1**, 1472 (1959).

22. A. V. Rzhanov, *Fiz. Tverd. Tela*, **3**, 822, 1718 (1961).

23. A. V. Rzhanov, *Proc. Intern. Conf. Semicond. Phys.*, Prague (1960), p. 503.

24. Yu. F. Novototskii-Vlasov and A. V. Rzhanov, *Surf. Sci.*, **2**, 93 (1964).

25. A. Many and D. Gerlich, *Phys. Rev.*, **107**, 404 (1957).

26. V. B. Sandomirskii, Dissertation, Institute of Physical Chemistry, Academy of Sciences of the USSR, Moscow (1955).

27. Th. Wolkenstein and V. B. Sandomirskii, *Probl. Kinet. Katal.*, **8**, 189 (1955).

28. V. B. Sandomirskii, *Izv. Akad. Nauk SSSR, Ser. Fiz.*, **21**, 211 (1957).

29. I. A. Myasnikov and S. Ya. Pshezhetskii, *Probl. Kinet. Katal.*, **8**, 175 (1955).

30. V. G. Litovchenko, V. I. Lyashenko, and O. S. Frolov, *Ukr. Fiz. Zh.*, **10**, 1334 (1965).

31. A. E. Bazhanova and Yu. A. Zarif'yants, *Vestn. Mosk. Gos. Univ., Ser. Fiz.*, No. 3, 355 (1972); *Vestn. Mosk. Gos. Univ.*, No. 3, 360 (1974).

32. E. N. Figurovskaya and V. F. Kiselev, *Dokl. Akad. Nauk SSSR*, **182**, 1365 (1968).

33. T. J. Gray, *Discuss. Faraday Soc.*, No. 8, 331 (1950).

34. V. I. Lyashenko, Doctoral Dissertation, Kiev State University (1955).

35. N. P. Keier and G. I. Chizhikova, *Dokl. Akad. Nauk SSSR*, **120**, 830 (1958).
36. J. N. Zemel and J. O. Varela, *J. Phys. Chem. Soc.*, 14, 142 (1960).
37. J. A. Dillon, Jr., and H. E. Farnsworth, *J. Appl. Phys.*, **28**, 174 (1957).
38. P. Handler and W. M. Portnov, *Phys. Rev.*, **116**, 516 (1959).
39. P. Kh. Burshtein, L. A. Larin, and G. F. Voronina, *Dokl. Akad. Nauk SSSR*, **133**, 148 (1960).
40. J. Stelzer, *J. Electron. Control*, Ser. I, **8**, 39 (1960).
41. C. Wagner, *J. Chem. Phys.*, **18**, 69 (1950).
42. W. E. Garner, T. J. Gray, and F. S. Stone, *Proc. R. Soc. London*, Ser. A, **197**, 296 (1949).
43. É. Kh. Enikeev, L. Ya. Margolis, and S. Z. Roginskii, *Dokl. Akad. Nauk SSSR*, **124**, 606 (1959).
44. R. L. Petritz, *Phys. Rev.*, **104**, 1508 (1956).
45. J. Bardeen, *Phys. Rev.*, **71**, 717 (1947).
46. E. T. Goodwin, *Proc. Cambridge Philos. Soc.*, **35**, 221, 232 (1939).
47. I. E. Tamm, *Z. Phys.*, **76**, 849 (1932).
48. A. V. Rzhanov, V. P. Migal, and N. N. Migal, *Fiz. Tekh. Poluprovodn.*, **3**, 1333 (1969).
49. V. G. Baru, *Kinet. Katal.*, **6**, 269 (1965).
50. V. G. Baru and Th. Wolkenstein, *Dokl. Akad. Nauk SSSR*, **167**, 1314 (1966).
51. Th. Wolkenstein and V. B. Sandomirskii, *Dokl. Akad. Nauk SSSR*, **118**, 980 (1958).
52. A. V. Rzhanov, Yu. F. Novototskii-Vlasov, and I. G. Neizvestnyi, *Zh. Tekh. Fiz.*, **17**, 2440 (1957).
53. A. V. Rzhanov, N. M. Pavlov, and M. A. Selezneva, *Zh. Tekh. Fiz.*, **28**, 2645 (1958).
54. A. V. Rzhanov, *Electronic Processes at Semiconductor Surfaces* [in Russian], Nauka, Moscow (1971).
55. V. I. Lyashenko, O. A. Serba, and I. I. Stepko, *Dokl. Akad. Nauk Ukr. SSR*, **3**, 350 (1962).
56. É. P. Mikheeva and N. P. Keier, *Kinet. Katal.*, **5**, 748 (1964).
57. L. I. Ivankiv, Dissertation, Lvov State University (1967).
58. S. A. Hoenig and J. R. Lane, *Surf. Sci.*, **11**, 163 (1968).
59. M. Constantinescu, E. Segal, and M. Vass, *Rev. Roum. Chim.*, **21**, 503 (1976).
60. M. Teodorescu, M. Vass, and E. Segal, *Zh. Fiz. Khim.*, **52**, 3071 (1978).
61. V. S. Kuznetsov and V. B. Sandomirskii, *Kinet. Katal.*, **3**, 724 (1962).
62. B. I. Boltaks, *Diffusion in Semiconductors* [in Russina], Fizmatgiz, Moscow (1961).
63. W. E. Garner, *Adv. Catal.*, **9**, 169 (1957).
64. T. I. Barry and F. S. Stone, *Proc. R. Soc. London, Ser. A*, **255**, 124 (1960).
65. O. Peshev and Th. Wolkenstein, *Zh. Fiz. Khim.*, **11**, 571 (1966); *Probl. Kinet. Katal.*, **12**, 144 (1967).
66. O. Peshev, *Kinet. Katal.*, **7**, 84 (1966).
67. D. Glemza and R. J. Kokes, *J. Phys. Chem.*, **66**, 566 (1962).

68. E. N. Figurovskaya, V. F. Kiselev, and Th. Wolkenstein, *Dokl. Akad. Nauk SSSR*, **161**, 1142 (1965).
69. L. E. Derlyukova, Dissertation, Institute of Physical Chemistry, Academy of Sciences of the USSR, Moscow (1970).
70. Th. Wolkenstein, V. S. Kuznetsov, and V. B. Sandomirskii, *Kinet. Katal.*, **4**, 24 (1963); in: *Surface Properties of Semiconductors* (A. N. Frumkin, ed.), Consultants Bureau, New York (1964), p. 136.
71. J. N. Butler, *J. Chem. Phys.*, **35**, 636 (1961).
72. N. D. Tomashov, *Corrosion and Protection of Metals* [in Russian], Izd-vo Akad. Nauk SSSR, Moscow (1959), p. 53.
73. B. Yu. Ainbinder and Th. Wolkenstein, *Zashch. Met.*, **7**, No. 3, 249 (1971).
74. N. F. Mott, *J. Chim. Phys.*, **44**, 172 (1947); *Trans. Faraday Soc.*, **43**, 429 (1947).
75. N. Cabrera and N. F. Mott, *Rep. Prog. Phys.*, **12**, 163 (1949).
76. A. T. Fromhold, *J. Phys. Chem. Solids*, **24**, 1081, 1309 (1963).
77. P. T. Landsberg, *J. Chem. Phys.*, **23**, 1079 (1955).
78. R. Hooke and T. P. Brody, *J. Chem. Phys.*, **42**, 4310 (1965).
79. T. B. Grimley and B. M. W. Trapnell, *Proc. R. Soc. London, Ser. A*, **234**, 405 (1956).
80. N. F. Mott and P. H. Fehlner, *Oxid. Met.*, **2**, 59 (1970).
81. D. D. Eley and P. W. Wilkinson, *Proc. R. Soc. London, Ser. A*, **254**, 327 (1960).
82. E. E. Huber and C. T. Kirk, *Surf. Sci.*, **5**, 447 (1966).
83. A. I. Loskutov, É. Kh. Enikeev, I. L. Rozenfel'd, and R. I. Nazarova, *Zashch. Met.*, **17**, No. 2, 163 (1981).
84. O. Kubaschewski and B. Hopkins, *Oxidation of Metals and Alloys*, Butterworths, London (1953).

CHAPTER 5

1. Th. Wolkenstein, *Probl. Kinet. Katal.*, **8**, 79 (1955).
2. V. V. Voevodskii, Th. Wolkenstein, and N. N. Semenov, in: *Problems of Chemical Kinetics, Catalysis, and Reactivity* [in Russian], Izd-vo Akad. Nauk SSSR, Moscow (1955), p. 423.
3. N. Thon and H. A. Taylor, *J. Am. Chem. Soc.*, **75**, 2447 (1953).
4. Th. Wolkenstein, *Zh. Fiz. Khim.*, **24**, 1068 (1950).
5. M. Boudart, *J. Am. Chem. Soc.*, **74**, 1531 (1952).
6. K. Hauffe, *Adv. Catal.*, **7**, 213 (1955).
7. K. Hauffe, *Dachema Monographien*, **26**, 222 (1956).
8. K. Hauffe, *Adv. Catal.*, **9**, 187 (1957).
9. G. K. Boreskov and V. V. Popovskii, *Probl. Kinet. Katal.*, **10**, 67 (1960).
10. I. A. Myasnikov and S. Ya. Pshezhetskii, *Probl. Kinet. Katal.*, **8**, 175 (1955).
11. O. V. Krylov, S. Z. Roginskii, and E. A. Fokina, *Izv. Akad. Nauk SSSR, Otd. Khim. Nauk*, No. 4, 421 (1957).
12. Th. Wolkenstein, *Probl. Kinet. Katal.*, **8**, 201 (1955).

13. G. M. Zhabrova, *Probl. Kinet. Katal.*, **8**, 209 (1955).
14. V. N. Vladimirova, É. Kh. Enikeev, G. M. Zhabrova, and L. Ya. Margolis, *Dokl. Akad. Nauk SSSR*, **131**, 342 (1960); O. M. Vinogradova, V. I. Vladimirova, and G. M. Zhabrova, *Dokl. Akad. Nauk SSSR*, **133**, 1375 (1960).
15. F. C. Whitmore, *J. Am. Chem. Soc*, **54**, 3274 (1932).
16. M. E. Winfield, *Catalysis*, **7**, 116 (1960).
17. V. I. Spitsyn, Th. Wolkenstein, I. E. Mikhailenko, and V. I. Pikaeva, *Izv. Akad. Nauk SSSR, Ser. Khim.*, No. 11, 2449 (1970).
18. T. Takaishi, *Z. Naturforsch., Teil A*, **11**, 297 (1956).
19. J. E. Germain, *Catalyse hétérogène*, Dunod, Paris (1959).
20. G. M. Schwab and J. Block, *Z. Phys. Chem. (BRD)*, **1**, 42 (1954).
21. G. Parravano, *J. Am. Chem. Soc.*, **74**, 1194 (1952), *J. Am. Chem. Soc.*, **75**, 1448, 1452 (1953).
22. N. P. Keier, S. Z. Roginskii, and I. S. Sazonova, *Dokl. Akad. Nauk SSSR*, **106**, 859 (1956); *Iv. Akad. Nauk. SSSR, Ser. Fiz.*, **21**, 183 (1957).
23. É. Kh. Enikeev, L. Ya. Margolis, S. Z. Roginskii, *Dokl. Akad. Nauk SSSR*, **130**, 807 (1960).
24. A. Bielanski and J. Deren, in: *Symposium on Electronic Phenomena in Chemisorption and Catalysis on Semiconductors* (K. Hauffe and Th. Wolkenstein, eds.), De Gruyter, Berlin (1969), p.156.
25. K. Hauffe, *Angew. Chem.*, **68**, 776 (1956).
26. D. A. Dowden, N. Mackenzie, and B. H. W. Trapnell, *Adv. Catal.*, **9**, 70 (1957).
27. Th. Wolkenstein, *Adv. Catal.*, **23**, 157 (1973).
28. E. Molinari and G. Parravano, *J. Am. Chem. Soc.*, **75**, 5233 (1953).
29. H. W. Kohn and E. H. Taylor, *J. Am. Chem. Soc.*, **79**, 252 (1957); *J. Phys. Chem.*, **63**, 967 (1959); *J. Catal.*, **2**, 32 (1963); *J. Catal.*, **2**, 208 (1963).
30. V. C. F. Holm and R. W. Clark, *Ind. Eng. Chem.*, **43**, 501 (1951); *Ind. Eng. Chem.*, **44**, 107 (1952).
31. L. F. Heckelsberg, A. Clark, and G. C. Bailey, *J. Phys. Chem.*, **60**, 559 (1956).
32. V. C. F. Holm and A. Clark, *J. Catal.*, **2**, 16 (1963).
33. S. E. Voltz and S. W. Weller, *J. Am. Chem. Soc.*, **75**, 5227, 5231 (1953); *J. Phys. Chem.*, **5**, 100 (1955).
34. W. E. Corner, *Adv. Catal.*, **9**, 169 (1957).
35. A. Cimino, *Naturwissenschaften*, **43**, 58 (1956).
36. A. Clark, *Ind. Eng. Chem.*, **45**, 1476 (1953); *J. Phys. Chem.*, **60**, 1506 (1956).
37. Y. L. Sundler and M. Gazith, *J. Phys. Chem.*, **63**, 1095 (1959).
38. S. Yu. Elovich and L. Ya. Margolis, *Izv. Akad. Nauk SSSR, Ser. Fiz.*, **21**, 206 (1957).
39. R. L. Petritz, *Phys. Rev.*, **104**, 1508 (1956).
40. Th. Wolkenstein, *Zh. Fiz. Khim.*, **25**, 1244 (1951).
41. I. S. Sazonova, N. P. Keier, T. P. Khokhlova, and I. L. Mikhailova, *Kinet. Katal.*, **11**, 447 (1970).

42. V. I. Spitsyn, Th. Wolkenstein, G. N. Pirogova, S. F. Timashev, R. I. Korosteleva, and A. A. Sonina, *Izv. Akad. Nauk SSSR, Ser. Khim.*, No. 4, 771 (1972).

43. G. K. Boreskov and K. I. Matveev, *Probl. Kinet. Katal.*, **8**, 165 (1955).

44. C. Rienäcker, *Chem. Technol.*, No. 5, 1 (1959).

45. R. P. Charman, R. H. Griffith, and J. D. F. Marsh, *Proc. R. Soc. London*, **224**, 419 (1954).

46. A. Bielanski, J. Deren, and J. Haber, *Bull. Acad. Pol. Sci. Cl. 3*, **3**, 491 (1955); *Bull. Acad. Pol. Sci. Cl. 3* **4**, 221 (1955); *Bull. Acad. Pol. Sci. Cl. 3* **4**, 523 (1956); *Bull. Acad. Pol. Sci. Cl. 3*, **5**, 197 (1957); *Probl. Kinet. Katal.*, **10**, 37 (1960).

47. V. I. Lyashenko and I. I. Stepko, *Probl. Kinet. Katal.*, **8**, 180 (1955).

48. V. I. Lyashenko and I. I. Stepko, *Zh. Fiz. Khim.*, **29**, 401 (1955); *Izv. Akad. Nauk SSSR, Ser. Fiz.*, **21**, 201 (1957).

49. I. I. Stepko, Dissertation, Kiev State University (1956).

50. V. I. Lyashenko, G. F. Romanova, and I. I. Stepko, *Probl. Kinet. Katal.*, **10**, 111 (1960).

51. O. V. Krylov, *Nonmetal Catalysis* [in Russian], Khimiya, Leningrad (1967).

52. Th. Wolkenstein, *Kinet. Katal.*, No. 3, 776 (1981).

53. Th. Wolkenstein, *Izv. Akad. Nauk SSSR, Otd. Khim. Nauk*, No. 2, 143 (1957).

54. V. G. Baru and Th. Wolkenstein, *Dokl. Akad. Nauk SSSR*, **167**, 1314 (1966).

55. Th. Wolkenstein and V. B. Sandomirskii, *Dokl. Akad. Nauk SSSR*, **118**, 980 (1958).

56. L. I. Ivankiv, M. V. Miliyanchuk, and A. K. Filatova, in: *All-Union Workshop on Deep Catalytic Reactions*, Moscow (1964), Digest of papers, p. 81.

57. V. P. Fentsik and P. M. Stadnik, in: *Surface and Contact Phenomena in Semiconductors* [in Russian], Tomsk State University Press, Tomsk (1964), p. 190.

58. N. P. Keier, É. P. Mikheeva, and L. M. Usol'tseva, *Kinet. Katal.*, **8**, 1199 (1967).

59. Th. Wolkenstein, V. S. Kuznetsov, and V. B. Sandomirskii, *Kinet. Katal.*, **4**, 24 (1963); in: *Surface Properties of Semiconductors* (A. N. Frumkin, ed.), Consultants Bureau, New York (1964), p. 136.

60. J. Deren and R. Russer, *J. Catal.*, **7**, 396 (1967).

61. J. Deren and J. Haber, in: *Studies on the Physicochemical and Surface Properties of Chromium Oxides*, Krakow (1969), p. 58.

62. Th. Wolkenstein, *Zh. Fiz. Khim.*, **22**, 311 (1948).

63. V. B. Sandomirskii, Dissertation, Institute of Physical Chemistry, Academy of Sciences of the USSR, Moscow (1955).

64. Th. Wolkenstein, *Usp. Fiz. Nauk*, **60**, 249 (1956).

65. G. M. Zhabrova, *Usp. Khim.*, **20**, 450 (1951).

66. N. I. Élement, Dissertation, Moscow Institute of Chemical Engineering (1949).

67. A. B. Shekhter and Yu. Sh. Moshkovskii, *Izv. Akad. Nauk SSSR, Otd. Khim. Nauk*, No. 4, 353 (1949).

68. N. P. Keier and G. I. Chizhikova, *Dokl. Akad. Nauk SSSR*, **120**, 830 (1958).
69. G. M. Schwab and J. Block., *Z. Elektrochem.*, **58**, 756 (1954).
70. G. M. Zhabrova and E. A. Fokina, *Izv. Akad. Nauk SSSR, Otd. Khim. Nauk*, No. 6, 963 (1955).
71. S. Z. Roginskii, *Probl. Kinet. Katal.*, **6**, 9 (1949).
72. Th. Wolkenstein and S. Z. Roginskii, *Zh. Fiz. Khim.*, **29**, 485 (1955).
73. L. Ya. Margolis and O. M. Todes, *Dokl. Akad. Nauk SSSR*, **58**, 421 (1947).
74. O. V. Krylov and L. Ya. Margolis, *Zh. Obshch. Khim.*, **20**, 1991 (1950).
75. F. H. Constable, *Proc. R. Soc. London, Ser. A*, **108**, 355 (1923).
76. W. Meyer and H. Neldel, *Phys. Z.*, **38**, 1014 (1937).
77. B. Claudel and J. Veron, *C. R. Acad. Sci., Ser. C*, **267**, 1195 (1968).
78. A. B. Shekhter and Yu. Sh. Moshkovskii, *Dokl. Akad. Nauk SSSR*, **72**, 339 (1950).
79. G. M. Schwab, *Z. Phys. Chem., Abt. A*, **144**, 243 (1929).
80. E. Cremer, *Adv. Catal.*, **7**, 75 (1955).
81. S. Z. Roginskii and Yu. L. Khait, *Dokl. Akad. Nauk SSSR*, **130**, 366 (1960).
82. V. P. Lebedev, *Probl. Kinet. Katal.*, **10**, 204 (1960).
83. G. I. Likhtenshtein, *Kinet. Katal.*, **4**, 35 (1963).
84. O. Exner, *Collect. Czech. Chem. Commun.*, **29**, 1094 (1964).
85. V. S. Kuznetsov, *Kinet. Katal.*, **5**, 277 (1964).
86. O. Peshev and G. Bliznakov, *J. Catal.*, **7**, 18 (1967).
87. O. Peshev, *Izv. Bolg. Akad. Nauk, Otd. Khim. Nauk*, **1**, 149 (1968).
88. O. Peshev, *Zh. Fiz. Khim.*, **44**, 370 (1970).
89. I. V. Nicolescu, M. Spinzi, and A. Suceveanu, *Chem. Technol.*, **15**, 226 (1963).
90. I. V. Nicolescu, M. Spinzi, and A. Suceveanu, *Rev. Chim.*, **20**, 337 (1969).

CHAPTER 6

1. S. Z. Roginskii, *Adsorption and Catalysis on Inhomogeneous Surfaces* [in Russian], Izd-vo Akad.Nauk SSSR, Moscow (1948).
2. V. I. Levin, *Usp. Khim.*, **17**, 174 (1948).
3. Th. Wolkenstein, *Zh. Fiz. Khim.*, **21**, 153 (1947).
4. J. K. Roberts, *Some Problems in Adsorption*, Cambridge University Press, London (1939).
5. J. Koutecky, *Trans. Faraday Soc.*, **54**, 1038 (1958).
6. V. L. Bonch-Bruevich and Th. Wolkenstein, *Zh. Fiz. Khim.*, **28**, 1219 (1954); *Probl. Kinet. Katal.*, **8**, 218 (1955).
7. O. Peshev, Dissertation, Moscow State University, Moscow (1965).
8. V. L. Bonch-Bruevich, *Zh. Fiz. Khim.*, **27**, 662, 960 (1953).
9. K. Hauffe, *Bull. Soc. Chim. Belg.*, **67**, 417 (1958).
10. Sh. M. Kogan and V. B. Sandomirskii, *Dokl. Akad. Nauk SSSR*, **127**, 377 (1959); *Izv. Akad. Nauk SSSR, Otd. Khim. Nauk*, No. 9, 1681 (1959).
11. E. Bauer and H. Staude, in: *Katalyze Bericht von der Hauptjahrestagung, 1958*, Chemische Gesellschaft in der DDR, Berlin (1959), p. 121.

12. A. V. Rzhanov, in: *Surface Properties of Semiconductors* (A. N. Frumkin, ed.), Consultants Bureau, New York (1964).

13. Yu. F. Novototskii-Vlasov and M. P. Sinyukov, in: *Surface Properties of Semiconductors* (A. N. Frumkin, ed.), Consultants Bureau, New York (1964); I. G. Neizvestnyi, in: *Surface Properties of Semiconductors* (A. N. Frumkin, ed.), Consultants Bureau, New York (1964).

14. N. M. Andreeva, V. V. Murina, and Yu. F. Novototskii-Vlasov, *Fiz. Tekh. Poluprovodn.*, 8, 788 (1974).

15. V. V. Murina, Yu. F. Novototskii-Vlasov, A. S. Petrov, R. V. Prudnikov, G. F. Golovanova, in: *Electronic Processes at Semiconductor Surfaces and the Semiconductor–Insulator Interface* [in Russian], Izd-vo Akad. Nauk SSSR, Novosibirsk (1974), p. 262.

16. N. I. Kobozev, *Acta Physicochim. URSS*, **13**, 469 (1940).

17. N. I. Kobozev, *Usp. Khim.*, **23**, 545 (1956).

18. Th. Wolkenstein, *Zh. Fiz. Khim.*, **23**, 917 (1949); *Probl. Kinet. Katal.*, **7**, 360 (1949).

19. J. H. de Boer, *Adv. Catal.*, **9**, 472 (1957).

20. I. M. Livshits and A. M. Kosevich, *Izv. Akad. Nauk SSSR, Ser. Fiz.*, **19**, 395 (1955).

21. V. B. Sandomirskii, *Radiotekh. Elektron.*, **7**, 1971 (1962).

22. O. Peshev, *C. R. Acad. Bulg. Sci.*, **29**, 201 (1976).

23. O. Peshev, *Kinet. Katal.*, **18**, 907 (1977).

24. O. Peshev, in: *Proc. 7th Inter. Vacuum Conf. and 3rd Inter. Conf. Solid Surfaces*, Vienna (1977), p. 1963.

25. O. Peshev, *Dispersed Semiconductors in Adsorption and Surface Reactions, Bulgarian Academy of Sciences*, Sofia (1980), Chap. 4.

26. Sh. M. Kogan, *Probl. Kinet. Katal.*, **10**, 52 (1960).

27. O. Peshev, *Zh. Fiz. Khim.*, **44**, 370 (1970).

28. O. Peshev and G. Bliznakov, *J. Catal.*, **7**, 18 (1967).

29. V. G. Baru, *Kinet. Katal.*, **6**, 269 (1965).

30. M. M. Egorov, K. G. Krasil'nikov, and V. F. Kiselev, *Zh. Fiz. Khim.*, **32**, 2448 (1958).

31. M. M. Egorov, T. S. Egorova, K. G. Krasil'nikov, and V. F. Kiselev, *Zh. Fiz. Khim.*, **32**, 2624 (1958).

32. O. Peshev, *Izv. Bulg. Akad. Nauk, Otd. Khim. Nauk*, **1**, 149 (1968).

33. O. Peshev, *Zh. Fiz. Khim.*, **52**, 3015 (1978).

34. O. Peshev, *J. Chim. Phys.*, **74**, 183 (1977).

35. I.-M. Herrman, P. Vergnon, and S.-I. Teichner, *J. Catal.*, **37**, 57 (1975).

36. I.-M. Herrman, P. Vergnon, and S.-I. Teichner, *React. Kinet. Catal. Lett.*, **2**, 199 (1975).

37. I.-M. Herman, *Kinet. Katal.*, **19**, 383 (1978).

38. P. Vergnon, I.-M. Herrman, and S.-I. Teichner, *Zh. Fiz. Khim.*, **52**, 3021 (1978).

39. G. Ya. Pikus and G. E. Chaika, *Ukr. Fiz. Zh.*, **18**, 931 (1978).

40. G. Ya. Pikus and G. E. Chaika, *Zh. Fiz. Khim.*, **52**, 3101 (1978).

41. G. Ya. Pikus, V. F. Shnyukov, and B. P. Nikonov, *Fiz. Tverd. Tela (Leningrad)*, **10**, 125 (1968).
42. G. Ya. Pikus, G. N. Tal'nova, and S. V. Tychkina, *Izv. Akad. Nauk SSSR, Neorg. Mater.*, **12**, 1955 (1976).
43. G. Ya. Pikus and V. P. Teterya, *Fiz. Tverd. Tela (Leningrad)*, **15**, 2098 (1973).
44. G. Ya. Pikus and G. N. Tal'nova, *Fiz. Tverd. Tela*, **18**, 2934 (1976).
45. G. N. Tal'nova and G. Ya. Pikus, *Zh. Fiz. Khim.*, **52**, 3107 (1978).
46. G. Ya. Pikus and V. P. Teterya, *Ukr. Fiz. Zh.*, **17**, 380 (1972).
47. G. Ya. Pikus and G. N. Tal'nova, *Izv. Akad. Nauk SSSR, Neorg. Mater.*, **13**, 433 (1977).
48. O. M. Braun, G. Ya. Pikus, and G. E. Chaika, *Ukr. Fiz. Zh.*, **21**, 742 (1976).
49. B. Yu. Aibinder and É. Kh. Enikeev, *VINITI Abstracts* [in Russian], No. 30 (1973).

CHAPTER 7

1. A. N. Terenin, *Probl. Kinet. Katal.*, **8**, 17 (1955).
2. Yu. P. Solonitsyn, *Zh. Fiz. Khim.*, **32**, 1241 (1958).
3. V. A. Kotel'nikov, *Kinet. Katal.*, **5**, 565 (1964).
4. J. Haber and F. S. Stone, *Trans. Faraday Soc.*, **59**, 192 (1963).
5. F. Romero-Rossi and F. S. Stone, in: Deuxième *Congrès International de Catalyse*, Paris (1960), Rapport No. 70.
6. F. S. Stone, in: *Coloquio Sobre Quimica Fisica de Processes on Superficies Solides*, Madrid (1965), p. 109.
7. T. Kwan, in: *Electronic Phenomena in Chemisorption and Catalysis on Semiconductors* (K. Hauffe and Th. Wolkenstein, eds.), De Gruyter, Berlin (1969), p. 184.
8. F. S. Stone, *Ipatieff Centenary Conf.*, Evanston (1967).
9. Y. Fujita and T. Kwan, *Bull. Chem. Soc. Jpn.*, **31**, 830 (1958).
10. T. I. Barry, in: *Deuxième Congrès International de Catalyse*, Paris (1960), Rapport No. 72.
11. A. N. Terenin and Yu. P. Solonitsyn, *Discuss. Faraday Soc.*, No. 28, 28 (1959).
12. D. Kennedy, M. Ritchil, and J. M. MacKenzie, *Trans. Faraday Soc.*, **54**, 119 (1958).
13. V. B. Kazanskii, O. V. Nikitina, G. B. Pariiskii, and V. F. Kiselev, *Dokl. Akad. Nauk SSSR*, **151**, 369 (1963).
14. J. Haber and A. Kowalska, *Bull. Acad. Pol. Sci., Ser. Sci. Chim.*, **13**, 463 (1965).
15. Yu. P. Solonitsyn, Dissertation, Leningrad State University (1960).
16. R. Coekelbergs, A. Crucq, J. Decot, L. Degols, M. Randoux, and L. Tommerman, *J. Phys. Chem. Soc.*, **26**, 1983 (1965).
17. Th. Wolkenstein (F. F. Vol'kenshtein), *The Electronic Theory of Catalysis on Semiconductors*, Pergamon Press, Oxford (1963).

18. Th. Wolkenstein, *Zh. Fiz. Khim.*, **26**, 1462 (1952).

19. Th. Wolkenstein and I. V. Karpenko, *Fiz. Tverd. Tela (Leningrad)*, **9**, 403 (1967).

20. Th. Wolkenstein and I. V. Karpenko, *J. Appl. Phys.*, Suppl., **33**, 460 (1962).

21. G. L. Bir, *Fiz. Tverd. Tela (Leningrad)*, **1**, 67 (1959).

22. Th. Wolkenstein and I. V. Karpenko, *Dokl. Akad. Nauk SSSR*, **165**, 1101 (1965).

23. H. W. Kohn and E. H. Taylor, in: *Deuxième Congrès International de Catalyse*, Paris (1960), Rapport No. 71.

24. H. W. Kohn, *Nature*, **184**, 630 (1959); *J. Chem. Phys.*, **33**, 1588 (1960); *J. Catal.*, **2**, 208 (1963).

25. G. K. Boreskov, V. B. Kazanskii, Yu. A. Mischenko, and G. B. Pariiskii, *Dokl. Akad. Nauk SSSR*, **175**, 384 (1964).

26. Yu. A. Mischenko, Dissertation, Moscow State University (1964).

27. G. M. Muha, *J. Phys. Chem.*, **70**, 1390 (1966).

28. D. N. Stamiers and J. Turkevich, *J. Am. Chem. Soc.*, **86**, 728, 757 (1964).

29. J. H. Lunsford and J. P. Jayne, *J. Phys. Chem*, **69**, 2183 (1965).

30. E. Bauer and H. Staude, in: *Katalyse, Bericht von der Hauptjahrestagung, 1958, Chemischen Gesellschaft in der DDR (1959)*, p. 121.

31. E. Kamke, *Differentialgleichungen, Lösungsmethoden und Lösungen*, Chelsea, New York (1948).

32. V. G. Baru and Th. Wolkenstein, *Zh. Fiz. Khim.*, **42**, 1317 (1968); *Surf. Sci.*, **13**, 294 (1969).

33. T. I. Barry and F. S. Stone, *Proc. R. Soc. London, Ser. A*, **255**, 124 (1960).

34. V. G. Baru and Th. Wolkenstein, *Usp. Khim.*, **37**, 1685 (1968).

35. F. Steinbach and R. Harborth, *Faraday Discuss. Chem. Soc.*, **58**, 143 (1974).

36. A. A. Lisachenko and F. I. Vilesov, in: *Photoadsorption and Photocatalytic Phenomena in Heterogeneous Systems* [in Russian], Novosibirsk (1974), p. 3.

37. T. T. Bykova, S. A. Komolov, and É. F. Lazneva, in: *Photoadsorption and Photocatalytic Phenomena in Heterogeneous Systems* [in Russian], Novosibirsk (1974), p. 102.

38. T. T. Bykova and É. F. Lazneva, *Fiz. Tekh. Poluprovodn.*, **7**, 1369 (1972).

39. R. I. Bickley and R. K. M. Jayanty, *Faraday Discuss. Chem. Soc.*, **58**, 935 (1974).

40. V. I. Gol'danskii, V. A. Namiot, and R. V. Khokhlov, *Zh. Éksp. Teor. Fiz.*, **70**, 2349 (1976).

41. A. Crucq and R. Coekelbergs, *Zh. Fiz. Khim.*, **53**, 2904 (1979).

42. E. Molinari, F. Cramarosa, and F. Paniccia, *J. Catal.*, **4**, 415 (1965).

43. K. Hauffe, *Reaktionen in und an festen Stoffen*, Springer, Berlin (1966).

44. Th. Wolkenstein, *Z. Phys. Chem. Neue Folge*, **108**, 97 (1977).

45. L. V. Lyashenko, in: *Photoadsorption and Photocatalytic Phenomena in Heterogeneous Systems* [in Russian], Novosibirsk (1974), p. 70.

46. A. A. Lisachenko and F. I. Vilesov, *Vestn. Leningr. Gos. Univ., Ser. Fiz. Khim.*, Issue 2, No. 10, 30 (1968).

47. I. S. McLintock and M. Ritchie, *Trans. Faraday Soc.*, **61**, 1007 (1965).

48. V.N. Filimonov, *Kinet. Katal.*, **7**, 512 (1966); *Dokl. Akad. Nauk SSSR*, **154**, 922 (1964); *Dokl. Akad. Nauk SSSR*, **158**, 1408 (1964).
49. G. M. Schwab, F. Steinbach, H. Noller, and M. Venugopalan, *Nature*, **193**, 774 (1962).
50. G.M. Schwab, H. Noller, F. Steinbach, and M. Venugopalan, *Z. Naturforsch.*, **19a**, 45, 145 (1964).
51. N. Djeghri, M. Formenti, F. Juvllet, and S. J. Techner, *Faraday Discuss. Chem. Soc.*, **58**, 185 (1974).
52. R.I. Bickley and R. K. M. Jayanty, *Faraday Discuss. Chem. Soc.*, **58**, 194 (1974).
53. A. A. Lisachenko and F. I. Vilesov, *Kinet. Katal.*, **9**, 935 (1968).
54. A. Bobrovskaya and V. E. Kholmogorov, *Zh. Teor. Éksp. Khim.*, **3**, 112 (1967).
55. L. V. Lyashenko and Ya. B. Gorokhvatskii, *Kinet. Katal.*, **8**, 694 (1967).
56. I. N. Borshchevskii and L. A. Nikolaev, *Zh. Fiz. Khim.*, **28**, 265, 2211 (1954); *Zh. Fiz. Khim.*, **33**, 1071 (1959); *Zh. Fiz. Khim.*, **36**, 249, 369 (1962).
57. H. W. Kohn and E. H. Taylor, *J. Catal.*, **2**, 32 (1963).
58. H. W. Kohn and E. H. Taylor, *J. Phys. Chem.*, **63**, 967 (1959).
59. J. Lunsford and T. Leland, *J. Phys. Chem.*, **66**, 2591 (1962).
60. G. F. Shipman, *J. Phys. Chem.*, **70**, 1120 (1962).
61. Th. Freund, *J. Catal.*, **2**, 289 (1964).
62. G. K. Boreskov, V. B. Kazanskii, Yu. A. Mishchenko, and G. P. Pariiskii, *Dokl. Akad. Nauk SSSR*, **157**, 384 (1964).
63. S. Yu. Elovich and L. Ya. Margolis, *Izv. Akad. Nauk SSSR, Ser. Fiz.*, **21**, 206 (1957).
64. Th. Wolkenstein, *Kinet. Katal.*, **2**, 481 (1961); *Discuss. Faraday Soc.*, **31**, 209 (1961).
65. Th. Wolkenstein and V. B. Nagaev, *Kinet. Katal.*, **16**, 381 (1975).
66. Th. Wolkenstein, *Adv. Catal.*, **23**, 137 (1973).
67. F. Romero-Rossi and F. Stone, in: *Deuxième Congrès International de Catalyse*, Paris (1960), Rapport No. 70.
68. F. Steinbach and K. Krieger, *Z. Phys. Chem., Neue Folge*, **58**, 290 (1968).
69. F. Steinbach, *Z. Phys. Chem., Neue Folge*, **60**, 126 (1968).
70. W. Dorfler and K. Hauffe, *J. Catal.*, **3**, 171 (1964).
71. W. Ritchey and J. Calvert, *J. Phys. Chem.*, **60**, 1465 (1960).
72. U. Fujita, *Catalyst*, **3**, 285 (1961).
73. G. A. Korsunskii, *Dokl. Akad. Nauk SSSR*, **113**, 853 (1957).
74. G. A. Korsunskii, *Zh. Fiz. Khim.*, **34**, 510 (1960).
75. G. A. Korsunskii and Yu. S. Lebedev, *Zh. Fiz. Khim.*, **35**, 1078 (1961).
76. G. A. Korsunskii, Doctoral Dissertation, State Optical Institute, Leningrad (1960).
77. R. E. Stephens, B. Ke, and P. Trivich, *J. Phys. Chem.*, **59**, 966 (1955).
78. T. S. Nagarjunan and J. Calvert, *J. Phys. Chem.*, **68**, 17 (1964).
79. F. Steinbach, *Nature*, **215**, 152 (1967).

80. V. S. Zakharenko, A. E. Cherkashin, N. P. Keier, and S. V. Kosheev, in: *Photoadsorption and Photocatalytic Phenomena in Heterogeneous Systems* [in Russian], Novosibirsk (1974), p. 54.

81. L. V. Lyashenko and Y. A. B. Gorokhovatskii, *Teor. Éksp. Khim.*, **3**, 218 (1967).

82. F. Steinbach and R. Harborth, *Faraday Discuss. Chem. Soc.*, **58**, 143 (1974).

83. V. S. Zakharenko, A. E. Cherkashin, and N. P. Keier, in: *Photoadsorption and Photocatalytic Phenomena in Heterogeneous Systems* [in Russian], Novosibirsk (1974), p. 48.

84. M. C. Marchem and K. J. Laidler, *J. Phys. Chem.*, **57**, 363 (1953).

85. V. I. Veselovskii and D. N. Shub, *Probl. Kinet. Katal.*, **8**, 43 (1955); *Zh. Fiz. Khim.*, **26**, 569 (1952).

86. V. I. Veselovskii, *Zh. Fiz. Khim.*, **21**, 983 (1947); *Zh. Fiz. Khim.*, **22**, 1302, 1427 (1948); *Zh. Fiz. Khim.*, **23**, 1096 (1949).

87. G. Calvert, K. Theurer, T. Rankin, and W. MacNevin, *J. Am. Chem. Soc.*, **75**, 2875 (1953).

88. G. M. Schwab, *Adv. Catal.*, **9**, 229 (1957).

89. Th. R. Rubin, G. Calvert, T. Rankin, and W. MacNevin, *J. Am. Chem. Soc.*, **75**, 2875 (1953).

90. E. Baur and C. Neuweller, *Helv. Chim. Acta*, **10**, 901 (1927).

91. L. I. Grossweiner, *J. Phys. Chem.*, **59**, 742 (1955).

92. Th. Wolkenstein and V. B. Nagaev, *Kinet. Katal.*, **16**, 388 (1975).

93. A. V. Pamfilov, Ya. S. Mazurkevich, and R. Ya. Mushchii, *Ukr. Khim. Zh.*, **25**, 453 (1959).

94. M. Constantinescu and E. Segal, *Rev. Roum. Chim.*, **17**, 83 (1972).

CHAPTER 8

1. V. A. Sokolov and A. N. Gorban, *Luminescence and Adsorption* [in Russian], Nauka, Moscow (1969).

2. Th. Wolkenstein, A. N. Gorban, and V. A. Sokolov, *Radical-Recombination Luminescence in Semiconductors* [in Russian], Nauka, Moscow (1976).

3. V. V. Malakhov, G. P. Peka, and Th. Wolkenstein, *J. Lumin.*, **5**, 252 (1972).

4. V. V. Malakhov, G. P. Peka, and Th. Wolkenstein, *J. Lumin.*, **5**, 261 (1972).

5. G. P. Peka and Yu. I. Karkhanin, *Dokl. Akad. Nauk SSSR*, **141**, 60 (1961).

6. I. B. Ermolovich, G. P. Peka, and M. K. Sheinkman, *Fiz. Tverd. Tela (Leningrad)*, **11**, 3002 (1969).

7. Th. Wolkenstein, V. V. Malakhov, and G. P. Peka, *Fiz. Tekh. Poluprovodn.*, **5**, 436 (1971).

8. O. S. Zinets, G. P. Peka, and Yu. P. Karkhanin, *Fiz. Tverd. Tela (Leningrad)*, **6**, 3516 (1964).

9. G. P. Peka and G. P. Zhdanov, *Fiz. Tverd. Tela (Leningrad)*, **11**, 1732 (1969).

10. G. P. Peka and A. S. Guzii, *Fiz. Tverd. Tela (Leningrad)*, **8**, 2293 (1966).

11. Yu. I. Karkhanin and V. E. Lashkarev, *Dokl. Akad. Nauk SSSR*, **97**, 1007 (1954).
12. D. T. Stevenson and R. I. Keyes, *Physica*, **20**, 1041 (1954).
13. M. K. Shteikman, I. B. Ermolovich, and G. L. Belen'kii, *Fiz. Tverd. Tela (Leningrad)*, **10**, 2628 (1968).
14. A. N. Gorban and V. A. Sokolov, *Opt. Spektrosk.*, **7**, 569 (1959).
15. V. A. Sokolov and A. N. Gorban, *Izv. Akad. Nauk SSSR, Ser. Fiz.*, **25**, 424 (1961).
16. K. M. Sancier, W. J. Frederics, and H. Wise, *J. Chem. Phys.*, **37**, 854 (1962).
17. V. V. Styrov and V. A. Sokolov, *Zh. Prikl. Spektrosk.*, **8**, 807 (1968).
18. D. P. Popov, Dissertation, Tomsk State University, Tomsk (1972).
19. A. I. Bazhin, V. V. Styrov, and V. A. Sokolov, *Zh. Fiz. Khim.*, **44**, 198 (1970).
20. V. A. Sokolov, V. V. Styrov, Yu. M. Naslednikov, V. D. Khoruzhii, G. A. Lubyanskii, B. G. Urusov, in: *Luminescent Materials and High Purity Materials* [in Russian], Proc. Stavropol University, No. 6 (1971), p. 88.
21. Yu. M. Naslednikov, Dissertation, Tomsk State University, Tomsk (1971).
22. A. I. Bazhin, V. V. Styrov, and V. A. Sokolov, *Izv. Vyssh. Uchebn. Zaved., Fiz.*, No. 7, 7 (1968).
23. V. V. Styrov and V. A. Sokolov, *Izv. Vyssh. Uchebn. Zaved., Fiz.*, No. 6, 135 (1967).
24. V. G. Kornich and A. N. Gorban, *Opt. Spektrosk.*, **21**, 234 (1966).
25. A.N. Gorban and V. G. Kornich, *Izv. Akad. Nauk SSSR, Ser. Fiz.*, **30**, 1424 (1966).
26. V. V. Styrov and A. I. Bazhin, *Izv. Vyssh. Uchebn. Zaved., Fiz.*, No. 12, 104 (1967).
27. W. Handle and H. Niermann, *Z. Naturforsch., Teil B*, **11a**, 395 (1956).
28. B. G. Urusov, Dissertation, Tomsk State University, Tomsk (1972).
29. V. V. Styrov, V. A. Sokolov, and Yu. P. Osipenko, *Izv. Vyssh. Uchebn. Zaved., Fiz.*, No. 11, 62 (1967).
30. K. F. Sommermeyer, *Z. Phys. Chem.*, **B41**, 433 (1939).
31. K. M. Sancier, W. J. Fredericks, and H. Wise, *J. Chem. Phys.*, **37**, 865 (1962).
32. Th. Wolkenstein, A. N. Gorban, and V. A. Sokolov, in: *Surface and Contact Phenomena in Semiconductors* [in Russian], Tomsk State University Press, Tomsk (1964), p. 457.
33. V. G. Kornich and A. N. Gorban, *Opt. Spektrosk.*, **21**, 390 (1966).
34. M. V. Fock, *Introduction to Kinetics of Luminescence of Phosphor Crystals* [in Russian], Nauka, Moscow (1964).
35. Th. Wolkenstein, A. N. Gorban, and V. A. Sokolov, *Kinet. Katal.*, **4**, 24 (1963).
36. Th. Wolkenstein, V. A. Sokolov, A. N. Gorban, and V. G. Kornich, in: *Proc. Inter. Conf. Luminescence, Budapest* (1966), p. 1433.
37. L. Reinecke, *Z. Phys.*, **135**, 361 (1953).
38. S. A. Krapivina, *Zh. Prikl. Spektrosk.*, **4**, 64 (1966).
39. P. J. Padlay, *Trans. Faraday Soc.*, **56**, 449 (1966).

40. Th. Wolkenstein and V. B. Sandomirskii, *Dokl. Akad. Nauk SSSR*, **118**, 980 (1960).
41. Th. Wolkenstein and V. G. Baru, *Dokl. Akad. Nauk SSSR* **167**, 1314 (1966).
42. Yu. N. Rufov, A. A. Kadushin, and S. Z. Roginskii, *Dokl. Akad. Nauk SSSR*, **171**, 905 (1966); S. Z. Roginskii and Yu. N. Rufov, *Kinet. Katal.*, **11**, 383 (1970).
43. A. A. Kadushin and Yu. N. Rufov, *Izv. Akad. Nauk SSSR, Otd. Khim. Nauk*, No. 12, 1245 (1966).
44. S. Z. Roginskii, in: *Symposium on Electronic Phenomena in Chemisorption and Catalysis on Semiconductors* (K. Hauffe and Th. Wolkenstein, eds.), De Gruyter, Berlin (1969), p. 212.
45. L. I. Ivankiv, V. P. Martynova, B. M. Palyukh, A. M. Pentsak, and Z. V. Solyanik, *Ukr. Fiz. Zh.*, **15**, 1208 (1971).
46. D. P. Popov and V. V. Styrov, in: *Collection of Scientific Papers of Noril'sk Evening Industrial Institute* [in Russian], No. 10, Krasnoyarsk (1971), p. 15; D. P. Popov, Dissertation, Tomsk State University, Tomsk (1972).
47. Th. Wolkenstein, V. A. Sokolov, G. P. Peka, V. V. Styrov, and V. V. Malakhov, *Izv. Akad. Nauk SSSR, Ser. Fiz.*, **37**, 855 (1973).
48. Th. Wolkenstein, V. A. Sokolov, D. P. Popov, and V. V. Styrov, *Kinet. Katal.*, **15**, 1250 (1974).
49. M. Breysse, L. Faure, B. Claudel, and J. Veron, *Prog. Vacuum Microbalance Tech.*, **2**, 229 (1973).
50. M. Breysse, L. Faure, B. Claudel, and J. Veron, *Le Vide*, Nos. 163–165, 24 (1973).
51. R. Roose and G. Offergald, *Electron. Fis. Apl.*, **16**, 47 (1973).
52. L. I. Ivankiv, Z. V. Solyanik, and A. M. Pentsak, *Ukr. Fiz. Zh.*, **16**, 1201 (1971).
53. L. I. Ivankiv, V. P. Martynova, B. M. Palyukh, B. M. Pentsak, and Z. V. Solyanik, *Ukr. Fiz. Zh.*, **16**, 1203 (1971).
54. V. V. Styrov, *Pis'ma Zh. Éksp. Teor. Fiz.*, **5**, 242 (1972).
55. I. A. Nikolaev, V. V. Styrov, and Yu. I. Tyurin, *Teor. Éksp. Khim.*, **16**, 67 (1980).
56. Th. Wolkenstein, *Zh. Fiz. Khim.*, No. 3, 773 (1976).
57. M. Breysse, B. Claudel, L. Faure, M. Culnin, R. T. T. Williams, and Th. Wolkenstein, *J. Catal.*, **45**, 137 (1976).

CHAPTER 9

1. S. R. Morrison, *The Chemical Physics of Surfaces*, Plenum Press, New York (1977).
2. Th. Wolkenstein, *Zh. Fiz. Khim.*, **21**, 1317 (1947).
3. V. L. Bonch-Bruevich, *Zh. Fiz. Khim.*, **25**, 1033 (1951).
4. A. Bielanski and J. Deren, in: *Symposium on Electronic Phenomena in Chemisorption and Catalysis on Semiconductors* (K. Hauffe and Th. Wolkenstein, eds.), De Gruyter, Berlin (1969), p. 156.

5.   N. P. Keier, in: *Proc. 4th Intern. Cong. Catal.* [in Russian], Nauka, Moscow
      (1969), p. 156.
6.   O. V. Krylov, *Probl. Kinet. Katal.*, **16**, 129 (1975).